Discrete Mathematics and Theoretical Computer Science

Springer

London
Berlin
Heidelberg
New York
Barcelona
Hong Kong
Milan
Paris
Singapore
Tokyo

Cristian S. Calude and Gheorghe Păun

Finite
Versus
Infinite

Contributions to an Eternal Dilemma

Springer

Cristian Calude
Department of Computer Science, University of Auckland, Private Bag 92019,
Auckland, New Zealand

Gheorghe Păun
Institute of Mathematics, Romanian Academy of Sciences, PO Box 1-764,
70700 Bucharest, Romania

ISBN 1-85233-251-4 Springer-Verlag Berlin Heidelberg New York

British Library Cataloguing in Publication Data
Finite versus infinite : contributions to an eternal
 Dilemma. – (Discrete mathematics and theoretical computer
 Science)
 1. Infinite 2. Finite, The 3. Computer science – Mathematics
 4. Mathematics - Philosophy 5. Mathematics – Data processing
 I. Calude, Christian S. II. Paun, Gheorghe
 510
ISBN 1852332514

Library of Congress Cataloging-in-Publication Data
Calude, Christian, 1952-
   Finite versus infinite : contributions to an eternal dilemma / Christian S. Calude and
 Gheorghe Paun.
     p. cm. – (Discrete mathematics and theoretical computer science)
   Includes bibliographical references.
   ISBN 1-85233-251-4 (alk. Paper)
   1. Processes, Infinite. 2. Infinite. I. Paun, Gheorghe, 1950- II. Title. III. Series.
 QA295.C25 2000
 515'.24—dc21   99-059492

Typesetting: Camera ready by authors
Printed and bound at the Athenæum Press Ltd., Gateshead, Tyne & Wear
34/3830-543210 Printed on acid-free paper SPIN 10749478

Preface

The finite – infinite interplay is central in human thinking, from ancient philosophers and mathematicians (Zeno, Pythagoras), to modern mathematics (Cantor, Hilbert) and computer science (Turing, Gödel). Recent developments in mathematics and computer science suggest a) radically new answers to classical questions (e.g., does infinity exist?, where does infinity come from?, how to reconcile the finiteness of the human brain with the infinity of ideas it produces?), b) new questions of debate (e.g., what is the role played by randomness?, are computers capable of handling the infinity through unconventional media of computation?, how can one approximate efficiently the finite by the infinite and, conversely, the infinite by finite?).

Distinguished authors from around the world, many of them architects of the mathematics and computer science for the new century, contribute to the volume. Papers are as varied as Professor Marcus' activity, to whom this volume is dedicated. They range from real analysis to DNA computing, from linguistics to logic, from combinatorics on words to symbolic dynamics, from automata theory to geography, and so on, plus an incursion into the old history of conceptions about infinity and a list of philosophical "open problems". They are mainly mathematical and theoretical computer science texts, but not all of them are purely mathematical. They deal directly with the finite – infinite interplay (one proves several times that paths from finite to infinite and conversely, from infinite to finite, are both inspiring and useful), or only implicitly (each grammar or automaton is a *finite* device meant to describe a possibly *infinite* object, a language; similarly, an axiomatic system is a finite construction aiming to cover an infinity of truths – whence the limits of such systems).

We emphasize the precaution taken in the title. The reader can find here only some *contributions* to an *eternal* debate. This is just a step in an infinite sequence of steps. We hope that this is a step forward...

*

Many thanks are due to all contributors, to Bev Ford and Rebecca

Mowat, from Springer, London, for the efficient and pleasant cooperation, as well as to our host institutions, JAIST, Ishikawa, Japan, and TUCS, Turku, Finland, for providing excellent environments for completing this book.

Cristian S. Calude
Gheorghe Păun
November 1999

Tribute to Professor Solomon Marcus

The present book is presented in honour of Professor Solomon Marcus, on the occasion of his 75th birthday (1st of March, 2000).

The co-editors of this volume have debated for a while what would be the most appropriate topic for such a work. The decision was, at the same time, difficult and easy.

Professor Marcus started by being a "pure mathematician", building for himself an international reputation in mathematical analysis and related areas; then, at the end of the fifties, he switched to an interesting career in mathematical linguistics (mainly analytical approaches to phonological, morphological, syntactic and semantic categories, but also automata and formal languages).[1] Although he was one of the most cited authors in this area, he has once again enlarged his interests – and definitely others' as well – by founding one more domain, mathematical poetics. Just *en passant*, in 1968, he introduced (Marcus) contextual grammars, a genuinely new generative device inspired from analytical models in the study of language.[2] This was not enough for Professor Marcus. In the same way as real analysis and mathematical linguistics provided to him tools for mathematical poetics, all these have been applied to semiotics, natural and social sciences.

Then, what to choose for a *Festscrift*? None of these alone would be enough, but all together are hard to be contained under the covers of a single book.[3]

A tempting idea was to look for major trends or meta-ideas in Professor Marcus' *oeuvre*. Again a difficult choice – at least for the present co-editors, although both of them were (and psychologically still are!) his students and had/have the privilege to be his close collaborators. Analogies/bridges between fields which, at the first sight, look remote? The finesse of study, no

---

[1] His book *Finite Grammars and Automata*, published unfortunately only in Romanian in 1964, was one of the earliest monographs on this topic.

[2] The domain has expanded beyond his expectations (as he confessed somewhere): a comprehensive book was published recently by Kluwer: *Marcus Contextual Grammars*.

[3] Another book in honour of Professor Marcus, including only papers in mathematical and computational linguistics, will be published by the Romanian Academy Publishing House.

matter what object of research? The attraction for counter-intuitive, even paradoxical facts, proven then to be common-frequent-beneficial? The visible passion for dichotomies? Which one? Artistic versus scientific? Finite physical existence versus infinite spiritual existence? Old versus young provocation? Discrete versus continuous? Analytical versus generative? Local versus global? Finite versus infinite?

That's it! It fits well with Professor Marcus' activity and personality, both of which leave the impression of infinity... About four hundred papers, more than forty authored or co-authored books (published in ten languages not including Romanian), twenty five edited books, hundreds of conferences attended, hundreds of lectures at universities, many domains of direct research, many more domains of general interest, an authoritative leader of schools, a respected cultural presence, a continuous promoter of the Romanian mathematical heritage and, at the same time, an energizer of the first steps of many young researchers, a perfect memory, a rigorous life style, active scientifically and physically at 75 (at a level which is tiring for many of his much younger colleagues) – all these have something to do with infinity...

Antonia, a 10 year old Romanian-Spanish girl from Tarragona, Spain, had a decisive influence in choosing the topic of the volume. Earlier than us, she called Professor Marcus *El infinito*, under circumstances pointing to his manner of approaching kids. In his opinion, a child should rediscover Zeno's paradoxes via a Socrates-type dialogue of the following kind: S.M.: "Look, Antonia, you have a bread, a usual one, and you eat today half of it. How much does it remain for tomorrow?" A.: "Half a bread, of course." S.M.: "Okay, but tomorrow you eat half of what you have got. Does some bread remain for after tomorrow?" A.: "Yes, sir, some bread still remains." S.M.: "Good, but after tomorrow you again eat half of the bread you have. Is it true that some bread still remains?" A. (already doubtful): "Yes, but"... After about five iterations, the two parts are always separated: A., or any other "victim", knowing for sure that (s)he can eat a *normal* bread in a day or so, hence any continuation of the dialogue is senseless (because it's breadless), S.M. looking desperately for the bright light of infinity in the others' eyes: eating every day half of the piece of bread you have got means securing bread for infinitely many days, isn't it?... In many cases, both sides are disappointed... In at least one case, S.M. got a surname: *El Infinito*...

Happy Birthday, *Profesore Infinito*!

Contents

Smoothing Data:
When Finite and Infinite Dimensional Help Each Other

Umberto Amato

Istituto per Applicazioni della Matematica CNR
Via Pietro Castellino 111, 80131 Napoli, Italy
E-mail: amato@iamna.iam.na.cnr.it

Dan Tudor Vuza

Institute of Mathematics of the Romanian Academy
P.O. Box 1-764, RO-70700 Bucharest, Romania
E-mail: dvuza@stoilow.imar.ro

1   Introduction

A huge number of applications require to solve the mathematical problem of removing noise from data. Therefore the problem has been widely considered and several methods have been developed for its solution under different points of view, e.g., statistics, approximation theory, inverse problems, just to cite some in the mathematical literature. Each field uses its own tools to solve the problem, i.e., nonparametric regression, approximation, regularization, respectively. The different points of view even gave rise to different denominations for the problem, e.g., smoothing data, de-noising (see in this respect the discussion in [13], where distinction between de-noising and smoothing data is made). In the present paper we shall be concerned with the problem of removing noise from data in the framework of inverse problems, where the problem is known as smoothing data, and indeed it is the simplest prototype.

The dilemma finite-infinite dimensional is classic in mathematics and obviously it also concerns smoothing data. Who could argue that infinite dimensional is more desirable? It is the dream of people to have a complete,

1

full representation of things, even the simplest (think of people who collect things, stamps, post-cards, and so on). It is mankind's dream translated into mathematical language, where it becomes a must. Most finite dimensional quantities rarely have an autonomous life and often live only as an approximation of infinite dimensional.

This is true also for smoothing data, where in addition we are generally faced with a real problem with real data. This has two peculiarities: a finite number of data, N, a finite (strictly greater than zero) noise. This means two things. First, the problem is finite-dimensional by its very nature, so that a finite-dimensional approach should be appropriate for its handling. Second, the type of convergence required is different from the one that is usually assumed in the general theory of inverse problems: we are not interested in the case when noise tends to zero, but in the convergence in some sense (maybe probabilistic) when the size of the sample increases.

Nevertheless, the analysis of the problem in the infinite dimensional setting has been an approach followed in the literature. There are at least two important reasons that go beyond the generality of the mathematicians' dream. First, there is a particular applicative interest when details finer than obtainable from the data only are required (e.g., when smooth curves are asked as in CAGD, or zooming in is needed in image processing). Second, exogenous information has to be introduced in the problem, often in an infinite dimensional setting, in order to remove the ambiguity due to the finite size of the sample, as currently done in discrete and semidiscrete inverse problems. A pioneer in this respect is the celebrated thin-plate spline, where exogenous information is given by the regularity of the solution expressed in terms of its p-th derivative. Then the (finite dimensional) thin-plate spline is the solution of the (infinite dimensional) problem

$$\min_{\tilde{f} \in H^p} \sum_{i=1}^{N} (\tilde{y}_i - y_i^\varepsilon)^2 + \lambda \|\tilde{f}^{(p)}\|_{L_2},$$

where H^p is the Sobolev space of regularity p, y_i^ε is a sample of an unknown function f in a proper grid of size N affected by noise and $\tilde{y}_i$ is a solution of the regularization problem collocated in the same grid.

Then we can say that, even though finite dimensional is the proper setting of this problem due to its experimental nature in general, however

infinite dimensional helps finite dimensional.

Recently wavelets have been considered as a tool for solving several mathematical problems, including noise removal. The interest is due to their properties to be localized both in time and frequency; moreover a fast algorithm can be developed for computing the wavelet transform, also suitable to be implemented on integrated circuits. What is interesting for noise removal is their capability to concentrate information in a limited amount of data; then pieces of not interesting information can be discarded without affecting too much the whole message whereas in this process a lot of the accompanying noise is also eliminated. This generates the asked noise suppression. Let us analyze in more detail this aspect.

Noise and signal are strongly linked and one knows indeed only their combination; then to reduce one implies reducing the other too, unfortunately. Luckily, noise and signal have very different characterizations for the majority of problems; in particular the signal has strong redundancies and correlations.

Therefore a suitable transformation of the signal gives very few, possibly independent, pieces of information and allows one to discard the redundant ones. Since the noise is intimately linked to the signal, the former is also discarded together with the redundant information and since noise does not exhibit the same redundancy as the signal, the part discarded with the latter is significant, giving rise to the noise removal. Then finite dimensional (more exactly, small finite dimensional) is the key for suppressing noise effectively. The question is how much finite? Of course it depends on the signal (the more the structures of signal and noise are different, the more noise removal is better), but in general we can say less, much less than the size of the sample. The ideal case would be fixed, not depending on the size of the sample at all. This would give high computational efficiency and best convergence rate (N^{-1}). However this (parametric) approach is not suitable when no enough information is available on the functional expression of the solution.

Therefore finite dimensional is something more than simply an obligation coming from the finite size nature of applications. It is the key for having effective noise removal. But surprisingly it is also the key for having effective infinite dimensional solutions. In fact we shall see that the infinite dimensional solution can be effectively defined only by a limit process of the finite dimensional problem when the size of the sample increases. An attempt to deal directly with the infinite dimensional setting gives rise to lack of convergence or discrepancies. Then we can say that

finite dimensional helps infinite dimensional

and the mutual help finite-infinite dimensional is complete.

2 Shrinking Functions and Inverse Problems

Let us introduce the noise removal problem

$$y_i^\varepsilon = y_i + \varepsilon_i, \ 1 \leq i \leq N, \tag{1}$$

where y is a sample of an unknown function f at some points and ε_i is a realization of noise having some suitable properties (initially we suppose white noise with variance σ^2). Then y^ε is the signal at our disposal, that is a sample of f affected by noise. In the following we suppose $N = 2^J$ and f sampled at equispaced points. Moreover we choose an orthogonal wavelet system and we indicate by $\{Y_{-1,0}, Y_{j,\ell}, 0 \leq j < J, 0 \leq \ell < 2^j\}$ the discrete wavelet transform (DWT) of $\{2^{-J/2}y_i\}_{i=1}^N$. For ease of notation we mainly indicate the DWT by $\{Y_k\}_{k=0}^{N-1}$, with k related to the scale and translation parameters by the relation $k = 0$ for $j = -1$ and $k = 2^j + \ell$ for $0 \leq j < J$, $0 \leq \ell < 2^j$. Y_k turns out to be an approximation of the continuous wavelet transform of f. Such an approximation is usual in all applications and justified by several papers [12], [20]; it is even better in the case of Coiflets basis [7]. Analogously $\underline{Y}^\varepsilon$ will denote the DWT of y^ε. Note that due to linearity of the model (1) and orthogonality of the transform, noise is white also in the wavelet domain (with variance $N^{-1}\sigma^2$).

Most methods for removing noise from signals based on the wavelet transform go through the following

Algorithm 2.1

1. Take the wavelet transform of the data $2^{-J/2}\underline{y}^\varepsilon$, yielding $\underline{Y}^\varepsilon$.

2. Shrink $\underline{Y}^\varepsilon$ according to

   $$\tilde{Y}_k = s(Y_k^\varepsilon; t)Y_k^\varepsilon, \ 0 \leq k < N, \tag{2}$$

   where $0 \leq s(Y_k^\varepsilon; t) \leq 1$ is a (generally nonlinear) shrinking function depending on a free parameter t.

3. Take the inverse wavelet transform $2^{J/2}\underline{\tilde{Y}}$, yielding $\tilde{\underline{y}}$, which is an estimate of the cleaned signal.

The reason why the association between the wavelet transform and Algorithm 2.1 became so popular in applications is provided by its computational speed and by the possibility to implement Algorithm 2.1 within integrated circuits. In the literature the number of shrinking functions developed by researchers is rapidly increasing; here we mention *hard, soft* [14],

[15], *DKLT* [11], *regularization* [6], [1], *firm* [16], *garrote* [8], to cite some. They were developed under different frameworks (statistics, approximation theory, inverse problems) with different purposes. All of them are nonlinear, except regularization that relies on tools taken from inverse problems methods (see the next section). Indeed, even though the shrinking functions were developed under such different frameworks, a unifying approach for them can be just regularization. In this respect in [2] it is proved that any shrinking method is the solution of a properly defined regularization problem: let us consider the prototype regularization problem

$$\min_{x \in \mathbf{R}} (x - y)^2 + 2g(x). \tag{3}$$

Then we have the following

Theorem 2.1 ([2], Th. 2.6) *Let* $r : \mathbf{R} \to \mathbf{R}$ *be an increasing antisymmetric function such that* $0 \le r(y) \le y$ *for* $y \in [0, \infty)$, $r(y) \to \infty$. *Then there is a continuous positive solid function* $g : \mathbf{R} \to \mathbf{R}$ *such that* $r(y)$ *is the unique solution of the problem* (3) *for every* y *at which* r *is continuous (a function* $f : X \to \mathbf{R} \cup \{\infty\}$ *is called solid if* $|x| \le |y| \Rightarrow f(x) \le f(y)$). *Moreover the expression of the function* g *underlying the regularization problem can be evaluated as*

$$g(x) = \int_0^x (s(t) - t)dt, \ x \ge 0,$$

with

$$s(x) = \sup\{y | r(y) \le x\}.$$

This means that the boundary among the several approaches to problem (1) broadens quite a lot, and that the approach pursued in the present paper, based on regularization, could provide a unifying framework for the problem of removing noise from data.

Finally we mention that convergence properties proved for the methods can be classified into two types, according to the purposes of the method; both consider the L_2-loss $L(\tilde{f}) := \|\tilde{f} - f\|^2 \simeq \|\tilde{y} - y\|^2 = \|\tilde{Y} - Y\|^2$ and the corresponding risk $\rho(\tilde{f}) := EL(\tilde{f})$, where E means taking the average value with respect to the noise distribution function. For regularization convergence holds when $L(\tilde{f}) \to 0$ asymptotically in some sense (in probability or in the average). For thresholding methods the convergence for an estimator $\tilde{f}$ is given in the framework of the minimax theory: let $\mathcal{F}[0, 1]$ be any of the function spaces $B_{p,q}^s[0, 1]$ (the Besov classes) or $F_{p,q}^s[0, 1]$ (the Triebel classes) and let $\mathcal{F}_c$ denote the ball of functions $\{f : \|f\|_{\mathcal{F}} \leq C\}$; moreover, let ψ be a wavelet system with basis elements in C^R and D vanishing moments so that $1/p < s < \min(R, D)$ holds. Then

Theorem 2.2 ([13], Th. 1.2) *For each ball $\mathcal{F}_c$ arising from an $\mathcal{F}[0, 1]$ as above, there is a constant $C_2(\mathcal{F}_c; \psi)$ not depending on N such that for all $N = 2^J$*

$$\sup_{f \in \mathcal{F}_c} \rho(\tilde{f}) \leq C_2 \log N \inf_{\hat{f}} \sup_{f \in \mathcal{F}_c} \rho(\hat{f}),$$

where $\hat{f}$ is any estimator.

3   Regularization

Let us consider the smoothing data problem (1), with $f \in H^p$. In [6], [1] a wavelet-based regularization method was introduced for solving the problem. The method was recently rediscovered in [9], [10]. Let f_N^ε be a function in the span of the first N wavelets and assuming the values y_k^ε at the design nodes. We consider the following penalization problem

$$\min_{\tilde{f} \in H^p} \|\tilde{f} - f_N^\varepsilon\|_{L_2}^2 + \lambda \|\tilde{f}\|_{H^p}^2, \tag{4}$$

with λ being the regularization parameter.

Considering equivalent norms in the wavelet space (see [18]), problem (4) can be written in the wavelet domain as

$$\min_{\tilde{Y}} \sum_{k=0}^{N-1} (\tilde{Y}_k - Y_k^\varepsilon)^2 + \lambda \sum_{k=0}^{N-1} a_{j(k)}^{2p} \tilde{Y}_k^2, \tag{5}$$

with

$$a_{j(k)} \equiv a_k := 2^{\max\{j(k),0\}} \tag{6}$$

and $\tilde{Y}_k = 0$ for $k \geq N$.

The solution of problem (5) is simply given by

$$\tilde{Y}_k = \frac{Y_k^\varepsilon}{1 + a_k^{2p}\lambda}. \tag{7}$$

Convergence of the regularized solution to the true one was proved according to the following

Theorem 3.1 ([6], Th. 3.1 and [1], Th. 3) *Let* $f \in H^p$, $p > \frac{1}{4}$. *Then*

$$
E\left[\|\tilde{f} - f\|^2_{L^2(0,1)}\right] \leq \left(\frac{\sigma^2}{N}\right)^{\frac{2p}{2p+1}} \left[D(p)\|f\|^{\frac{2}{2p+1}}_{H^p} + \left(\frac{\sigma^2}{N}\right)^{\frac{1}{2p+1}}\right] + \frac{\|f\|^2_{H^p}}{N^{2p}},
$$

for $\lambda = C(p)\left(\frac{\sigma^2}{\|f\|^2_{H^p}N}\right)^{\frac{2p}{2p+1}}$, *where* $C(p)$ *and* $D(p)$ *do not depend on* N.

Note that the rate of convergence $(N^{-2p/(2p+1)})$ is optimal in the framework of nonparametric regression [17].

Two different criteria were considered for the choice of the regularization parameter based on the available noise data only. In [1] an adaption of the GCV criterion [21] to the wavelet framework was considered: λ is chosen as the minimizer of $V_N(\lambda)$, where

$$
V_N(\lambda) = \frac{\|(I - R_N(\lambda))\,\underline{Y}^\varepsilon\|^2}{\left[\frac{1}{N-1}\mathrm{Tr}\,(I - R_N(\lambda))\right]^2},
$$

with $R_N(\lambda)$ being a diagonal matrix whose diagonal elements are given by $(1 + a_k^{2p}\lambda)^{-1}$. For this criterion the following asymptotic efficiency theorem has been proved.

Theorem 3.2 *Let* $f \in H^p$, $p > 1/2$. *Then the choice of* λ *provided by the GCV is asymptotically optimal in the sense that if* λ_N *is any minimizer of* $EV_N(\lambda)$, *then*

$$
\lim_{N \to \infty} \frac{\rho_{\lambda_N}(\tilde{f})}{\min_{\lambda \geq 0}\rho_\lambda(\tilde{f})} = 1,
$$

with $\rho_\lambda(\tilde{f})$ *being the risk of the regularized solution corresponding to the regularization parameter* λ.

In [6] the regularization parameter is chosen by the Mallows criterion endowed with an estimate of variance, σ^2_{MS}, based on finite differences of the sample $\underline{y}^\varepsilon$ [19]: the expression to be minimized by λ is

$$
\tilde{\Lambda}(\lambda) = \frac{\sigma^2_{\mathrm{MS}}}{N}\left(\sum_{j=-1}^{J-1} a_j \frac{1 - a_j^{2p}\lambda}{1 + a_j^{2p}\lambda}\right) + \sum_{k=1}^{N-1}\left[\left(\frac{a_k^{2p}\lambda}{1 + a_k^{2p}\lambda}\right)^2 (Y_k^\varepsilon)^2\right].
$$

For this criterion the following convergence theorem has been proved:

Theorem 3.3 *Let* $f \in H^p$, $p > 1/2$. *Then*

$$\tilde{\Lambda}(\lambda) = \rho_\lambda(\tilde{f}) + \left(\frac{\log N}{N^{2p}}\right).$$

From Theorem 3.1 it is clear that the optimal regularization parameter goes to zero with N. In the infinite dimensional case $\lambda = 0$ and the regularized solution (7) is just the sample affected by noise; of course this does not converge to the true function (recall that we are interested in convergence when noise is finite). Therefore the infinite dimensional solution of the regularization problem can be properly defined only as the limit of the finite dimensional solution when the sample grows. A similar behaviour also happens in shrinking methods based on thresholds (e.g., *hard, soft*), where the asymptotic solution is defined as the limit when the threshold goes to 0 and it is different from the value corresponding to the threshold value $t = 0$.

In [5] it is proved that convergence of the regularized solution endowed with the GCV criterion also holds in the H^q-norm, $0 \le q < p$, that is

Theorem 3.4 ([5], Proposition 3.2) *If* $f \in H^{p+q} \setminus \{0\}$, $0 \le p < q$, *then*

$$\lim_N E\left[\|\tilde{f} - f\|^2_{H^{p+q}}\right] \to 0.$$

Therefore convergence of the regularized solution to the true function holds, but in addition the same happens also for some of its derivatives, which is interesting for those applications where derivatives are useful for extracting features from a signal.

The wavelet regularization method has been generalized to the case of correlated noise in [4]. The shrinking function is modified there as

$$\tilde{s}(Y_k^\varepsilon; \lambda) = \frac{1}{1 + a_k^{2p} v_k^2 \lambda}, \qquad 0 \le k < N, \tag{8}$$

with v_k^2 being the diagonal elements of the wavelet transform of the input error covariance matrix. In order to estimate the regularization parameter a modification of the GCV criterion has been introduced in [4], CGCV, as

$$\mathrm{CV}_N(\lambda) = \frac{\|(I - R_N(\lambda))\,\underline{Y}^\varepsilon\|_2^2}{\left\{\frac{1}{N}\mathrm{Tr}\left[C\left(I - R_N(\lambda)\right)\right]\right\}^2}$$

with C being a diagonal matrix whose non null elements are given by v_k^2.

In the case of correlated noise analog theorems on convergence (at the optimal rate) and asymptotic efficiency of the CGCV criterion have been stated. In [4] an algorithm that computes v_k^2 (that is the diagonal part of the transformed covariance matrix) in $O(N)$ operations, instead of the $O(N^2)$ required to compute the whole matrix, is introduced.

A further generalization has also been proposed in [3], aimed at making the wavelet regularization method fully adaptive. Indeed the shrinking function should assume high values when the noise affecting a coefficient is small with respect to the true coefficient itself (signal to noise ratio) and low values in the opposite situation. In the regularization shrinking function this is accomplished through the regularization parameter λ (that depends on the overall variance of noise affecting data, see Theorem 3.1), the coefficients v_k^2 (that account for noise correlation), and the values a_k^{2p} (that in some sense account for a certain decrease of the true wavelet coefficients according to the regularity of the underlying function). It is just this dependence on the true function that is not taken into account completely: for example the coefficients a_k^{2p} only depend on the scale but not on the translation parameter, so that any dependence of the shrinking function on the latter is completely missing. Therefore the following shrinking function has been introduced in [3] that generalizes (7) and (8):

$$\tilde{s}(Y_k^\varepsilon; \lambda) = \frac{1}{1 + a_k^{2p} t_k \lambda}, \qquad 0 \le k < N, \tag{9}$$

with t_k satisfying $0 < c_1 \le t_k \le c_2 < \infty$ for any k and N.

The choice of t_k based on the data only is an open problem; the optimal t_k gives rise to the known *oracle*

$$s^{\text{oracle}}(Y_k) = \frac{1}{1 + \frac{v_k^2 \sigma^2 / N}{Y_k^2}} \tag{10}$$

that depends on the (true) signal to noise ratio for each coefficient. As such, Eq. (10) can be used only for theoretical purposes with synthetic data. However, when the true (unknown) coefficients Y_k^2 are approximated in some way, Eq. (10) can yield plenty of shrinking functions, including known and possibly new ones.

4 Numerical Experiments

In the present section we show the results of numerical experiments based on synthetic data. We shall pursue the following purposes: to validate the GCV regularization method basing on significant examples; to understand a possible improvement of the method achievable by introducing fully adaptability of the coefficients; to compare the performance of wavelet regularization vs. Fourier regularization.

Function	Expression
Blocks	$\sum_{k=1}^{11} \alpha_k(1 + \text{sgn}(x - \beta_k))$
Bumps	$\sum_{k=1}^{11} \gamma_k(1 + \zeta_k\lvert x - \eta_k\rvert)^{-4}$
Heavysine	$4\sin(4\pi x) - \text{sgn}(x - 0.3) - \text{sgn}(0.72 - x)$
Doppler	$\sqrt{x(1 - x)}\sin(2.1\pi/(x + 0.05))$
Ramp	$x,\ x \leq x_0;\ x - 1$ otherwise
Cusp	$\sqrt{\lvert x - x_0\rvert}$
Sing	$\lvert x - x_0\rvert^{-1}$
HiSine	$\sin(0.6902N\pi x)$
LoSine	$\sin(0.3333N\pi x)$
Leopold	$\delta(x - x_0)$
Riemann	$\Re(\text{IFFT}(\underline{z}));\ z_{i^2} = 1/i,\ 1 \leq i \leq N;\ z_i = 0$ otherwise
Piece-regular	$\begin{array}{ll} -70\exp(-(x - 0.5)^2/2\sigma^2), & 0 \leq x \leq 1/7 \\ -35\exp(-(x - 0.5)^2/2\sigma^2), & 1/7 < x \leq 1/5 \\ -70\exp(-(x - 0.5)^2/2\sigma^2), & 1/5 < x \leq 1/3 \\ -15\text{Bumps}(x), & 1/3 < x \leq 1/2 \\ -\exp(4x), & 1/2 < x \leq 7/12 \\ -\exp(5 - 4x), & 7/12 < x \leq 2/3 \\ 25, & 11/15 < x \leq 49/60 \\ \exp(4x) - \exp(4), & 49/60 < x \leq 403/420 \\ 0, & \text{otherwise} \end{array}$
Piece-polynomial	$\begin{array}{ll} 20(x^3 + x^2 + 4), & 0 \leq x \leq 0.05 \\ 10, & 0.05 < x \leq 0.15 \\ 20(x^3 + x^2 + 4), & 0.15 < x \leq 0.2 \\ 10(0.6 - x)^3 + 45, & 0.2 < x \leq 0.4 \\ 40(2x^3 + x) + 100, & 0.4 < x \leq 0.6 \\ 16x^2 + 8x + 16, & 0.6 < x \leq 0.8 \\ 36 - 20x, & 0.8 < x \leq 0.9 \\ 150, & 0.9 < x \leq 0.95 \\ 36 - 20x, & 0.95 < x \leq 1 \end{array}$

Table 1: List of considered functions, $f(x)$, $0 \leq x \leq 1$, and corresponding expressions. In the table $\underline{\alpha}$, $\underline{\beta}$, $\underline{\gamma}$, $\underline{\zeta}$, $\underline{\eta}$ are suitable vectors, $x_0 = 0.37$, $\delta(x)$ is the Dirac function, IFFT is inverse Discrete Fourier transform, $\sigma = 6/40$.

It is clear from the paper that wavelet regularization is introduced for *smooth* functions ($f \in H^p$, $p > 1/4$). Nevertheless, many interesting signals in applied sciences do not share such a smoothness feature, in general. Therefore, in order that regularization be a candidate for removing noise from real signals, its performance has be to evaluated on non smooth functions. In the present paper we consider several of them, summarized in Table 1. They have been extensively used in the current literature or implemented in wavelet software.

We sampled each test function in $N = 2^J$ equispaced points according to the model (1), with variance of noise affecting data such that the signal to noise ratio is 7.

In order to evaluate the performance of the regularization method, we introduce the index I,

$$I = \sqrt{\frac{\sum_{k=0}^{N-1}(\tilde{Y}_k - Y_k)^2}{\sum_{k=0}^{N-1} Y_k^2}}. \tag{11}$$

Index I is an estimate of the relative (square root) L_2-loss and therefore it is less sensitive to the particular function. In order to damp influence of the random noise, 100 different realizations of noise were generated and index I^2 was averaged over them (estimate of the L_2-risk). Of course for a fair comparison all sets of tests were run on the same noise sequence.

Another interesting index considered for evaluating performance is the estimate of the convergence rate, γ, of the L_2-risk, ρ; it is computed by least squares fitting the experimental estimate of the risk as a function of the sample size according to the law $\rho = \alpha N^{-\gamma}$. Note that for *smooth* functions γ should be related to the smoothness index of a function, p_f (generally unknown), as $\gamma = 2p_f/(2p_f+1)$, provided that the wavelet system is regular enough and the actual p used in regularization satisfies $p \geq p_f$. The fit (and then convergence rate) is significant only when convergence holds, which, as we shall see, is not always the case.

Table 2 shows the indexes I for a sample size $N = 2^{12}$ and γ (asymptotic value) for the functions of Table 1. The values of I shown correspond to regularization endowed with the Generalized Cross Validation criterion for estimating the regularization parameter. This criterion, even though defined on an asymptotic basis, is highly efficient even for $N \geq 2^{11}$; the differences between the digits shown and optimal regularization (i.e., when the regularization parameter is chosen optimally, which is possible in simulation) are negligible. Moreover $p = 2$ was used in regularization.

Function	REGULARIZATION		ORACLE	
	I $(N = 10^{12})$	γ	I $(N = 10^{12})$	γ
Blocks	0.077	0.50	0.034	0.74
Bumps	0.076	0.79	0.035	0.82
Heavysine	0.034	0.50	0.016	0.79
Doppler	0.054	0.78	0.023	0.84
Ramp	0.066	0.50	0.015	0.83
Cusp	0.020	0.56	0.013	0.78
Sing	N.A.	N.A.	0.017	0.83
HiSine	N.A.	N.A.	N.A.	N.A.
LoSine	N.A.	N.A.	N.A.	N.A.
Leopold	N.A.	N.A.	0.016	0.84
Riemann	0.12	0.30	0.086	0.30
Piece-regular	0.063	0.52	0.029	0.79
Piece-polynomial	0.083	0.49	0.032	0.77

Table 2: Numerical experiments obtained by wavelet system. For several test functions the error index (Eq. (11)) is shown for $N = 2^{12}$ together with its convergence rate, ρ, when applicable. Lack of convergence is indicated by Not Applicable (N.A.). Values are shown for GCV regularization and for the *oracle* solution.

First we note from the table that convergence indeed holds even when the functions are not smooth, at a convergence rate that strongly depends on the function itself, of course. Paradoxically, the most interesting indications come from failures of the method, that occur with highly oscillating functions (HiSine and LoSine) and highly peaked ones (Sing and Leopold). A natural question is whether failures depend on the wavelet regularization method chosen (and accordingly whether the L_2-risk and the convergence rate can be improved). To this purpose we estimated the analog indexes I and γ for the *oracle* solution (Eq. (10), also shown in Table 2). They are to be intended as a trend in a possible improvement achievable by wavelet methods for removing noise from data based on other (nonlinear) shrinking functions (e.g., adaptive regularization, Hard-Thresholding, Soft-Thresholding). Comparison of left and right columns of Table 2 shows that this is the case indeed. Convergence now holds also for highly peaked functions (but again not for highly oscillating functions) and the rate is much better in general. We also make the interesting observation that most functions exhibit a smooth-like convergence rate of about 0.8, corresponding to a smoothness-like parameter of about 2. In some sense it is an experimental evidence that to consider regularization with parameter $p = 2$ is a good choice (this also applies to spline smoothing).

Wavelet based methods are only a class for solving the problem of noise removal from data. Before their astonishing success, the methods described or cited in the present paper were earlier employed in conjunction with the Fourier transform instead of the wavelet transform. Debate is active in the scientific community concerning the performance of wavelet based methods against Fourier based ones and selecting the applications for which a definite answer concerning the best performing one can be given. In the present section we attempt an experimental answer, by performing the same numerical experiments shown in Table 2 by means of a transform in the Fourier domain. Results are shown in Table 3.

Function	REGULARIZATION		ORACLE	
	$I\ (N = 10^{12})$	γ	$I\ (N = 10^{12})$	γ
Blocks	0.079	0.51	0.066	0.51
Bumps	0.079	0.81	0.064	0.81
Heavysine	0.034	0.51	0.029	0.51
Doppler	0.052	0.81	0.047	0.81
Ramp	0.064	0.52	0.056	0.52
Cusp	0.015	0.68	0.013	0.68
Sing	N.A.	N.A.	N.A.	N.A.
HiSine	N.A.	N.A.	0.047	0.47
LoSine	N.A.	N.A.	0.047	0.47
Leopold	N.A.	N.A.	N.A.	N.A.
Riemann	0.12	0.33	0.040	0.48
Piece-regular	0.064	0.53	0.054	0.53
Piece-polynomial	0.083	0.53	0.067	0.53

Table 3: Numerical experiments obtained by Fourier system. For several test functions the error index (Eq. (11)) is shown for $N = 2^{12}$ together with its convergence rate, ρ, when applicable. Lack of convergence is indicated by Not Applicable (N.A.). Values are shown for GCV regularization and for the *oracle* solution.

The first (positive) remark for Fourier regularization is that the method is very powerful among other nonlinear shrinking functions, since the estimated risk approaches the ideal one within 20% in general. Also the convergence rate is the same. There are no appreciable differences between wavelet and Fourier regularization (compare the left columns of Tables 2 and 3). However comparison of the right columns of the same tables clearly shows that Fourier methods are much less performing than wavelet ones in removing noise, both in terms of risk and convergence rate. This immediately means that there are significant improvements that are possible by nonlinear methods (e.g., adaptive regularization, thresholding methods) only in the wavelet domain. This confirms experimentally that the association between the wavelet transform and shrinking functions can be a good candidate for noise removal.

Finally we note that the Fourier methods are not applicable in the case of highly peaked functions (where wavelet regularization failed), even though linear regularization fails in cleaning noise from the signal.

We also remark that several other functions were considered for the tests, not shown here, mainly concerned with different kinds of chirps. For them full failure was detected with both wavelet and Fourier methods, so that specific methods have to be developed for such signals.

5   Conclusions

The present paper dealt with the problem of removing noise from signals. By its very nature, the problem is strictly finite dimensional. However a quite intriguing mutual aid finite-infinite dimensional is established in solving the problem. A wavelet based regularization method that removes noise from data is discussed in the paper. Convergence properties are given together with a data driven criterion for estimating the optimal regularization parameter. Moreover a generalization of the method to the case of correlated noise and to adaptive regularization is presented. Wavelet methods directly compete with analogous methods that can be developed by relying on a transform in the Fourier domain. So far we do not know about theoretical results concerning a comparison of accuracy between the two methods (e.g., examples can be worked out where Fourier regularization works better than wavelet regularization). A numerical comparison is shown in the present paper on the basis of a wide set of test functions. The main conclusions from the experiments are that wavelet and Fourier regularization behave quite similarly, as far as both accuracy and convergence rate are concerned. However, when adaptive methods are considered, wavelet methods show a much bigger potential for improvement with respect to Fourier methods, both in accuracy and convergence rate. Moreover wavelet methods are able to clean noise also from highly peaked signals, whereas Fourier methods are to be preferred for pure trigonometric signals, as could be expected.

References

[1] U. Amato, D. T. Vuza, Wavelet approximation of a function from samples affected by noise, *Rev. Roumaine Math. Pures Appl.*, **42** (1997), 481–493.

[2] U. Amato, D. T. Vuza, Besov regularization, thresholding and wavelets for smoothing data, *Numer. Funct. Anal. and Optimiz.*, **18** (1997), 461–493.

[3] U. Amato, M. R. Occorsio, D. T. Vuza, Fully adaptive wavelet regularization for smoothing data, *Rend. Circ. Matem. Palermo Serie II*, **52** (1998), 207–222.

[4] U. Amato, D. T. Vuza, Smoothing data with correlated noise by wavelet regularization, *Rev. Roumaine Math. Pures Appl.*, **43** (1998), 253–275.

[5] U. Amato, D. T. Vuza, Wavelet simultaneous approximation from samples affected by noise, *Computers & Mathematics with Applications*, **36** (1998).

[6] A. Antoniadis, Smoothing noisy data with tapered Coiflet series, *Scand. J. Stat.*, **23** (1996), 313–330.

[7] G. Beylkin, R. Coifmann, V. Rokhlin, The fast wavelet transform and numerical algorithms, *Comm. Pure Appl. Math.*, **44** (1991), 141–163.

[8] L. Breiman, Better subset regression using the nonnegative garrote, *Technometrics*, **37** (1995), 373–384.

[9] L. T. Dechevsky, S. I. Penev, *On Penalized Wavelet Estimation*, Report No. S98-15, Department of Statistics, School of Mathematics, The University of New South Wales, 1998.

[10] L. T. Dechevsky, S. I. Penev, *Weak Penalized Least Squares Wavelet Regression Estimation*, Report No. S99-1, Department of Statistics, School of Mathematics, The University of New South Wales, 1999.

[11] R. A. DeVore, G. Kyriazis, D. Leviatan, V. M. Tikhomirov, Compression and nonlinear n-widths, *J. Adv. Comp. Math.*, **1** (1993), 197–214.

[12] D. L. Donoho, *Interpolating Wavelet Transforms*, Technical Report 480, Department of Statistics, Stanford University, Stanford, 1992.

[13] D. L. Donoho, De-noising by soft-thresholding, *IEEE Trans. Inf. Th.*, **41** (1995), 613–627.

[14] D. L. Donoho, I. M. Johnstone, Ideal spatial adaption by wavelet shrinkage, *Biometrika*, **81** (1994), 425–455.

[15] D. L. Donoho, I. M. Johnstone, Adapting to unknown smoothness via wavelet shrinkage, *J. Amer. Statist. Assoc.*, **90** (1995), 1200–1224.

[16] H. Y. Gao, A. G. Bruce, WaveShrink with firm shrinkage, *Stat. Sinica*, **7** (1997), 855–874.

[17] I. A. Ibragimov, R. Z. Khasminskii, Nonparametric regression estimation, *Dokl. Akad. Nauk SSSR*, **252** (1980), 780–784.

[18] Y. Meyer, *Ondelettes et opérateurs*, Hermann, Paris, 1990.

[19] H. G. Muller, U. Stadtmuller, Estimation of heteroscedasticity in regression analysis, *Ann. Statist.*, **15** (1987), 610–625.

[20] W. Sweldens, R. Piessens, Quadrature formulae and asymptotic error expansion for wavelet approximations of smooth functions. *SIAM J. Numer. Anal.*, **31** (1994), 1240–1264.

[21] G. Wahba, *Spline Models for Observational Data*, SIAM, Philadelphia, 1990.

Rearrangements of Real Functions Derived from the Combinatorics of Young Tableaux

Hari Bercovici

Mathematics Department
Indiana University
Bloomington, IN 47405, USA
E-mail: bercovic@indiana.edu

Abstract. The relationship between Young tableaux and longest increasing subsequences is extended to measurable functions on an interval. We also extend Knuth equivalence to this context, and we show that every function is Knuth equivalent to a function associated with a continuous Young tableau.

1 Introduction

Young tableaux and their rich combinatorics are related to diverse areas of mathematics, including representation theory, symmetric functions, the geometry and topology of Grassmannians, and module theory. More recently it was noted in [1] that a continuous version of Young tableaux is the key to the study of certain lattices of invariant subspaces of operators on Hilbert space. What played a major role in [1] was an extension to continuous diagrams of the classical Littlewood–Richardson rule. Our purpose in this paper is to find continuous analogues of other aspects of the combinatorics of Young tableaux. We will extend to this context Knuth equivalence and a version of the Schensted bumping algorithm. We will recall in Section 2 the basics of Young tableaux, and develop our continuous analogue in Section 3. It is worthwhile to note that, while the statements in Section 3 involve real-valued functions, the proofs are done by reducing to the combinatorial case. It may be interesting to search for direct proofs.

2   Knuth Equivalence and Young Tableaux

Our basic reference on Young tableaux is the book [2] where all the results in this section are proved. Some of the original sources are [3] and [4]. In this section we will consider words $w = w_1 w_2 \cdots w_n$ over the alphabet of real numbers, i.e., $w_1, w_2, \ldots, w_n \in \mathbf{R}$. As usual, uv indicates the concatenation of u and v. Knuth equivalence $\equiv_K$ on words is the smallest equivalence relation with the following two properties:

(a) if $x, y, z \in \mathbf{R}$, $x \leq z < y$, and u, v are words, then $uxyzv \equiv_K uyxzv$;

(b) if $x, y, z \in \mathbf{R}$, $z < x \leq y$, and u, v are words, then $uxyzv \equiv_K uxzyv$.

     A basic result establishes a complete set of representatives of equivalence classes of words relative to Knuth equivalence. These representatives are words associated with Young tableaux. To explain how these are constructed, recall that a diagram is a left-justified collection of boxes placed in rows with $\lambda_1, \lambda_2, \ldots, \lambda_k$ boxes, where $\lambda_1 \geq \lambda_2 \geq \cdots \geq \lambda_k$. A (column strict) Young tableau is obtained by placing a number in each box of a diagram in such a way that the rows are increasing and the columns are strictly increasing. Thus, each number in the tableau is strictly less than the one above (if any), and at least equal to the one to its left (if any). Given a tableau T over a diagram with k rows, let us call w_j the word obtained by reading the elements in the jth row from left to right. The word $w(T)$ of T is defined as

$$w(T) = w_k w_{k-1} \cdots w_1.$$

Theorem 2.1 *For every word w there exists a unique Young tableau T such that $w \equiv_K w(T)$.*

     There are several algorithms for calculating T from w in the above theorem. One is given by Schensted bumping and it can be done fairly quickly by hand. Schensted was actually interested in another aspect of words w, namely in their increasing subsequences. Assume that the tableau T has row lengths $\lambda_1, \lambda_2, \ldots, \lambda_k$, and $w \equiv_K w(T)$. Then w has increasing subsequences of length λ_1 and no longer increasing subsequences. More generally, denote by $L_j(w)$ the largest sum of the form $n_1 + n_2 + \cdots + n_j$ such that w has disjoint increasing subsequences of lengths $n_1, n_2, \ldots, n_j$. Then

$$L_j(w) = \lambda_1 + \lambda_2 + \cdots + \lambda_j.$$

This gives a method for calculating the diagram of T. In order to calculate T itself we need an additional observation. For a fixed real number α denote by w^α the word obtained from w by deleting all the letters which are greater than or equal to α.

Proposition 2.1 *If $w \equiv_K w'$, then $w^\alpha \equiv_K w'^\alpha$ for every real number α.*

This, combined with the fact that w^α is a tableau word if w is one, can now be used to determine the tableau T with $w \equiv_K w(T)$. It will be convenient to set $L_0(w) = 0$ for all w.

Proposition 2.2 *Let w be a word and T a Young tableau such that $w \equiv_K w(T)$, say $w(T) = w_k w_{k-1} \cdots w_1$, where w_j are the row words of T. For $j = 1, 2, \ldots, k$ and real α, w_j has precisely $L_j(w^\alpha) - L_{j-1}(w^\alpha)$ elements strictly less than α.*

To see how this works, take the word $w = 236145$ (each digit is a letter) and note that $L_1(w) = 4$ and $L_2(w) = 6$, which gives $\lambda_1 = 4$ and $\lambda_2 = 2$. Look next at $w^6 = 23145$ for which $L_1 = 4$ and $L_2 = 5$. This indicates that 6 belongs in the last box of the second row of T. Continuing in this way we find that the top row of T is 1345 and the bottom row 26, so that $w(T) = 261345 \equiv_K 236145$.

This method of calculating T is not particularly efficient, but I chose to present it because it is the one which is most easily extended to measurable functions.

3 Knuth Equivalence of Measurable Functions

In this section the discrete words of the preceding section will be replaced by "continuous" words. Such a continuous word is simply a real valued, Lebesgue measurable function f with bounded domain $D(f) \subset \mathbf{R}$. Just as words are entirely determined by their letters, the relevant aspect of f will be determined by its values, not by the parametrization. Thus, two functions f and g will be considered identical if there exists an essentially invertible, measure preserving map $\varphi : D(f) \to D(g)$ such that $f = g \circ \varphi$ almost everywhere. Thus any function f can be replaced by another one defined on the interval $(0, \ell)$, where $\ell = |D(f)|$ (we denote by $|\sigma|$ the Lebesgue measure of σ). Indeed, the function $\varphi : D(f) \to (0, \ell)$ defined by $\varphi(t) = |D(f) \cap (-\infty, t)|$ is essentially invertible and measure preserving.

Concatenation is defined in the following way. A function f is the concatenation of g and h if $D(f) = D(g) \cup D(h)$, $D(g)$ lies to the left of $D(h)$ (i.e., $x < y$ if $x \in D(g)$ and $y \in D(h)$), $f|D(g) = g$, and $f|D(h) = h$. We will indicate this by writing $f = g \sqcup h$. Given functions g and h, one can always form $g \sqcup h$; one might have to shift the domains so that the domain of h is on the right. More generally, given any family $(f_i)_{i \in I}$ of functions indexed by a totally ordered countable set I, one can form the concatenation $\bigsqcup_{i \in I} f_i$ provided that $\sum_{i \in I} |D(f_i)| < \infty$. In this construction $D(f_i)$ must precede $D(f_j)$ if $i < j$. In this paper we will only use infinite concatenations indexed by the natural numbers with the reverse ordering, i.e., concatenations of the form

$$\bigsqcup_{n=\infty}^{1} f_n = \cdots \sqcup f_n \sqcup \cdots \sqcup f_2 \sqcup f_1.$$

We now define the continuous version of Young tableaux. Consider a sequence $T = (f_n)_{n=1}^{\infty}$ of measurable functions, with $D(f_n) = (0, \lambda_n)$, where $\lambda_n \geq 0$ for $n = 1, 2, \ldots$ Then T will be called a (column strict) Young tableau if the following conditions are satisfied:

(i) $\lambda_1 \geq \lambda_2 \geq \cdots$;

(ii) $\sum_{n=1}^{\infty} \lambda_n < \infty$;

(iii) each f_n is increasing (not necessarily strictly);

(iv) $f_n(x) \leq f_{n+1}(x)$, for all $x \in (0, \lambda_{n+1})$; and

(v) if $x < y < \lambda_{n+1}$, then $f_n(x) < f_{n+1}(y)$.

A few comments are in order. Increasing functions have at most countably many discontinuities, so it is always possible to assume that the functions f_n in a Young tableau are continuous from the right or from the left. To visualize conditions (iii) and (iv), let us consider the extended graph of f_n. This is a continuous curve consisting of all points (x, y) such that $x \in (0, \lambda_n)$ and $\lim_{t \uparrow x} f_n(t) \leq y \leq \lim_{t \downarrow x} f_n(t)$. If f_n is bounded below, one also adds the points of coordinates $(0, y)$ with $y \leq \lim_{t \downarrow 0} f_n(t)$. Similarly, if f_n is bounded above, the points (λ_n, y) with $y \geq \lim_{t \uparrow \lambda_n} f_n(t)$ are added to the extended graph of f_n. (In the language of monotone operators, the extended graph of f_n is a maximal monotone extension of f_n with the smallest possible domain. The range of this extension is the whole real line.) Condition (iii) indicates that the extended graph of f_{n+1} lies above that of f_n, while condition (iv) requires that these two graphs have no horizontal segment in common. These conditions can be expressed in terms of the distribution functions of the f_n. The distribution function μ_n of f_n is simply defined by $\mu_n(\alpha) = |\{t \in (0, \lambda_n) : f_n(t) < \alpha\}|$. The function μ_n is defined on the whole line, it is increasing and continuous from the left, and its extended graph is the reflection of the extended graph of f_n in the line $x = y$. Thus the conditions on f_n correspond with the fact that the extended graph of μ_n lies above the extended graph of μ_{n+1}, and these extended graphs have no vertical segments in common. In other words,

(a) $\mu_n(\alpha) \geq \mu_{n+1}(\alpha)$ for all α; and

(b) $\mu_n(\alpha) \geq \lim_{\beta \downarrow \alpha} \mu_{n+1}(\beta)$ for all α.

The alert reader will have noticed that condition (a) is in fact a consequence of (b) since the functions in question are increasing and continuous from the left. Similarly, condition (iii) on the functions f_n follows from (iv) if the functions f_n are assumed continuous from the left. In the following, the sequence $(\lambda_n)_{n=1}^\infty$ will be called the shape of the tableau T.

Given a Young tableau $T = (f_n)_{n=1}^\infty$, one can form the tableau function $f_T = \bigsqcup_{n=\infty}^1 f_n$. We can always think of f_T as being defined on the interval $(0, \ell)$, with $\ell = \sum_{n=1}^\infty \lambda_n$. If we set $\ell_n = \sum_{j=n}^\infty \lambda_j$, we have $f_T(t) = f_n(t - \ell_{n+1})$ for $t \in (\ell_{n+1}, \ell_n)$.

Given a function f and a real number α we will denote by f^α the restriction of f to $\{t \in D(f) : f(t) < \alpha\}$. If $T = (f_n)_{n=1}^\infty$, we also set $T^\alpha = (f_n^\alpha)_{n=1}^\infty$. The following result is left as an exercise for the reader.

Lemma 3.1 *If T is a Young tableau then T^α is also a Young tableau.*

Fix now a measurable function f and a natural number n. We will denote by $L_n(f)$ the least upper bound of all sums of the form $|\sigma_1| + |\sigma_2| + \cdots + |\sigma_n|$, where $\sigma_1, \sigma_2, \ldots, \sigma_n$ are measurable, pairwise disjoint subsets of $D(f)$ such that $f|\sigma_j$ is increasing for every j. For convenience we set $L_0(f) = 0$. Observe that it is possible that $L_n(f) = 0$ for all n, even though $|D(f)| \neq 0$. For instance, f could be strictly decreasing.

Lemma 3.2 *If T is a Young tableau of shape $(\lambda_n)_{n=1}^\infty$, then*

$$L_n(f_T) = \lambda_1 + \lambda_2 + \cdots + \lambda_n.$$

Proof. The inequality $L_n(f_T) \geq \lambda_1 + \lambda_2 + \cdots + \lambda_n$ is obvious. With the notation used in the definition of f_T, let $\sigma \subset (0, \ell)$ be a measurable set such that $f_T | \sigma$ is increasing. Up to a countable set we have

$$\sigma = \bigcup_{k=1}^{\infty} [\sigma \cap (\ell_{k+1}, \ell_k)] = \bigcup_{k=1}^{\infty} [\sigma^k + \ell_{k+1}],$$

where $\sigma^k \subset (0, \lambda_k)$. We claim that the sets σ^k are essentially pairwise disjoint. In fact σ^k and $\sigma^{k'}$ cannot have more than one point in common if $k \neq k'$. Indeed, if $x, y \in \sigma^k \cap \sigma^{k'}$ with $x < y$ and $k < k'$ we must have

$$f_{k+1}(y) \leq f_{k'}(y) = f_T(y + \ell_{k'+1}) \leq f_T(x + \ell_{k+1}) = f_k(x),$$

where we used $k' \geq k + 1$, the definition of f_T, the inequality $x + \ell_{k+1} \geq y + \ell_{k'+1}$, and the fact that f_T is increasing on σ. This inequality is not possible by property (iv) of Young tableaux. It follows that $|\sigma| = |\sigma'|$, where $\sigma' = \bigcup_{k=1}^{\infty} \sigma^k \subset (0, \lambda_1)$. Note that $\sigma' = u(\sigma)$, where $u : (0, \ell_1) \to (0, \lambda_1)$ is defined by $u(t) = t - \ell_{k+1}$ for $t \in (\ell_{k+1}, \ell_k)$. The above argument shows that $u | \sigma$ is essentially one-to-one on σ if $f_T | \sigma$ is increasing. Note furthermore that $u^{-1}(t)$ contains exactly j points if $t \in (\lambda_{j-1}, \lambda_j)$.

Consider now pairwise disjoint sets $\sigma_1, \sigma_2, \ldots, \sigma_n \subset (0, \ell_1)$ such that $f | \sigma_j$ is increasing for $j = 1, 2, \ldots, n$. The above argument shows that

$$\chi_{u(\sigma_1)}(t) + \chi_{u(\sigma_2)}(t) + \cdots + \chi_{u(\sigma_n)}(t) \leq j,$$

for $t \in (\lambda_{j-1}, \lambda_j)$, and therefore

$$\chi_{u(\sigma_1)} + \chi_{u(\sigma_2)} + \cdots + \chi_{u(\sigma_n)} \leq \chi_{(0,\lambda_1)} + \chi_{(0,\lambda_2)} + \cdots + \chi_{(0,\lambda_n)}$$

almost everywhere. Therefore

$$\sum_{j=1}^{n} |\sigma_j| = \sum_{j=1}^{n} |u(\sigma_j)| = \int \sum_{j=1}^{n} \chi_{u(\sigma_j)} \leq \int \sum_{j=1}^{n} \chi_{(0,\lambda_j)} = \sum_{j=1}^{n} \lambda_j,$$

and this implies the desired inequality $L_n(f) \leq \sum_{j=1}^{n} \lambda_j$. $\square$

It will be useful to know that the definition of $L_n(f)$ is related with the corresponding L_n defined for words. The relevant result is as follows.

Lemma 3.3 *Let $w = w_1 w_2 \cdots w_k$ be a word, and let f be a function defined on the interval $(0, \ell)$ such that $f(t) = w_j$ for $t \in ((j-1)\ell/k, j\ell/k)$. Then $L_n(f) = (\ell/k) L_n(w)$ for every natural number n.*

Proof. Any increasing subsequence of length j of w gives rise to set σ with $|\sigma| = (\ell/k)j$ such that $f|\sigma$ is increasing; σ simply consists of a union of j intervals of length ℓ/k. Moreover, disjoint subsequences determine disjoint subsets. This proves the inequality $L_n(f) \geq (\ell/k)L_n(w)$. For the opposite inequality, assume that $\sigma_1, \sigma_2, \ldots, \sigma_n$ are pairwise disjoint sets such that the restrictions $f|\sigma_i$ are increasing. If a point of one of the intervals $I_j = ((j-1)\ell/k, j\ell/k)$ belongs to one of the sets σ_i, we can add the whole interval I_j to σ_i, and f will still be increasing on the resulting set. In order to keep the sets σ_i disjoint, the parts that are added to a set σ_i must be deleted from the other sets; this does not decrease the sum of the measures of these sets. Applying this procedure at most k times we arrive at a new collection $\tilde{\sigma}_1, \tilde{\sigma}_2, \ldots, \tilde{\sigma}_n$ of pairwise disjoint sets, each one arising from an increasing subsequence of w, and with $\sum_{j=1}^{n} |\tilde{\sigma}_j| \geq \sum_{j=1}^{n} |\sigma_j|$. This proves the opposite inequality $L_n(f) \geq (\ell/k)L_n(w)$.   □

For an arbitrary function f let us denote

$$\lambda_n(f) = L_n(f) - L_{n-1}(f), \quad n = 1, 2, \ldots$$

Lemma 3.2 says simply that the numbers $\lambda_n(f_T)$ give precisely the shape of the tableau T.

Theorem 3.1 *For any measurable function f, the sequence $(\lambda_n(f))_{n=1}^{\infty}$ is decreasing.*

Proof. We need to prove that $2L_n(f) \geq L_{n-1}(f) + L_{n+1}(f)$ for $n \geq 1$. Fix a positive number η and choose sets $\sigma_1, \sigma_2, \ldots, \sigma_{n-1}, \sigma'_1, \sigma'_2, \ldots, \sigma'_{n+1}$ such that

(i) the restrictions $f|\sigma_i$ and $f|\sigma'_j$ are increasing;

(ii) $\sigma_i \cap \sigma_j = \emptyset$ for $i \neq j$, $\sigma'_i \cap \sigma'_j = \emptyset$ for $i \neq j$;

(iii) $\sum_{i=1}^{n-1} |\sigma_i| \geq L_{n-1}(f) - \eta$ and $\sum_{j=1}^{n+1} |\sigma'_j| \geq L_{n+1}(f) - \eta$.

Let us set $g = f|\sigma$, where $\sigma = \left(\bigcup_{i=1}^{n-1} \sigma_i\right) \cup \left(\bigcup_{j=1}^{n+1} \sigma'_j\right)$. Since $L_n(f) \geq L_n(g)$, while clearly $L_{n\pm1}(g) \geq L_{n\pm1}(f) - \eta$, we have

$$2L_n(f) - L_{n-1}(f) - L_{n+1}(f) \geq 2L_n(g) - L_{n-1}(g) - L_{n+1}(g) - 2\eta.$$

If the theorem were true for functions of the type of g, the result would follow for f as well because η is arbitrary. Therefore it suffices to prove the theorem for functions f for which there exist measurable sets $\omega_1, \omega_2, \ldots, \omega_m$ with $D(f) = \bigcup_{j=1}^m \omega_j$, and the restrictions $f|\omega_j$ are increasing. Assume therefore that f satisfies this property, and observe that the sets ω_j can be assumed to be pairwise disjoint; one just needs to replace ω_j by $\omega_j \setminus \left(\bigcup_{i<j} \omega_i\right)$ for $j > 1$. Fix again a positive number η and select compact sets $K_j \subset \omega_j$ such that $\sum_{j=1}^m |\omega_j \setminus K_j| < \eta$. Define then $K = \bigcup_{j=1}^m K_j$ and $h = f|K$. Since $|D(f)| - |D(h)| < \eta$ it is clear that $L_k(h) \leq L_k(f) \leq L_k(h) + \eta$ for all k. Therefore

$$2L_n(f) - L_{n-1}(f) - L_{n+1}(f) \geq 2L_n(h) - L_{n-1}(h) - L_{n+1}(h) - 2\eta.$$

Again, since η is arbitrary, it follows that it suffices to prove the theorem for functions whose domain is a disjoint union of compact sets $K_1, K_2, \ldots, K_m$ such that the restrictions $f|K_j$ are increasing. Assume therefore that f has this property. There exists then a positive number ε such that $|x - y| > \varepsilon$ whenever $x \in K_i$, $y \in K_j$, and $i \neq j$. None of the intervals $[q\varepsilon, (q+1)\varepsilon]$ (with q an integer) will intersect more than one of the sets K_j, and a finite number of these intervals, say $I_1, I_2, \ldots, I_p$, will cover $D(f)$. We may assume that $I_j = [q_j\varepsilon, (q_j+1)\varepsilon]$ with $q_1 < q_2 < \cdots < q_p$. If we set now $f_j = f|(I_j \cap D(f))$, we see that $f_1, f_2, \ldots, f_p$ are increasing and $f = f_1 \sqcup f_2 \sqcup \ldots \sqcup f_p$. At this point it will be convenient to make a measure preserving change of variable so that f is defined on an interval $(0, \alpha_p)$, where $0 = \alpha_0 < \alpha_1 < \cdots < \alpha_p$ and $f|(\alpha_{j-1}, \alpha_j)$ is increasing for $j = 1, 2, \ldots, p$. Once more fix a positive number η and set $= N = [\alpha_p/\eta]$. Define next a function v which is constant and equal to $\inf_{t\in((j-1)\eta, j\eta)} f(t)$ on the interval $((j-1)\eta, j\eta)$ for $1 \leq j \leq N$. Note that at most $p-1$ of these intervals contains one of the points α_i; call ω the union of these intervals. If $f|\sigma$ is increasing then clearly $v|\sigma \setminus \omega$ is increasing and $|\sigma \setminus \omega| \geq |\sigma| - (p-1)\eta$. We conclude that $L_k(v) \geq L_k(f) - (p-1)\eta$ for all k. If, on the other hand, σ is a set such that $v|\sigma$ is increasing, then $f|\sigma$ is not necessarily increasing, but f will be increasing on the smaller set obtained by removing from $\sigma \setminus \omega$ the intervals of the form $((j-1)\eta, j\eta)$ which intersect σ, and are closest to the left and right of one of the points α_i, $1 \leq i < p$. This way we obtain the inequalities $L_k(v) \leq L_k(f) + 3k(p-1)\eta$. As in the preceding reductions, it is easy to see now that it suffices to prove the theorem for the function v. However, v is derived from a word like in Lemma 3.3, and the desired inequality follows from that lemma and from the facts presented in Section 2.   □

Let us note that the functions $\alpha \mapsto L_n(f^\alpha)$ are increasing and left-continuous; this follows from the fact that $\lim_{\varepsilon \downarrow 0} |\{t : f(t) \in (\alpha - \varepsilon, \alpha)\}| = 0$.

The limits $\lim_{\beta \downarrow \alpha} L_n(f^\beta)$ can also be calculated. Indeed, if $\varepsilon > 0$ is small, and $\sigma \subset D(f^{\alpha+\varepsilon})$ is such that $f|\sigma$ is increasing then $f|\sigma'$, $\sigma' = \sigma \cap \{t : f(t) \leq \alpha\}$, is also increasing and $|\sigma| - |\sigma'| \leq |\{t : f(t) \in (\alpha, \alpha + \varepsilon)\}|$, and this quantity tends to zero as $\varepsilon \to 0$. We conclude that $\lim_{\beta \downarrow \alpha} L_n(f^\beta) = L_n(f^{\alpha+})$, where $D(f^{\alpha+}) = \{t : f(t) \leq \alpha\}$.

We can now improve the preceding theorem.

Theorem 3.2 *Let f be a measurable function, and set $\mu_n(\alpha) = L_n(f^\alpha) - L_{n-1}(f^\alpha)$ for $\alpha \in \mathbf{R}$ and $n = 1, 2, \ldots$*

(1) *For each α, the sequence $(\mu_n(\alpha))_{n=1}^\infty$ is decreasing.*

(2) *For each n, $\mu_n(\alpha)$ is increasing and continuous from the left.*

(3) *For each n, $\lim_{\alpha \to -\infty} \mu_n(\alpha) = 0$ and $\lim_{\alpha \to \infty} \mu_n(\alpha) = \lambda_n(f)$.*

(4) *For each n and α we have $\mu_n(\alpha) \geq \lim_{\beta \downarrow \alpha} \mu_{n+1}(\beta)$.*

Proof. Statement (1), and the left continuity in (2) follow from the preceding theorem and the discussion above. The equalities in (3) follow easily from the fact that

$$\lim_{\alpha \to -\infty} |\{t : f(t) < \alpha\}| = \lim_{\beta \to \infty} |\{t : f(t) \geq \beta\}| = 0.$$

To verify (4) and the remainder of (2) fix a positive number η. As in the proof of the preceding theorem, we can find a restriction g of f which is a finite concatenation of increasing functions, such that $\lim_{n \to \infty} L_n(f) - |D(g)| < \eta$. Since $L_n(g^\alpha) \leq L_n(f^\alpha) \leq L_n(g^\alpha) + \eta$, we must also have $|\lambda_n(g^\alpha) - \lambda_n(f^\alpha)| < 2\eta$ for every n and α. Thus it suffices to prove (4) and (2) in case f is a finite concatenation of increasing functions. Similarly, as in the conclusion of the proof of Theorem 3.1, we can further reduce to the case when f is associated with a word as in Lemma 3.3. For these functions the theorem follows from the results described in Section 2. $\qquad\square$

Corollary 3.1 *For every measurable function f on a bounded set there exists an essentially unique Young tableau T such that $L_n(f^\alpha) = L_n(f_T^\alpha)$ for every $\alpha \in \mathbf{R}$ and every natural number n.*

Proof. Let μ_n be as in the preceding result and define f_n to be the unique left-continuous function which has distribution function μ_n. Then $T =$

$(f_n)_{n=1}^{\infty}$ is a Young tableau by properties (2) and (4) of the preceding result. Clearly T satisfies the requirements of the corollary.   □

Consider the class $\mathcal{F}$ of measurable functions with the property that $\lim_{n\to\infty} L_n(f) = |D(f)|$. The natural extension of Knuth equivalence to the class $\mathcal{F}$ is as follows: $f, g \in \mathcal{F}$ are Knuth equivalent if $L_n(f^{\alpha}) = L_n(g^{\alpha})$ for every $\alpha \in \mathbf{R}$ and every $n = 1, 2, \ldots$ We indicate this equivalence by $f \equiv_K g$. Observe that the functions f^{α} belong to $\mathcal{F}$ if $f \in \mathcal{F}$, and therefore Knuth equivalent functions have the same distribution, that is $|\{t : f(t) < \alpha\}| = |\{t : g(t) < \alpha\}|$. In particular, if $f \in \mathcal{F}$ and T is the tableau associated with f by the preceding corollary, the function f_T can be regarded as a rearrangement of f. As in the combinatorial situation, the data for performing this rearrangement can be codified in a second Young tableau (a standard one), thus giving rise to an extension of the Robinson–Schensted correspondence. We plan to pursue these matters in future work. Let us just note that a Young tableau T is standard if the tableau function f_T is an essentially invertible, measure preserving transformation of its domain $(0, \ell)$.

It is instructive to look at a simple example. Consider the function $f : (0, 2) \to \mathbf{R}$ defined by $f(t) = t^2$ for $t \in (0, 1]$ and $f(t) = t - 1$ for $t \in (1, 2)$. This is not a Young tableau and one can calculate $L_1(f) = 5/4$, $L_2(f) = 3/4$. The longest set on which f is increasing is $(0, 1/2] \cup [5/4, 2)$. Observe that the break points $1/2$ and $5/4$ have the property that $f(1/2) = f(5/4)$ and $f'(1/2) = f'(5/4)$; this is not an accident. The reader should have no difficulty finding the Young tableau associated to f.

References

[1] H. Bercovici, W. S. Li, T. Smotzer, A continuous analogue of the Littlewood-Ridchardson rule and its application to invariant subspaces, *Advances in Mathematics*, **134** (1998), 278–293.

[2] W. Fulton, *Young Tableaux, With Applications to Representation Theory and Geometry*, Cambridge University Press, Cambridge, 1997.

[3] D. E. Knuth, Permutations, matrices, and generalized Young tableaux, *Pacific Journal of Mathematics*, **34** (1970), 709–727.

[4] C. Schensted, Longest increasing and decreasing subsequences, *Canadian Journal of Mathematics*, **13** (1961), 179–191.

A Possibilistic Distance
for Sequences of Equal and Unequal Length

Massimo Borelli[1], Andrea Sgarro[2]

[1,2]Department of Mathematical Sciences
University of Trieste, 34100 Trieste, Italy

[1]Scuola Media Superiore Italiana di Rovigno
Rovinj, Croatia
E-mail: sgarro@univ.trieste.it

Abstract. We generalize Hamming distances to sequences of possibilistic letters: this corresponds to passing from ordinary binary logics to an infinite-valued logical setting. Our proposal is validated on a possibilistic model of a noisy communication channel, as opposed to the random-noise models which are current in information theory; it might prove to be a basis for "soft decoding" of noisy data. We take into account both synchronous and non-synchronous channels; the latter require a further generalization to sequences of unequal length in the spirit of Levenštejn distance; by so doing, we re-take a problem which Prof. Solomon Marcus had suggested to the second author when both were even younger than they are nowadays.

1 History and Introduction

It has been very early understood that a binary (black and white) logical apparatus was wide of the mark if one wanted to cope with the complexity of human thought; even Aristotelian logics is *ternary*, having at its disposal three logical values, *false* = 0, *true* = 1, and *possible* = $\frac{1}{2}$ (the numerical encoding is of course modern). A much bolder step has been taken in this century by such giants of modern logics as Łukasiewicz or Moisil; in multi-valued logics the range of (numerically encoded) logical values can cover

27

the whole of the interval $[0, 1]$ and so spreads over a continuum; the basic logical operations, *disjunction, conjunction* and *negation* are (usually) implemented by the mathematical operations of *maximum, minimum* and *complementation to* 1. Nowadays, quite a variety of "soft logics" are available, say *fuzzy logics*, or *possibility theory*; the relationship between these approaches is clear at the *formal* level, but unfortunately formal coincidence or inclusion is not a proof that a similar relationship holds also at the level of contents, and so philosophers are offered a chance of unending quarreling. It should be pointed out that these theories are not just a matter of philosophical discussions: soft (flexible, nuanced) logics have proved to be extremely advantageous for industrial applications. Actually, and rather unexpectedly, the motto of these sophisticated tools when used in the "real world" appears to be: cheap, quick, and effective. The inspiring idea is trying to achieve the "unreasonable success" of human thought, which is imprecise, vague and ambiguous, but in practice quite reliable and surprisingly effective. When one of us is driving, one does not solve differential equations to avoid crashing onto the guardrail: now, why should a robot driver be obliged to do so, when he has to perform the very same task? Therefore, the sophisticated and abstract meditations of Lukasiewicz or Moisil have in the end produced a whole range of very concrete objects, which are supposed to make our life more pleasant (and which do make a few of us richer).

A compelling generalization of black-and-white Hamming distances between strings to a "soft" setting was investigated in [1]; the starting point was a problem in the classification of languages, more precisely Romance languages, due to the well-known Croatian romanist Ž. Muljačić, who stressed the role of the (extinct) Dalmatic language as a "bridge" between the western group and the eastern group (i.e., Romanian in its four variants, Dacoromanian, Aromanian, Meglenoromanian and Istroromanian, as spoken south of Trieste). Now, some of the linguistic features were fuzzy. At that time Prof. Solomon Marcus pointed out to the author of [1] the interest of further extending fuzzy Hamming distances to strings of unequal lengths; the thing was put forward so clearly that presumably Prof. Marcus knew very well how to solve the problem he was stating. Unfortunately, the extension to unequal lengths given by the inexperienced author of [1] was of a very limited scope only. In this paper we shall re-take Professor Marcus' request, though we have to frankly acknowledge a deplorable delay of more than twenty years.

We shall have to extend the notions of [1]; actually the distance in [1] involves real (fuzzy) logical values rather than possibilistic letters, as dealt with below; so, it is again a moot point whether our *formal* generalization has also some "philosophical" content. We shall validate our proposal on a possibilistic model of a noisy communication channel, as opposed to the random-noise models which are current in information theory. Actually, our approach may prove to be the basis for "soft decoding" of noisy data.

A masterly and classical reference to non-classical logics is [2]. As a standard reference to fuzzy sets we suggest [3], where several similarity and dissimilarity measures are listed; as a standard reference to possibility theory cf., e.g., [4]. A comprehensive introduction to "unorthodox" theories of uncertainty representation and management is [5].

2 Possibilistic Letters and Their Transmission

We recall that a *possibility distribution* Π over the finite set, or *alphabet*, $\mathcal{A} - \{u_1, \ldots, a_K\}$, $K \geq 2$, is defined by the *possibility vector* $(x_1, \ldots, x_K)$ whose components are the possibilities of the K singletons: $\Pi(a_i) = x_i$; $0 \leq x_i \leq 1, 1 \leq i \leq K$. For each subset $A \subseteq \mathcal{A}$ one sets $\Pi(A) = \max_{a_i \in A} x_i$, $\Pi(\emptyset) = 0$. Observe that we do not even require $\Pi(\mathcal{A}) = 1$, i.e., $\max_i x_i = 1$. So, we allow for *incomplete* possibility vectors; incomplete theories of uncertainty management are quite popular nowadays, but actually go back to Rényi's incomplete probabilities; cf. [6]. Nowadays incompleteness is rather associated with the representation of "self-contradictory" states of knowledge; cf., e.g., [7,8].

Consider a noisy medium through which letter $a_j \in \mathcal{A}$ is sent. We shall describe the noisy medium by means of a *possibilistic channel*. This means that at the output a possibilistic letter Y is received which is described through a possibility vector $(y_1, \ldots, y_K)$, $0 \leq y_j \leq 1$; we shall write directly $Y = (y_1, \ldots, y_K)$. Also the input letter a_j can be thought of as a possibilistic letter X; just set $x_j = 1$, else $x_i = 0$. Actually, we shall not in any way distinguish between such a possibilistic letter X and the corresponding alphabet symbol a_j; we shall usually write $X = a_j$ rather than $X = (0, \ldots, 0, 1, 0, \ldots, 0)$. Should one assume that the possibilistic channel increases the possibility vector which describes the input letter, then $y_j = 1$; however, this is never required below. For the time being we confine ourselves to *observation channels*, i.e., to the transmission of single letters, rather than to the transmission of whole sequences; as for these cf. Section 4. What we need is a sort of "distance" between possibilistic letters to be put to good use in *minimum distance decoding*, as explained in Section 4; we stress that the term "distance" is used in this paper in a purely heuristic sense.

Example. Let $\mathcal{A}$ be the standard stock of printed letters; say capital letter E has been transmitted and its lower part is erased during transmission; the possibilistic letter Ξ output by the possibilistic channel has 1 in correspondence to E and F, else zero.

We define some terms. The possibilistic letter X is a *deterministic letter*, when for a specified letter $a_j \in \mathcal{A}$ one has $x_j = 1$, while $x_i = 0$ whenever $i \neq j$ (deterministic letters should not be confused with *crisp letters*, a more general notion when all the logical values are either zero or one). In our context we find it convenient to say that X is a (partial) *erasure* to mean that at least two of its logical values are equal to 1 while all the others are equal to zero. In a *total erasure* X all the logical values are equal to 1; we stress that a total erasure is *located*, in the sense that, when it is received at the output of a possibilistic channel, we know that an input letter must have been transmitted; in Section 4 we shall also deal with erasures which are not located. X is a *random letter* when the sum of its possibilistic values x_i is equal to 1: $\sum_i x_i = 1$; the binary case $K = 2$ will be of special interest to us, as dealt with below (we stress that the term *random* should be understood in a purely formal sense, our context being *not* a probabilistic one). Deterministic letters and erasures are *normal*, or *complete*, in the sense that $\max_i x_i$ is equal to 1. Instead, random letters are usually *incomplete*, being complete only in the deterministic case. A possibilistic letter X is *concentrated* on a_i when x_i is the only positive possibility component; deterministic letters are both concentrated and complete. More generally, the *support* of a possibilistic letter is the subset of the alphabet $\mathcal{A}$ which corresponds to possibility components which are strictly positive. The extreme case of incompleteness is the all-zero possibilistic letter, whose support is void: we shall call it the (totally) *inconsistent* letter.

We define the *incompleteness*, or *inconsistency*, $\iota(X)$ of the possibilistic letter X by setting:

$$\iota(X) = 1 - \max_i x_i. \tag{1}$$

One has: $0 \leq \iota(X) \leq 1$, with $\iota(X) = 0$ iff X is normal, $\iota(X) = 1$ iff X is totally inconsistent. One defines a partial ordering between possibilistic letters by setting:

$$X \leq Y \iff x_i \leq y_i \text{ for all } i. \tag{2}$$

In this ordering the first element is the totally inconsistent letter for which all the possibilities are zero, while the last element is the total erasure for which all the possibilities are equal to 1. One has:

$$X \leq Y \implies \iota(X) \geq \iota(Y). \tag{3}$$

Observe that, in our application to channels, we might have considered only distances between an input letter X and an output letter Y where the input X is constrained to be deterministic; this restriction is never used in the section to follow, however, since it would greatly hamper the flexibility of the model.

We need one more technicality. If $\mathcal{I}$ is a set (finite in our case), we shall say that a symmetric function $\delta : \mathcal{I}^2 \to \Re^+$ is a *pseudometric* when it satisfies, for all $x, y, z \in \mathcal{I}$:

$$0 \leq \delta(x, x) \leq \delta(x, y) \leq \delta(x, z) + \delta(z, y).$$

The inequality on the right is called the *triangle inequality*. Observe that we do not even require $\delta(x, x) = 0$. To achieve this without jeopardizing the triangle inequality, it would be enough to set $\delta'(x, y) = 0$ if $x = y$, else $\delta'(x, y) = \delta(x, y)$; this change would be of no use in our context, though. Actually, our "distance" as defined below will verify the triangle inequality only when the "triangulating" element z is constrained to belong to a subset to be specified.

3 A Possibilistic Non-Metric Distance

Let $X = (x_1, \ldots, x_K)$ and $Y = (y_1, \ldots, y_K)$ be two possibilistic letters described by the corresponding possibility vectors over the alphabet $\mathcal{A}$ of K deterministic letters. It is *possible* that X and Y are equal when *there exists* a deterministic letter a_j such that it is possible *both* that $X = a_j$ and $Y = a_j$. Consequently, we tentatively set:

$$\rho(X, Y) = \max_{1 \leq i \leq K} \min(x_i, y_i). \tag{4}$$

to measure the *similarity* between the two possibilistic letters X and Y. We stress the question was *not* "are the two letters equal?" but rather "is it possible that they are equal?" Since we need a *dissimilarity* index, rather than a similarity one, we (still tentatively) define the *possibilistic distance* between the two possibilistic letters X and Y simply as:

$$d(X, Y) = 1 - \rho(X, Y)$$

Example. Let $X = E$ be deterministically equal to capital letter E, $Y = F$ be deterministically equal to capital letter F, let Ξ be the E/F erasure as in the example above, let Ω be a total erasure as described by the all-one possibility vector. Then: $d(E, F) = 1$, $d(E, \Xi) = d(F, \Xi) = 0$, $d(E, \Omega) = d(F, \Omega) = d(\Xi, \Omega) = 0$. All this is as it should be.

As for the formal properties of $d(X, Y)$, which is obviously symmetric, we have:

$$0 \le d(X, X) \le d(X, Y) \le 1 \tag{5}$$

as is soon checked. The "distance" $d(X, Y)$ takes on the minimum value 0 iff $x_i = y_i = 1$ for some i, i.e., when $X = a_i$ and $Y = a_i$, are both possible options (which are not necessarily true); $d(X, Y)$ takes on its maximum value 1 iff the supports of X and Y are disjoint. In particular $d(X, X)$ coincides with incompleteness as defined in (1) above:

$$d(X, X) = \iota(X)$$

and so is zero iff X is normal, and is 1 iff X is totally inconsistent. Therefore, incomplete letters are characterized by a positive "self-distance", and this does make sense. Actually, the lattice structure of possibilistic letters allows one to *always* re-express the distance $d(X, Y)$ as an incompleteness, more precisely as the incompleteness of the "lattice meet" $X \wedge Y$, whose possibilities are equal to $\min(x_i, y_i)$:

$$d(X, Y) = \iota(X \wedge Y) \tag{6}$$

(recall that the lattice meet is the "highest" element which precedes both arguments).

The case of the "usual" (deterministic) Hamming distances and the fuzzy case of [1] are both re-found; the latter, which is rather more complex, will be dealt with in an addendum at the end of this section.

Deterministic letters. Say X and Y are deterministic: $X = a_i$, $Y = a_j$. Then $d(X, Y)$ is equal to 0 or 1 according to whether $i = j$ or $i \ne j$, as in the usual Hamming case. More generally:

$$X = a_i \text{ is deterministic} \implies d(X, Y) = 1 - y_i. \tag{7}$$

Still more generally, if X is concentrated on a_i, then $d(X, Y) = 1 - \min(x_i, y_i)$. As for the triangle property, the following example shows that unfortunately it does not hold even if two of the letters involved are constrained to be deterministic.

Example. Take $X = E$, $Y = F$, both deterministic, and Ξ as in the examples above. Then $d(E, \Xi) + d(\Xi, F) = 0 < d(E, F) = 1$. Actually, it is enough to take Ξ concentrated on $\{E, F\}$ and such that $\xi_E + \xi_F > 1$ to have $d(E, \Xi) + d(\Xi, F) = 2 - (\xi_E + \xi_F) < d(E, F)$.

However the triangle property *does hold* when it is the "triangulating" letter Z which is constrained to be deterministic:

Theorem 3.1 *If* $Z = a_j$ *is deterministic then* $d(X, Z) + d(Z, Y) \geq d(X, Y)$. *A criterion for equality is that at least one of the following conditions* i) *or* ii) *holds true:* i) $d(X, Z) = 0$ *and* $d(Y, Z) = d(Y, Y)$; ii) *symmetrically,* $d(Y, Z) = 0$ *and* $d(X, Z) = d(X, X)$.

Proof. Because of (7) the first side is $(1 - x_j) + (1 - y_j)$; so $d(X, Z) + d(Z, Y) \geq \max\left[(1 - x_j), (1 - y_j)\right] \geq \min_i\left[\max\left[(1 - x_i), (1 - y_i)\right]\right] = d(X, Y)$. To have equality in the first inequality either $x_j = 1$ or $y_j = 1$, i.e., either $d(X, Z) = 0$ or $d(Y, Z) = 0$. Suppose that $d(X, Z) = 0$; then to have equality also in the second inequality one must have $y_j = \max_i y_i$, i.e., $d(Y, Z) = d(Y, Y)$. $\square$

Theorem 3.1 is useful in view of the applications to channels, where one needs distances of the type $d(X, Y)$ with X deterministic and variable, Y fixed. We find it meaningful to determine the set of all possibilistic letters Z that "universally" verify the triangle inequality; in so doing we generalize the inequality part of Theorem 3.1.

Theorem 3.2 *Let* Δ *be the set of possibilistic letters* Z *such that* $d(X, Z) + d(Z, Y) \geq d(X, Y)$, $\forall X$, $\forall Y$; *let* $\Delta^* \supseteq \Delta$ *be the set of possibilistic letters* Z *such that* $d(X, Z) + d(Z, Y) \geq d(X, Y)$, *for all* X *and* Y *deterministic; let* Δ^{**} *be the set of possibilistic letters* Z *whose two highest possibility components,* z_M *and* z_m, *say, sum at most to* 1: $z_M + z_m \leq 1$. *Then* $\Delta = \Delta^* = \Delta^{**}$.

Proof. The last example shows that a necessary condition for Z to belong to Δ^*, and so to Δ, is $Z \in \Delta^{**}$. We go to sufficiency and prove that $Z \in \Delta^{**}$ implies $Z \in \Delta$. In terms of $\rho(X, Y)$, which was defined in (4), the triangle inequality is re-written as $\rho(X, Z) + \rho(Z, Y) \leq 1 + \rho(X, Y)$. We distinguish the case when $d(X, Z)$ and $d(Z, Y)$, or, equivalently, $\rho(X, Z)$ and $\rho(Z, Y)$ are achieved in the same position and the case when it is not so. Actually, the triangle inequality *always* holds when $\rho(X, Z)$ and $\rho(Z, Y)$, are achieved in the same position, i, say, since one has $\rho(X, Z) + \rho(Z, Y) = \min(x_i, z_i) + \min(z_i, y_i) \leq x_i + y_i = \max(x_i, y_i) + \min(x_i, y_i) \leq$

$1+\min(x_i, y_i) \leq 1+\rho(X, Y)$. Assume now that the positions where $\rho(X, Z)$ and $\rho(Z, Y)$ are achieved are distinct, $i \neq j$, say. Then $\rho(X, Z)+\rho(Z, Y) = \min(x_i, z_i) + \min(z_j, y_j) \leq z_i + z_j \leq z_M + z_m \leq 1 \leq 1 + \rho(X, Y)$. □

As a remark, formula (6) soon implies that the triangle inequality always holds when the three letters involved are ordered; cf. (2). Assume that Z is the "triangulating" sequence and assume, without real restriction, $X \leq Y$. If $X \leq Y \leq Z$, then the triangle inequality becomes $\iota(Y) \geq 0$ and so holds with equality iff Y is normal (in this case, if Y is normal, then also Z is such). If $X \leq Z \leq Y$, then the triangle inequality becomes $\iota(Z) \geq 0$ and so holds with equality iff Z is normal. If $Z \leq X \leq Y$, then the triangle inequality becomes $2\iota(Z) \geq \iota(X)$ and so holds with equality iff Z is normal (use (3) with $Z \leq X$). Note that in all three cases a necessary condition to have equality is that the triangulating letter Z should be normal.

Addendum: Fuzzy Features

Assume $K = 2$; assume that the second logical value is bound to negate, i.e., complement, the first, i.e., $X = (x, 1 - x)$, $Y = (y, 1 - y)$. Formally, we are dealing with a binary random letter, but the meaning is rather the presence/absence of a fuzzy feature. Then $d(X, Y)$ is precisely the pseudometric $\delta(x, y)$ as given in [1], as we now argue.

Example. Say a Romance language is being studied and X denotes the presence/absence of rotacism in that language; if this linguistic feature is only weakly present in the language one can set $X = (\frac{1}{2}, \frac{1}{2})$, as the linguist Muljačić used to do in similar cases.

If $x_1, \ldots x_i, \ldots$ are logical values, then we set:

$$\sigma(x_1, \ldots, x_i, \ldots) = \min(x_1, \ldots, x_i, \ldots, 1 - x_1, \ldots, 1 - x_i, \ldots).$$

One has $0 \leq \sigma \leq \frac{1}{2}$; $\sigma = 0$ if at least one of the logical values is crisp (i.e., either zero or one), $\sigma = \frac{1}{2}$ if all the logical values are equal to $\frac{1}{2}$, i.e., if they are all "totally uncrisp". In [1] the following pseudometric distance between logical values was investigated:

$$\delta(x, y) = \max\left[\min(x, 1 - y), \min(1 - x, y)\right].$$

So $\delta(x, y)$ is the logical value of the proposition: "either the first logical value is true and the second is false, or the first logical value is false and the second is true" (i.e., "the two logical values are different"). Notice that

$\delta(x,x) = \sigma(x)$, and so is zero iff the logical value x is crisp, as it should be. The maximum value $\delta(x,y) = 1$ is reached when x and y are crisp and distinct. The equivalence with the alternative definition (8) as given below is checked in [1]; (8) stresses the relation between $\delta(x,y)$ and is the "usual" Euclidean distance $|x - y|$:

$$\delta(x,y) = |x - y| + \sigma(x,y). \tag{8}$$

Actually, we also have:

$$\delta(x,y) = \frac{1 + |x - y| - |1 - x - y|}{2}. \tag{9}$$

Proof. Because of (8), it is enough to check that:

$$\sigma(x,y) = \frac{1 - |x - y| - |1 - x - y|}{2}.$$

Assume $x = \sigma(x,y)$ (else use the transformations $x \leftrightarrow y$ and/or $x \leftrightarrow 1 - x$, $y \leftrightarrow 1 - y$, which leave both sides unchanged). Then also the second side is equal to x. □

Now, for $X = (x, 1 - x)$, $Y = (y, 1 - y)$, $\rho(X,Y) = \delta(x, 1 - y)$ and $d(X,Y) = 1 - \delta(x, 1 - y) = \delta(x,y)$; the last equality is soon checked by means of (9).

4 Sequence Distances and Possibilistic Channel Decoding

We now extend possibilistic distances from letters to sequences; our proposal is validated by means of two models of possibilistic channels, the first synchronous, the second asynchronous.

Synchronous channel (channel with located erasures)

In this case the channel is used n times; n is the length (number of possibilistic letters) both of the input and the output sequence. Only some sequences, called *codewords*, are admissible as input sequences; the output sequence will be decoded to a codeword which is "nearest". This case ties up with usual channel decoding when one minimizes the usual (deterministic) Hamming distance; accordingly, we define the Hamming distance between possibilistic n-sequences as the sum of the Hamming distances of its letters, i.e., as the "number of positions" (in a possibilistic sense) where they differ:

$$d(\underline{X}, \underline{Y}) = \sum_{1 \leq \ell \leq n} d(X_\ell, Y_\ell). \tag{10}$$

Here X_ℓ and Y_ℓ are the ℓ-th possibilistic letter of the input n-sequence $\underline{X}$ and the output n-sequence $\underline{Y}$, respectively. Properties for $d(\underline{X}, \underline{Y})$ are soon obtained from (5); in particular: $0 \leq d(\underline{X}, \underline{Y}) \leq n$. Because of Theorem 3.1, if all the letters of sequence $\underline{Z}$ are deterministic, one has $d(\underline{X}, \underline{Z}) + d(\underline{Z}, \underline{Y}) \geq d(\underline{X}, \underline{Y})$.

The fuzzy Hamming distance of [1] is re-found when the letters $X_1, \ldots, X_n$ which make up the sequence $\underline{X}$ are all constrained to be binary and "random". In this case the possibilistic n-sequence $\underline{X}$ can be re-interpreted as a fuzzy subset $F_{\underline{X}} \subseteq \{1, 2, \ldots, n\}$ described through its indicator function $(x_1, \ldots, x_n)$, $X_\ell = (x_\ell, 1 - x_\ell)$. As an example, the linguist Muljačić used to characterize a language through n linguistic fuzzy features, the logical value x_ℓ telling whether the ℓ-th feature, e.g., rotacism, was present in the language; the Hamming distance between two languages expressed how far apart the two languages were with respect to the selected features. Notice that the fuzzy Hamming distance $d(\underline{X}, \underline{Y})$ as in [1] is nothing else but the fuzzy cardinality $|F_{\underline{X}} \triangle F_{\underline{Y}}|$ of the symmetric difference between the corresponding fuzzy subsets, $F_{\underline{X}}$ and $F_{\underline{Y}}$.

Example. Say the synchronous channel is used $n = 5$ times; the possible codewords to be used as input sequences are all the meaningful English word of length 5. Let Ξ be the E/F erasure, as above. The output sequence $\Xi\Xi$AST is decoded to FEAST, since EEAST, EFAST, FFAST are not possible inputs; $d(\Xi\Xi$AST, FEAST$) = 0$, $d(\Xi\Xi$AST, BEAST$) = 1$, BEAST being a possible input, i.e., a meaningful English word of length 5.

Asynchronous channel (channel with located and unlocated erasures)

In this case the channel is used n times; n is the length of the input sequence, while the length of the output sequence can be smaller than n, m say, because *unlocated* erasures may have occurred; the channel can erase letters in such a way that there is no way to know where this has taken place. This case ties up with the usual Levenštejn distance, rather than Hamming distance; it is a "bounded" Levenštejn distance, though, n being the bound. We observe that to come closer to a genuine "unbounded" Levenštejn distance, we should have considered also unlocated *insertions*, and not only unlocated erasures; we think however that our bounded case is enough to cover quite a meaningful class of noisy communication media.

We can try to compensate for unlocated erasures by re-insertions of letters, these re-insertions having their cost, however. We shall take the easy way out and assume, as is standard, that each re-insertion has cost one, the same incurred when the decoder confuses two (distinct) deterministic letters. To express this fact we find it convenient to add a new "letter" $\aleph$ which has the property that $d(X,\aleph) = d(\aleph, X) = d(\aleph,\aleph) = 1$ whatever the possibilistic letter X may be. Observe that (5) still holds when the letters X and/or Y are allowed to be equal to $\aleph$; the triangle inequality always holds when at least one of the three letters involved is equal to $\aleph$; compare with Theorem 3.2. Now, we shall *extend* the output sequence $\underline{Y}$ to a sequence $\underline{U}$ of length n, by inserting $n - m$ times the letter $\aleph$; this can be done in $\binom{n}{m}$ ways. We shall write $\underline{Y} \dashv \underline{U}$. We generalize (10) and set:

$$d(\underline{X},\underline{Y}) = \min_{\underline{U}:\underline{Y}\dashv\underline{U}} d(\underline{X},\underline{U}).$$

Clearly:

$$n - m \ \leq \ d(\underline{X},\underline{Y}) \ \leq \ n.$$

Moreover, if one assume that $\underline{X}$ and $\underline{Z}$ have both length n (are two possible input sequences, or codewords), and $\underline{Z}$ is deterministic, then the triangle inequality $d(\underline{X},\underline{Z}) + d(\underline{Z},\underline{Y}) \geq d(\underline{X},\underline{Y})$ does hold. A proof of this fact follows. Say $\underline{U}$ achieves $d(\underline{Z},\underline{Y})$: $\underline{Y} \dashv \underline{U}$, $d(\underline{Z},\underline{Y}) = d(\underline{Z},\underline{U})$; since $\underline{U}$ belongs also to the minimization set in the definition of $d(\underline{X},\underline{Y})$ one has $d(\underline{X},\underline{Y}) \leq d(\underline{X},\underline{U})$. Then the triangle inequality for sequences of length n (cf. Theorem 3.1) gives $d(\underline{X},\underline{Z})+d(\underline{Z},\underline{Y}) = d(\underline{X},\underline{Z})+d(\underline{Z},\underline{U}) \geq d(\underline{X},\underline{U}) \geq d(\underline{X},\underline{Y})$.

Example. Say the asynchronous channel is used $n = 11$ times and the output sequence is $\Phi\Psi$NVERSY; for the example's sake, here Φ is a possibilistic letter which has 1 in correspondence to vowels, and $\frac{1}{2}$ in correspondence to consonants, while Ψ has 1 in correspondence to consonants, and $\frac{1}{2}$ in correspondence to vowels. There are $n - m = 3$ unlocated erasures, and so the distance has to be at least equal to 3. Ω denotes a (located) total erasure, as above. The decoded sequence is ANNIVERSARY; actually, by taking $\underline{U} = \Phi\Psi$N$\aleph$VERS$\aleph\aleph$Y one has $d(\text{ANNIVERSARY}, \Phi\Psi\text{NVERSY}) = 3$; instead $d(\text{CONTROVERSY}, \Phi\Psi\text{NVERSY}) = 4$; take $\underline{U} = \Phi\Psi$N$\aleph\aleph\aleph$VERSY.

References

[1] A. Sgarro, A fuzzy Hamming distance, *Bulletin Mathématique*, **21**, 1-2 (1977).

[2] Gr. C. Moisil, *Încercări vechi şi noi de logică neclasică*, Editura Ştiinţifică, Bucureşti, 1965.

[3] D. Dubois, H. Prade, *Fuzzy Sets and Systems: Theory and Applications*, Academic Press, 1980.

[4] D. Dubois, H. Prade, *Possibility Theory*, Plenum Press, 1987.

[5] G. J. Klir, T. A. Folger, *Fuzzy Sets, Uncertainty and Information*, Prentice-Hall, 1988.

[6] A. Rényi, On measures of entropy anf information, now in *Selected Papers of Alfréd Rényi*, **2**, Akadémiai Kiadó, Budapest, 1976, 565–579.

[7] A. Sgarro, An open-frame theory of incomplete interval probabilities, *International Journal of Uncertainty, Fuzziness and Knowledge-Based Systems*, **6**, 6 (1998).

[8] P. Castellan, A. Sgarro, Open-frame Dempster conditioning for incomplete interval probabilities, *Proc. First International Symposium on Imprecise Probabilities and their Applications*, Ghent, Belgium, June 1999.

Solving Problems with Finite Test Sets[1]

Cristian S. Calude

Department of Computer Science, The University of Auckland
Private Bag 92019, Auckland, New Zealand
E-mail: cristian@cs.auckland.ac.nz

Helmut Jürgensen

Department of Computer Science, The University of Western Ontario
London, Ontario, Canada N6A 5B7, and

Institut für Informatik, Universität Potsdam
Am Neuen Palais 10, D-14469, Potsdam, Germany
E-mail: helmut@uwo.ca.

Shane Legg

Department of Mathematics, The University of Auckland
Private Bag 92019, Auckland, New Zealand

Abstract. *Every finite and every co-finite set of non-negative integers is decidable.* This is true and it is not, depending on whether the set is given constructively. A similar constraint is applicable in language theory and many other fields. The constraint is usually understood and, hence, omitted.

The phenomenon of a set being finite, but possibly undecidable, is, of course, a consequence of allowing non-constructive arguments in proofs. In this note we discuss a few ramifications of this fact. We start out with showing that every number theoretic statement that can be expressed in first-order logic can be reduced to a finite set, to be called a test set. Thus, if one knew the test set, one could determine the truth of the statement. The crucial point is, of course, that we may not be able

---

[1]The research reported in this paper was partially supported by Auckland University, Research Grant A18/XXXXX/62090/3414050, and by the Natural Sciences and Engineering Council of Canada, Grant OGP0000243.

to know what the finite test set is. Using problems in the class Π_1 of the arithmetic hierarchy as an example, we establish that the bound on the size of the test set is Turing-complete and that it is upper-bounded by the busy-beaver function.

This re-enforces the fact that there is a vast difference between finiteness and constructive finiteness. In the context of the present re-opened discussion about the notion of computability – possibly extending its realm through new computational models derived from physics – the constraint of constructivity of the model itself may add another twist.

1   Introduction

In the early days of decidability theory and also of theoretical computer science it was not uncommon to find statements like *every finite and every co-finite set of non-negative integers is decidable* in the research literature and in textbooks, and to find "proofs" of this using the argument that a decision algorithm could use table look-up; moreover, such statements themselves would be used in proofs of the decidability of other problems via reduction to finite or co-finite sets.[2] Of course every finite or co-finite set is decidable, but only – as is well-known – if it is given constructively. Similar constraints are applicable in language theory and many other fields. The constraint is, of course, usually understood and, hence, omitted. For example, in the case of the D0L equivalence problem[3] it was known for quite some time that this problem could be reduced to the problem of deciding whether two regular languages are equal. Unfortunately, this reduction was not constructive and a constructive one eluded researchers for several years.

The phenomenon of a set being finite, but possibly undecidable, is, of course, a consequence of allowing non-constructive arguments in proofs. In this note we discuss a few ramifications of this fact. We start out with showing that every number theoretic statement that can be expressed in first-order logic can be reduced to a finite set, to be called a test set. Thus, if one knew the test set, one could determine the truth of the statement. This

---

[2]We refrain from giving references, because pointing to past mistakes is not the aim of this paper. However, the interested reader is likely to find such statements by just perusing a few older books.

[3]Given a finitely generated free monoid X^* with set X of generators, elements $u, v \in X^*$ and endomorphisms g and h, the D0L equivalence problem is to decide whether the sets $\{u, g(u), g(g(u)), \ldots\}$ and $\{v, h(v), h(h(v)), \ldots\}$ are equal.

rather simple result models what is sometimes referred to as experimental mathematics: simply stated, if the statement is true we don't need to do anything and if it is false we find the smallest counter-example by computer. We then show how several classical problems fall into this category. The crucial point is, of course, that we may not be able to know what the finite test set is. Using problems in the class Π_1 of the arithmetic hierarchy as an example, we establish that the bound on the size of the test set is Turing-complete and that it is upper-bounded by the busy-beaver function.

This re-enforces the fact that there is a vast difference between finiteness and constructive finiteness. In the context of the present re-opened discussion about the notion of computability – possibly extending its realm through new computational models derived from physics – the constraint of constructivity of the model itself may add another twist.

Let $\mathbf{N}$ denote the set of positive integers, let $\mathbf{N_0} = \mathbf{N} \cup \{0\}$, and, for $k \in \mathbf{N}$, consider a k-ary predicate P on $\mathbf{N}$, that is, a mapping of $\mathbf{N}^k$ into the set $\mathbf{B} = \{0, 1\}$ of truth values. Consider the formula

$$f = Q_1 n_1 \, Q_2 n_2 \, \ldots \, Q_k n_k \, P(n_1, n_2, \ldots, n_k),$$

where $Q_1, Q_2, \ldots, Q_k \in \{\forall, \exists\}$ are quantifier symbols. In analogy to the arithmetic classes, we say that f is in the class $\hat{\Pi}_s$ or $\hat{\Sigma}_s$ if the quantifier prefix of f starts with $\forall$ or $\exists$, respectively, and contains $s - 1$ alternations of quantifier symbols. When P is computable, then f is in Π_s or Σ_s, respectively.[4] It is sufficient to consider only such formulæ f in which no two consecutive quantifier symbols are the same; in the following we make this assumption without special mention. With f as above, one has $s = k$.

As usual in logic, we write $P(n_1, \ldots, n_k)$ instead of $P(n_1, \ldots, n_k) = 1$ when $n_1, \ldots, n_k$ are elements of $\mathbf{N}$. Thus, $\neg P(n_1, \ldots, n_k)$ if and only if $P(n_1, \ldots, n_k) = 0$. Moreover, since we consider variable symbols only in the domain $\mathbf{N}$, if f is any formula in first-order logic, we write f *is true* instead of f *is true in* $\mathbf{N}$.

Let Γ_s be one of the classes $\hat{\Pi}_s$, $\hat{\Sigma}_s$, Π_s, and Σ_s. We refer to the task of proving or refuting a first-order logic formula as a *problem* and especially, to problems expressed by formulæ in Γ_s as Γ_s-*problems*.

We say that a problem is being *solved* if the corresponding formula is proved or disproved to be true, that is, if the truth value of the formula is determined. A problem is said to be *finitely solvable* if it can be solved by examining finitely many cases.[5]

---

[4] See [32] for general background on arithmetic classes.
[5] A rigorous definition of this notion is given in Section 3 below.

42   C. S. Calude, H. Jürgensen, S. Legg

For example, consider the predicate

$$P(n) = \begin{cases} 1, & \text{if } n \text{ is even or } n = 1 \text{ or } n \text{ is a prime,} \\ 0, & \text{otherwise,} \end{cases}$$

that is, $P(n) = 0$ if and only if n is an odd number greater than 1 which is not a prime. Then the problem expressed by the formula $\forall n\, P(n)$ is finitely solvable;[6] indeed, it is sufficient to check all n up to and including 9.

In this paper, we mainly consider $\hat{\Pi}_1$-problems and Π_1-problems. For example, *Goldbach's conjecture* is a Π_1-problem. It states that *every even* $n \in \mathbf{N}$ *is the sum of two primes.*[7] To express this in the terminology as introduced, let $P_G : \mathbf{N} \to \mathbf{B}$ be such that

$$P_G(n) = \begin{cases} 1, & \text{if } n \text{ is odd or } n \text{ is the sum of two primes,} \\ 0, & \text{otherwise.} \end{cases}$$

Thus, $f_G = \forall n\, P_G(n)$ is true if and only if Goldbach's conjecture is true.

Similarly, *Riemann's hypothesis* is a Π_1 problem.[8] Consider the complex function

$$\zeta(s) = \frac{1}{1 - 2^{1-s}} \cdot \sum_{n=1}^{\infty} \frac{(-1)^{n-1}}{n^s},$$

where $s = \sigma + \mathrm{i}\,t$, $\sigma, t \in \mathbf{R}$, $\sigma > 0$, and $s \neq 1$. Riemann conjectured that all zeroes $s_0 = \sigma_0 + \mathrm{i}\,t_0$ of ζ satisfy $\sigma_0 = \frac{1}{2}$ and are simple [30].

By a result of [14], Riemann's hypothesis can be expressed in terms of the function $\delta_R : \mathbf{N} \to \mathbf{R}$ defined by

$$\delta_R(k) = \prod_{n<k} \prod_{j \leq n} \eta_R(j),$$

where

$$\eta_R(j) = \begin{cases} p, & \text{if } j = p^r \text{ for some prime } p \text{ and some } r \in \mathbf{N}, \\ 1, & \text{otherwise.} \end{cases}$$

---

[6]This example is based on a folklore joke on induction proofs: to prove that *all odd natural numbers greater than 2 are primes* one proceeds as follows: 3 is a prime; 5 is a prime; 7 is a prime; 9 is a measuring error; 11 is prime; 13 is a prime; this is enough evidence.

[7]The conjecture was stated in 1742 by Goldbach in a letter to Euler [17]. According to [22], in 1980 the Goldbach conjecture was known to be true for all $n \leq 10^8$; in [35] of December 1994, it is claimed that no counter-example exists up to $2 \cdot 10^{10}$. Hardy states that the Goldbach problem is "probably as difficult as any of the unsolved problems in mathematics" [19]. See also [26] and [34].

[8]The problem is first proposed in [30]; see also [31].

Riemann's hypothesis is equivalent with the assertion that

$$\left(\sum_{k \leq \delta_R(n)} \frac{1}{k} - \frac{n^2}{2}\right)^2 < 36n^3,$$

for all $n \in \mathbf{N}$, see [14].[9] Hence, let

$$P_R(n) = \begin{cases} 1, & \text{if } \left(\sum_{k \leq \delta_R(n)} \frac{1}{k} - \frac{n^2}{2}\right)^2 < 36n^3, \\ 0, & \text{otherwise.} \end{cases}$$

Thus, $f_R = \forall n\, P_R(n)$ is true if and only if the Riemann hypothesis is true. Clearly, P_R is decidable. Therefore, Riemann's hypothesis is a Π_1-problem.

As in the case of the Goldbach conjecture, also for the Riemann hypothesis huge computations have been performed to search for a counterexample – or to increase the confidence [3], [4], [5], [6], [25].

Of course, not every mathematical statement is a Π_1-problem. For instance, the conjecture stating the existence of infinitely many twin primes, that is, consecutive odd primes such as 857 and 859, is not a Π_1-problem. With

$$P_T(n, m) = \begin{cases} 1, & m > n \text{ and } m \text{ and } m + 2 \text{ are primes,} \\ 0, & \text{otherwise,} \end{cases}$$

this conjecture can be stated as

$$f_T = \forall n \exists m\, P_T(n, m).$$

The formula f_T is in the class Π_2. Bennett claims that most mathematical conjectures can be settled indirectly by proving *stronger* Π_1-problems, see [2]. For the twin-prime conjecture such a stronger Π_1-problem is obtained as follows. Consider the predicate

$$P_T'(n) = \begin{cases} 1, & \text{if there is } m \text{ with } 10^{n-1} \leq m \leq 10^n, m \text{ and } m + 2 \text{ primes,} \\ 0, & \text{otherwise.} \end{cases}$$

Let $f_T' = \forall n P_T'(n)$. Thus, f_T' gives rise to a Π_1-problem and, if f_T' is true, then f_T is also true.

In this paper we discuss the fact – surprising (only) at first thought – that every $\hat{\Pi}_s$-problem and every $\hat{\Sigma}_s$-problem has a finite test set. Of course, there cannot be a constructive proof of this statement. Moreover, already for $s = 1$ the size of the test sets behaves as badly as the busy beaver.

---

[9]For another proof see [23], pp. 117–122.

2   Notation and Basic Notions

In this section we briefly review some basic notions and introduce some
notation. Let X be a non-trivial alphabet, that is, a non-empty, finite set
with at least 2 elements. Then X^* is the set of all *words* over X. A *(formal)
language over* X is a subset of X^*.

We assume that the reader is familiar with the theory of computable
functions on integers or strings (see [32], [8]). If U is a universal Turing
machine which maps strings over X to non-negative integers, and π is a
program for U then $U(\pi)$ denotes the result of applying U to π and an
empty input tape. In particular, we write $U(\pi) = \infty$ when U does not halt
on π.

3   Finite Solvability

For $s \in \mathbf{N}$, let $\hat{\Gamma}_s$ denote any of $\hat{\Pi}_s$ and $\hat{\Sigma}_s$, and let Γ_s denote any of Π_s
and Σ_s.

Definition 3.1 *Let*

$$f = Q_1 n_1\, Q_2 n_2\, \ldots Q_s n_s\, P(n_1, n_2, \ldots, n_s),$$

with $s \in \mathbf{N}$, where $Q_1, Q_2, \ldots, Q_s$ are alternating quantifier symbols.

  1. *A* test set *for f is set $T \subseteq \mathbf{N}^s$ such that f is true in $\mathbf{N}^s$ if and only
     if it is true in T.*

  2. *The* problem *of f is* finitely solvable *if there is a finite test set for f.*

Theorem 3.1 *Let $s \in \mathbf{N}$. Every $f \in \hat{\Gamma}_s$ is finitely solvable.*

Proof. Let
$$f = Q_1 n_1\, Q_2 n_2\, \ldots Q_s n_s\, P(n_1, n_2, \ldots, n_s),$$

with $s \in \mathbf{N}$, where $Q_1, Q_2, \ldots, Q_s$ are alternating quantifier symbols.
We determine a sequence $N_1, N_2, \ldots, N_s$ of finite sets with $N_i \subseteq \mathbf{N}^i$
such that the problem posed by f can be solved by checking all s-tuples
$(n_1, n_2, \ldots, n_s) \in N_s$.

We define the sets N_i by induction on i. For this purpose, let

$$f_i(m_1, \ldots, m_{i-1}) = Q_i n_i\, \ldots Q_s n_s\, P(m_1, \ldots, m_{i-1}, n_i, \ldots, n_s),$$

where $m_1, \ldots, m_{i-1} \in \mathbf{N}$. In particular, $f_1(\ ) = f$ and $f_{s+1}(m_1, \ldots, m_s) = P(m_1, \ldots, m_s)$.

For $i = 1$, if $Q_1 = \forall$, let

$$\nu_1 = 1 \quad \text{if } f = f_1(\ ) \text{ is true,}$$

and

$$\nu_1 = \min\{m_1 \mid m_1 \in \mathbf{N}, \neg f_2(m_1)\} \quad \text{otherwise;}$$

if $Q_1 = \exists$, let

$$\nu_1 = 1 \quad \text{if } f = f_1(\ ) \text{ is not true,}$$

and

$$\nu_1 = \min\{m_1 \mid m_1 \in \mathbf{N}, f_2(m_1)\} \quad \text{otherwise.}$$

Let $N_1 = \{(m_1) \mid m_1 \in \mathbf{N}, m_1 \leq \nu_1\}$.

Now, suppose N_{i-1} has been defined and $i \leq s$. For each $(m_1, \ldots, m_{i-1}) \subset N_{i-1}$, define $\nu_i(m_1, \ldots, m_{i-1}) \in \mathbf{N}_0$ as follows. If $Q_i = \forall$, let

$$\nu_i(m_1, \ldots, m_{i-1}) = 1 \quad \text{if } f_i(m_1, \ldots, m_{i-1}) \text{ is true,}$$

and

$$\nu_i(m_1, \ldots, m_{i-1}) = \min\{m_i \mid m_i \in \mathbf{N}, \neg f_{i+1}(m_1, \ldots, m_{i-1}, m_i)\} \quad \text{otherwise;}$$

if $Q_i = \exists$, let

$$\nu_i(m_1, \ldots, m_{i-1}) = 1 \quad \text{if } f_i(m_1, \ldots, m_{i-1}) \text{ is not true,}$$

and

$$\nu_i(m_1, \ldots, m_{i-1}) = \min\{m_i \mid m_i \in \mathbf{N}, f_{i+1}(m_1, \ldots, m_{i-1}, m_i)\} \quad \text{otherwise.}$$

Let

$$N_i = \{(m_1, \ldots, m_i) \mid (m_1, \ldots, m_{i-1}) \in N_{i-1}, m_i \in \mathbf{N}, \\ m_i \leq \nu(m_1, \ldots, m_{i-1})\}.$$

We now prove,[10] by induction on i, that each set $T_i = N_i \times \mathbf{N}^{s-i}$ is a test set for f. Then, in particular, N_s is a finite test set for f.

---

[10]We decided to include this rather straight-forward proof as it was only by this proof that we discovered some subtle traps in the construction of the test sets.

Consider $i = 1$. Suppose first that $Q_1 = \forall$. The set N_1 is $\{(1)\}$ and, clearly, the set T_1 is a test set[11] for f. When f is false the set N_1 consist of all positive integers up to the first counter-example for the first variable of P. Hence, again, T_1 is a test set for f. On the other hand, suppose that $Q_1 = \exists$. Then $N_1 = \{(1)\}$ when f is false. Clearly T_1 is a test set[12] for f. When f is true the set N_1 consists of all positive integers up to the first witness for the first variable of P. Again T_1 is a test set for f.

Now consider $i > 1$ and assume that T_{i-1} is a test set for f. First suppose that $Q_i = \forall$. Consider $(m_1, \ldots, m_{i-1}) \in N_{i-1}$. If $f_i(m_1, \ldots, m_{i-1})$ is true then $\nu_i(m_1, \ldots, m_{i-1}) = 1$. As T_{i-1} is a test set for f, to test whether f is true on $\{(m_1, \ldots, m_{i-1})\} \times \mathbf{N}^{s-i+1}$ it suffices to test on $\{(m_1, \ldots, m_{i_1}, 1)\} \times N^{s-i}$, and $(m_1, \ldots, m_{i-1}, 1) \in N_i$. If $f_i(m_1, \ldots, m_{i-1})$ is false, then N_i contains all the i-tuples $(m_1, \ldots, m_{i-1}, m_i)$ with m_i ranging from 1 to the smallest counter-example. Hence, as T_{i-1} is a test set for f so is T_i.

Now suppose that $Q_i = \exists$. If $f_i(m_1, \ldots, m_{i-1})$ is false then $\nu_i(m_1, \ldots, m_{i-1}) = 1$. As T_{i-1} is a test set for f, to test whether f is true on $\{(m_1, \ldots, m_{i-1})\} \times \mathbf{N}^{s-i+1}$ it suffices to test on $\{(m_1, \ldots, m_{i_1}, 1)\} \times N^{s-i}$, and $(m_1, \ldots, m_{i-1}, 1) \in N_i$. If $f_i(m_1, \ldots, m_{i-1})$ is true then N_i contains all the i-tuples $(m_1, \ldots, m_{i-1}, m_i)$ with m_i ranging from 1 to the smallest witness. Hence, as T_{i-1} is a test set for f so is T_i. $\square$

The proof of Theorem 3.1 is non-constructive and this remains so even when P is decidable. Thus, from this proof we do not learn anything about the number of cases one needs to check in order to prove or disprove the truth of f. It is clear from the theories of arithmetic classes and degrees of unsolvability that, in general, finite test sets cannot be *constructed* for this type of problems even when the predicate is computable. We try to shed some light, from a different perspective, on some of the reasons why this cannot be done.

The proof of Theorem 3.1 highlights a typical pitfall in proofs in computability theory when the reasoning of classical logic is used. The proof and the statement proved are computationally meaningless as neither helps with *actually solving the $\hat{\Gamma}_s$-problem*. The "construction" of the sets N_i in the proof disguises the fact that none of these finite sets may be computable.

---

[11] In fact, the empty set would be a test set for f. However, if one uses this idea, that is sets ν_1 to 0 rather than 1 – and similarly for ν_i in general – then the "construction" seems to break down.

[12] Again the empty set could have been used were it not for problems with the subsequent steps of the "construction".

See, for example, the formula f_G expressing Goldbach's conjecture.

The statement of Theorem 3.1 has some similarity with the *Test Set Theorem* in formal language theory. This theorem, originally known as Ehrenfeucht's conjecture, can be stated as follows: *Let X and Y be alphabets, and let $L \subseteq X^*$. There exists a finite subset F of L, a test set, such that, for any two morphisms f, g from X^* to Y^*, $f(u) = g(u)$ for all $u \in L$ whenever $f(u) = g(u)$ for all $u \in F$.* This was proved independently in [1] and [18].[13] In [7] and also [11] it is pointed out that the existence of the test sets is not constructive.[14] In the statement of the Test Set Theorem for languages, the order of the quantifiers, that is, $\forall L \exists F \forall f \forall g$, is very important. The modified order $\forall L \forall f \forall g \exists F$ results in a far simpler statement, for which a proof can be given using the same ideas as in the proof of Theorem 3.1.

In the following, for $f \in \hat{\Gamma}_s$, let $N(f) = N_s$ with N_s as in the proof of Theorem 3.1. In particular, when $s = 1$, then $N(f)$ is the set $\{(n_1) \mid n_1 \in \mathbf{N}, n_1 \leq \nu_1\}$. For this case, we define $\nu(f) = \nu_1$.

4 Π_1-**Problems**

In this section, we analyze the case of Π_1-problems in greater detail. Let X be an arbitrary but fixed alphabet. We use X as the alphabet for programs of universal Turing machines. We also fix a computable bijective function $\langle , \rangle : X^* \times \mathbf{N}_0 \to X^*$. Consider $f = \forall n\, P(n)$ where P is a computable predicate on $\mathbf{N}$. We assume that P is given as a program for an arbitrary, but fixed universal Turing machine U. Thus P is given as a word $\pi_P \in X^*$ such that $U(\langle \pi_P, n \rangle) = P(n)$ for all $n \in \mathbf{N}$. One can, therefore, consider ν as a partial function of X^* into $\mathbf{N}_0$, that is, $\nu(\pi_P) = \nu(f)$ with f as above. We first determine an upper bound on $\nu(f)$ for $f \in \Pi_1$.

The *busy beaver function* $\sigma : \mathbf{N} \to \mathbf{N}$ ([29]; see also [15], [16], Chapter 39) is defined as follows:

$$\sigma(n) = \max\{U(x) \mid x \text{ is a program of length } n \text{ for } U$$
$$\text{and } U(x) \text{ halts on } x\}.$$

Let P be a computable unary predicate on $\mathbf{N}$, let $f = \forall n\, P(n)$, hence $f \in \Pi_1$. Consider a program p_f for U such that

$$U(p_f) = \min\{n \mid \neg P(n)\},$$

---

[13]Explanations of the proofs are given in [27] and [33]. For further information see [13].

[14]Under special assumptions on L like regularity, test sets can be effectively constructed [20], [21]; see also [13].

if f is not true, and such that U runs forever on p_f if f is true. Such a program always exists because the program, which tries $P(1)$, $P(2)$, ... and halts with the first n such that $\neg P(n)$, has the required properties. Let $m_f = |p_f|$. If f is not true, then U halts on p_f with output $\nu(f)$. Hence $\nu(f) \leq \sigma(m_f)$. If f is true, then $\nu(f) = 0$. This proves the following statement.

Proposition 4.1 *For every $f \in \Pi_1$, $\nu(f) \leq \sigma(m_f)$.*

By Theorem 4.1, to solve the problem of f one only needs to check the truth value of $P(n)$ for all n not exceeding $\sigma(m_f)$. This could be very useful if σ were computable. However, σ grows faster than any computable function. Hence, the bound $\nu(f) \leq \sigma(m_f)$ does not help in the actual solution of the problem of f. In fact, no computable bound exists! Here is the argument. For any $\pi \in X^*$, define the predicate P_π on **N** by

$$P_\pi(n) = \begin{cases} 1, & U(\pi) \text{ does not halt within } n \text{ steps,} \\ 0, & \text{otherwise.} \end{cases}$$

Clearly, the predicate is computable. Let $f_\pi = \forall n \, P_\pi(n)$. Then f_π is true if and only if $U(\pi)$ does not halt.

Assume now that there is a program to compute an upper bound of $\nu(f)$ for any $f \in \Pi_1$; this program takes, as input, a program ρ computing the predicate P^ρ and computes as output an integer $\nu'(\rho)$ such that $\nu(f^\rho) \leq \nu'(\rho)$, where $f^\rho = \forall n \, P^\rho(n)$. We show that this assumption implies the existence of an algorithm deciding the halting problem for Turing machines. Indeed, consider $\pi \in X^*$. To decide whether $U(\pi)$ halts, first compute a program p_π computing P_π. Next compute $\nu'(p_\pi)$. As $f_\pi = f^{p_\pi}$, one has $\nu(f_\pi) \leq \nu'(p_\pi)$. Hence, to determine whether f_π is true, it is sufficient to determine whether $P_\pi(n)$ for all $n \leq \nu'(p_\pi)$. If so, then $U(\pi)$ halts; otherwise it doesn't.

Theorem 4.1 *The upper bound ν is Turing-complete.*

Proof. We already showed that an oracle for ν or an upper bound on ν allows one to decide the halting problem. The converse follows from Proposition 4.1. □

Corollary 4.1 *There is no constructive proof showing that every $f \in \Pi_1$ has a finite test set.*

With appropriate modifications, a statement similar to Corollary 4.1 can be proved for Σ_1. In fact, for any $s \in \mathbf{N}$ and any Γ_s, there is no constructive proof of the fact that every $f \in \Gamma_s$ has a finite test set.

5 Conclusions

Many true Π_1-problems are undecidable, hence independent with respect to a given sufficiently rich, sound, and computably axiomatizable theory. The analysis above can help us in understanding this phenomenon. Knowing that P is false can be used to get a proof that "P is false": we keep computing $P(n)$ for large enough n until we get an n such that $\neg P(n)$. But this situation is not symmetric: if we know that P is true we might not be able to prove that "P is true", and this case is quite frequent [10]. Indeed, even when we "have" the proof, that is, we have successfully checked that $P(n) \neq 0$, for all $n \leq \nu((\forall n)P(n))$, we might not be able to "realize" that we have achieved the necessary bound.

The correspondence $P \mapsto \nu((\forall n)P(n))$ exists and is perfectly legitimate from a classical point of view, but has no constructive "meaning". To a large extent the mathematical activity can be regarded as a gigantic, collective effort to compute individual instances of the function $\nu((\forall n)P(n))$. This point of view is consistent with Post's description of mathematical creativity [28]: "Every symbolic logic is incomplete and extendible relative to the class of propositions constituting K_0. The conclusion is inescapable that even for such a fixed, well defined body of mathematical propositions, *mathematical thinking is, and must remain, essentially creative.*" [15] It also gives support to the "quasi-empirical" view of mathematics, which sustains that although mathematics and physics are different, it is more a matter of degree than black and white [12, 9]; see also [24].

In essence, the seemingly paradoxical situation arises from the fact that, in classical logic, it may happen that only finite resources are needed for defining a finite object while finite resources will not suffice to determine the same object constructively. The finite "character" of a problem may nevertheless rule out – in a very fundamental way – that its solution can be obtained by finite means.

---

[15] As usual, K_0 means the halting problem in this quote.

Acknowledgment

We thank Douglas Bridges, Greg Chaitin and Solomon Marcus for stimulating discussions.

References

[1] M. H. Albert, J. Lawrence, A proof of Ehrenfeucht's conjecture, *Theoret. Comput. Sci.*, **41** (1985), 121–123.

[2] C. H. Bennett, Chaitin's Omega, in *Fractal Music, Hypercards, and More* ... (M. Gardner, ed.), W. H. Freeman, New York, 1992, 307–319.

[3] R. P. Brent, J. v. d. Lune, H. J. J. t. Riele, D. T. Winter, On the zeros of the Riemann zeta function in the critical strip I, *Math. Comp.*, **33** (1979), 1361–1372.

[4] R. P. Brent, J. v. d. Lune, H. J. J. t. Riele, D. T. Winter, On the zeros of the Riemann zeta function in the critical strip II, *Math. Comp.*, **39** (1982), 681–688.

[5] R. P. Brent, J. v. d. Lune, H. J. J. t. Riele, D. T. Winter, On the zeros of the Riemann zeta function in the critical strip III, *Math. Comp.*, **41** (1983), 759–767.

[6] R. P. Brent, J. v. d. Lune, H. J. J. t. Riele, D. T. Winter, On the zeros of the Riemann zeta function in the critical strip IV, *Math. Comp.*, **46** (1986), 667–681.

[7] C. Calude, Note on Ehrenfeucht's conjecture and Hilbert's basis theorem, *Bull. EATCS*, **29** (1986), 18–22.

[8] C. Calude, *Theories of Computational Complexities*, North-Holland, Amsterdam, 1988.

[9] C. S. Calude, G. J. Chaitin, Randomness everywhere. *Nature*, 400, **22** July (1999), 319–320.

[10] C. Calude, H. Jürgensen, M. Zimand, Is independence an exception? *Applied Mathematics and Computation*, **66** (1994), 63–76.

[11] C. Calude, D. Vaida, Ehrenfeucht test set theorem and Hilbert basis theorem: A constructive glimpse, in *Mathematical Foundations of*

Computer Science, 1989 (A. Kreczmar, G. Mirkowska, eds.), *Lecture Notes in Computer Science* **379**, Springer-Verlag, Berlin, 1989, 177–184.

[12] G. J. Chaitin, *The Unknowable*, Springer-Verlag, Singapore, 1999.

[13] C. Choffrut, J. Karhumäki, Combinatorics on words, in *Handbook of Formal Language Theory* (G. Rozenberg, A. Salomaa, eds.), Vol. 1, Springer-Verlag, Berlin, 1987, 329–438.

[14] M. Davis, Y. V. Matijasevič, J. Robinson, Hilbert's tenth problem. Diophantine equations: Positive aspects of a negative solution, in *Mathematical Developments Arising from Hilbert Problems* (F. E. Browder, ed.), American Mathematical Society, Providence, RI, 1976, 323–378.

[15] A. K. Dewdney, A computer trap for the busy beaver, the hardest-working Turing machine, *Scientific American*, **251**(8) (1984), 19–23.

[16] A. K. Dewdney, *The New Turing Omnibus*, Computer Science Press, New York, 1993.

[17] L. E. Dickson, *History of the Theory of Numbers*, Carnegie Institute, Washington, 1919, 1920, 1923, 3 volumes.

[18] V. S. Guha, The equivalence of infinite systems of equations in free groups and semigroups, *Mat. Zametki*, **40** (1986), 321–324 (in Russian).

[19] G. H. Hardy, Goldbach's theorem, *Mat. Tid. B*, **1** (1922), 1–16. Reprinted in *Collected Papers of G. H. Hardy*, vol. 1, Oxford University Press, Oxford, 1966, 545–560.

[20] J. Karhumäki, W. Rytter, S. Jarominek, Efficient constructions of test sets for regular and context-free languages, *Theoret. Comput. Sci.*, **116** (1993), 305–316.

[21] J. Karhumäki, W. Plandowski, W. Rytter, Polynomial size test sets for context-free languages, *J. Comput. System Sci.*, **50** (1995), 11–19.

[22] W. A. Light, T. J. Forres, N. Hammond, S. Roe, A note on the Goldbach conjecture, *BIT*, **20** (1980), 525.

[23] Y. V. Matijasevič, *Hilbert's Tenth Problem*, MIT Press, Cambridge, MA, 1993, 117–122.

[24] S. Marcus, Bridging linguistics and computer science, via mathematics, in *People and Ideas in Theoretical Computer Science* (C. S. Calude, ed.), Springer-Verlag, Singapore, 1998, 163–176.

[25] A. M. Odlyzko, Tables of zeros of the Riemann zeta function, http://www. research.att.com/~amo/zeta_tables/index.html.

[26] C.-T. Pan, *Goldbach Conjecture*, Science Press, Beijing, 1992.

[27] D. Perrin, On the solution of Ehrenfeucht's conjecture, *Bull. EATCS*, **27** (1985), 68–70.

[28] E. L. Post, Recursively enumerable sets of positive integers and their decision problems, *Bull. (New Series) Amer. Math. Soc.*, **50** (1944), 284–316.

[29] T. Rado, On non-computable numbers, *Bell System Tech. J.*, **3** (1962), 977–884.

[30] B. Riemann, Über die Anzahl der Primzahlen unter einer gegebenen Größe, in *Gesammelte mathematische Werke und, wissenchaftlicher Nachlaß*, Springer-Verlag, Berlin, 1990, 177–185.

[31] H. Riesel, *Prime Numbers and Computer Methods for Factorization*, Birkhäuser, Boston, second ed., 1994.

[32] H. Rogers, *Theory of Recursive Functions and Effective Computability*, McGraw-Hill, New York, 1967.

[33] A. Salomaa, The Ehrenfeucht conjecture: A proof for language theorists, *Bull. EATCS*, **27** (1985), 71–82.

[34] W. Yuan, ed., *Goldbach Conjecture*, World Scientific, Singapore, 1984.

[35] Names of large numbers and unsolved problems, http://www. smartpages.com/faqs/sci-math-faq/unsolvedproblems/faq.html, December 1994.

State Complexity of Regular Languages: Finite versus Infinite[1]

Cezar Câmpeanu

Faculty of Mathematics
University of Bucharest
Str. Academiei 14, 70109 Bucureşti, Romania
E-mail: cezar@funinf.math.unibuc.ro

Kai Salomaa, Sheng Yu

Department of Computer Science
University of Western Ontario
London, Ontario, Canada N6A 5B7
E-mail: {ksalomaa, syu}@csd.uwo.ca

Abstract. We consider the state complexity of regular languages and their operations. Especially, we compare the state complexity results on finite languages and general regular languages. The similarity relation $\sim_L$ and the equivalence relation $\equiv_L$ over Σ^* are also compared. Their applications on minimization of deterministic finite cover automata and deterministic finite automata, respectively, are investigated.

1 Introduction

There are many ways to measure the complexity of a deterministic finite automaton (DFA): (1) the number of states, (2) the number of transitions, or (3) both the number of states and the number of transitions. Although (3) gives more complete information, (1) is both simpler in presentation and cleaner and purer for investigation. The number of states also gives a

---

[1]This research is supported by the Natural Sciences and Engineering Research Council of Canada grants OGP0041630 and OGP0147224.

linear bound on the number of transitions. In the case of a complete DFA, i.e., a DFA whose transition function is defined for each state and each letter in the alphabet, the number of transitions is totally determined by the number of states when the alphabet is given. Therefore, the number of states is a natural complexity measure for a DFA as well as for the language it accepts. In the following, by a DFA we always mean a complete DFA.

A regular language is accepted by infinitely many different DFA. However, there is one that has the minimal number of states and it is unique up to a renaming of the states. We use the number of states of the minimal automaton to measure the complexity of the given language. Thus, by the *state complexity* of a regular language L, denoted $C(L)$, we mean the number of states in the minimal DFA that accepts L. By the state complexity of a class $\mathcal{L}$ of regular languages, denoted $C(\mathcal{L})$, we mean the maximum among all $C(L)$, $L \in \mathcal{L}$. When we speak about the state complexity of an operation on regular languages, we mean the state complexity of the resulting languages from the operation. For example, we say that the state complexity of the intersection of an m-state DFA language, i.e., a language accepted by an m-state complete DFA, and an n-state DFA language is exactly mn. This means that mn is the state complexity of the class of languages each of which is the intersection of an m-state DFA language and an n-state DFA language. In other words, there exist two regular languages that are accepted by an m-state DFA and an n-state DFA, respectively, such that the intersection of them is accepted by a minimal DFA of mn states, and this is the worst case. So, in a certain sense, state complexity is a worst-case complexity.

The state complexity of a regular-language operation gives a lower-bound for the space as well as the time complexity of the same operation. In many cases, the bounds given are tight. For example, the state complexity of the union of an m-state DFA language and an n-state DFA language is exactly mn. This also gives both the space and time complexity of the union operation within a constant factor.

State complexity is a complexity measure only for regular languages. However, it can be extended to cover other families of languages as well. For example, the automaticity studied by Shallit et al. [17] can be considered as an extension of the state complexity. We will not consider any extension of state complexity in this article.

Examining the state complexity results on the basic operations (e.g., catenation, union, intersection, and complementation) on regular languages in [19], one would notice that all the worst cases are given by using infinite languages only. This observation raises the question: are finite languages significantly different from (infinite) regular languages in state complexity of their operations? For example, would the state complexity of the union of two finite languages accepted by an m-state and an n-state DFA, respectively, be $O(m + n)$ instead of mn? We will investigate these questions in this article.

Finite languages are, perhaps, one of the most often used but least studied classes of languages. Finite languages are exactly the languages accepted by acyclic finite automata. It has been shown that there is a linear (time) algorithm for the minimization of an acyclic DFA by Revuz in 1992 [13]. However, for the minimization of a general DFA, the best known algorithm has a time complexity $O(n \log n)$ by Hopcroft in 1971 [6].

In this article, we compare the state complexity results for finite and infinite regular languages. We first consider the relatively simple cases, i.e., the operations on languages over a one-letter alphabet. Then we consider the general cases. In the one-letter cases, most of the operations on finite languages have a much lower state complexity than the corresponding operations on regular languages. However, in the general cases, only the catenation of two finite languages, when the first language is accepted by a DFA with a constant number of final states, has a much lower state complexity than its regular language counterpart.

Due to the not-so-positive results in the general cases, we resort to a different concept to try to reduce the number of states for DFA accepting finite languages. The concept of cover automata for finite languages is described in the last section of this article. In many cases, cover automata are a much more concise representation than DFA for finite languages.

2 Preliminaries

A deterministic finite automaton (DFA) is denoted by a quintuple $(Q, \Sigma, \delta, q_0, F)$, where Q is the finite set of states, Σ is the finite alphabet, $\delta : Q \times \Sigma \to Q$ is the transition function, $q_0 \in Q$ is the start state, and $F \subseteq Q$ is the set of final states. In this paper, *all the DFAs are assumed to be complete DFAs*. By a complete DFA we mean that there is a transition defined for each letter of the alphabet from each state, i.e., δ is a total function. In contrast, a DFA is called an incomplete DFA if its transition function is a partial function.

For any $x \in \Sigma^*$, we use $\#(x)$ to denote the length of x and $\#_a(x)$ for some $a \in \Sigma$ to denote the number of appearances of a in x. The empty word is denoted by ε.

The transition function δ of a DFA is extended to $\hat{\delta} : Q \times \Sigma^* \to Q$ by setting $\hat{\delta}(q, \varepsilon) = q$ and $\hat{\delta}(q, ax) = \hat{\delta}(\delta(q, a), x)$ for $q \in Q$, $a \in \Sigma$, and $x \in \Sigma^*$. In the following, we simply use δ to denote $\hat{\delta}$ if there is no confusion.

A word $w \in \Sigma^*$ is accepted by a DFA $A = (Q, \Sigma, \delta, q_0, F)$ if $\delta(q_0, w) \in F$. The language accepted by A, denoted $L(A)$, is the set $\{w \in \Sigma^* \mid \delta(q_0, w) \in F\}$. Two DFA are said to be equivalent if they accept the same language.

An incomplete DFA can be transformed to an equivalent complete DFA by adding a 'sink state' and transitions, which were undefined before, to the 'sink state', as well as transitions from the 'sink state' to the 'sink state'.

Let $A = (Q, \Sigma, \delta, s, F)$ be a DFA. Then

a) a state q is said to be accessible if there exists $w \in \Sigma^*$ such that $\delta(s, w) = q$;

b) a state q is said to be useful if there exists $w \in \Sigma^*$ such that $\delta(q, w) \in F$.

It is clear that for every DFA A there exists an equivalent DFA A' such that every state of A' is accessible and at most one state is useless (the 'sink state'). A DFA A' as above is called a *reduced* DFA. We will use only reduced DFA in the following.

A nondeterministic finite automaton (NFA) is denoted also by a quintuple $(Q, \Sigma, \eta, q_0, F)$ where $\eta \subseteq Q \times (\Sigma \cup \{\varepsilon\}) \times Q$ is a transition relation rather than a function, and Q, Σ, q_0, and F are defined similarly as in a DFA. The words and languages accepted by NFA are defined similarly as for DFA.

For a set s, we use $|s|$ to denote the cardinality of s. For a language L, we define $L^{\leq l} = \bigcup_{i=0}^{l} L^i$.

For $L \subseteq \Sigma^*$, we define a relation $\equiv_L \subseteq \Sigma^* \times \Sigma^*$ by

$$x \equiv_L y \text{ iff } xz \in L \Leftrightarrow yz \in L \text{ for all } z \in \Sigma^*.$$

Clearly, $\equiv_L$ is an equivalence relation, which partitions Σ^* into equivalence classes. The number of equivalence classes of $\equiv_L$ is called the *index* of $\equiv_L$. The Myhill-Nerode Theorem [7] states that L is regular if and only if $\equiv_L$ has a finite index and the minimal number of states of a complete DFA that accepts L is equal to the index of $\equiv_L$.

For a rather complete background knowledge in automata theory, the reader may refer to [7], [15].

The following lemmas will be used in the subsequent sections. They can be proved rather easily. Thus, we omit the proofs to concentrate on our main results.

Lemma 2.1 *Let $R \subseteq \Sigma^*$ be a regular language. If there exists an integer n such that*

$$max\{\#(w) \mid w \in \Sigma^* \ \& \ w \notin R\} = n,$$

then any DFA accepting R needs at least $n + 2$ states. In particular, if Σ is a singleton, the minimal DFA accepting R uses exactly $n + 2$ states.

Lemma 2.2 *Let $m, n > 0$ be two arbitrary integers such that $(m, n) = 1$ (m and n are relatively prime).*

(i) *The largest integer that cannot be presented as $cm + dn$ for any integers $c, d > 0$ is mn.*

(ii) *The largest integer that cannot be presented as $cm + dn$ for any integers $c > 0$ and $d > 0$ is $(m-1)n$.*

(iii) *The largest integer that cannot be presented as $cm + dn$ for any integers $c, d \geq 0$ is $mn - (m+n)$.*

3 Finite Versus Regular Languages Over a One-Letter Alphabet

As we have mentioned in the introduction, we start our comparison of the state complexity of operations on regular and finite languages from the relatively easy cases, i.e., the languages over a one-letter alphabet.

We first list the basic results below and then give detailed explanations for some of the operations.

We assume that L_1 is an m-state DFA language and L_2 an n-state DFA language, $\Sigma = \{a\}$, and $m, n > 1$.

	Finite	Regular	
$L_1 \cup L_2$	$max(m, n)$	mn,	for $(m, n) = 1$
$L_1 \cap L_2$	$min(m, n)$	mn,	for $(m, n) = 1$
$\Sigma^* - L_1$	m	m	
$L_1 L_2$	$m + n - 1$	mn,	for $(m, n) = 1$
L_1^R	m	m	
L_1^*	$m^2 - 7m + 13$, for $m > 4$	$(m-1)^2 - 1$	

Note that for *finite languages*, the state complexity for each of the union, intersection, and catenation operations is linear, while it is quadratic for infinite regular languages.

In the above table, all results for finite languages are relatively trivial except for L_1^*. We give an informal proof in the following. Let L_1 be accepted by an m-state DFA A_1 and A is a minimal DFA accepting L_1^*. It is clear that the length of the longest word accepted by A_1 is $m - 2$. (Note that the m states include a 'sink state'.) We consider the following three cases: (1) A_1 has one final state; (2) A_1 has two final states; or (3) A_1 has three or more final states. If (1), then A has $m - 1$ states. For (2), the worst case is given by $L = \{a^{m-2}, a^{m-3}\}$. By (iii) of Lemma 2.2, the length of the longest word that is not in L_1^* is

$$(m - 2)(m - 3) - (2m - 5) = m^2 - 7m + 11.$$

Then A has exactly $m^2 - 7m + 13$ states. In case (3), it is easy to see that A cannot have more than $m^2 - 7m + 13$ states.

For *regular languages*, we give a more detailed discussion below.

For the *union* operation, it is clear that mn states are sufficient for the resulting minimal DFA. To show that mn states are necessary, it suffices to show that there are at least mn distinct equivalence classes of the relation $\equiv_{L_1 \cup L_2}$. Let $L_1 = (a^m)^*$ and $L_2 = (a^n)^*$, $m, n > 1$ and $(m, n) = 1$. For positive integers p and q, let $m_p = p \bmod m$, $n_p = p \bmod n$, $m_q = q \bmod m$, and $n_q = q \bmod n$, $0 \le m_p, m_q < m$, $0 \le n_p, n_q < n$. It turns out that if $m_p \ne m_q$ or $n_p \ne n_q$, then a^p and a^q are not equivalent. However, this is not immediately clear for some cases. For example, let $m_p = -2 \bmod m$, $n_p = -1 \bmod n$, $m_q = -1 \bmod m$, and $n_q = -2 \bmod n$. (In order to explain easily, we use the negative numbers.) Then neither $m_p = m_q$ nor $n_p = n_q$, but $m_p = n_q$ and $n_p = m_q$. So, both $a^p a^2, a^q a^2 \in L_1 \cup L_2$ and $a^p a, a^q a \in L_1 \cup L_2$. It then appears that $a^p \equiv_{L_1 \cup L_2} a^q$. However, this is not true because it can be proved that $a^p a^{2+m} \in L_1 \cup L_2$, but $a^q a^{2+m} \notin L_1 \cup L_2$ assuming that $m < n$.

The state complexity result for the *intersection* of two regular languages can be similarly proved.

The result for the *catenation* of two regular languages is more involved. We outline a proof in the following. A more detailed proof can be found in [19].

We first give a general example of an m-state DFA language and an n-state DFA language, $(m, n) = 1$, such that mn states are necessary for any

DFA that accepts the catenation of the two languages. Let $L_1 = a^{m-1}(a^m)^*$ and $L_2 = a^{n-1}(a^n)^*$. Obviously, L_1 and L_2 can be accepted by an m-state DFA and an n-state DFA, respectively, and $L = L_1 L_2 = \{a^i \mid i = (m-1)+(n-1)+cm+dn$ for some integers $c, d \geq 0\}$. By (iii) of Lemma 2.2, for $(m, n) = 1$, the largest number that cannot be represented by $cm + dn$, $c, d \geq 0$, is $mn - (m + n)$. Then the largest i such that $a^i \notin L$ is $mn - 2$. So, the minimal DFA that accepts L has at least mn states. We show that mn states are sufficient in the following theorem.

Theorem 3.1 *For any integers $m, n \geq 1$, let A and B be an m-state DFA and an n-state DFA, respectively, over a one-letter alphabet. Then there exists a DFA of at most mn states that accepts $L(A)L(B)$.*

Proof. The cases when $m = 1$ or $n = 1$ are trivial. We assume that $m, n \geq 2$ in the following. Let $A = (Q_A, \{a\}, \delta_A, s_A, F_A)$ and $B = (Q_B, \{a\}, \delta_B, s_B, F_B)$. By a variation of the subset construction, we know that $L(A)L(B)$ is accepted by the DFA $C = (Q_C, \{a\}, \delta_C, s_C, F_C)$ where

$Q_C = \{\langle q, P \rangle \mid q \in Q_A \,\&\, P \subseteq Q_B\}$;
$s_C = \langle s_A, \emptyset \rangle$ if $s_A \notin F_A$ and $s_C = \langle s_A, \{s_B\} \rangle$ if $s_A \in F_A$;
$\delta_C(\langle q, P \rangle, a) = \langle q', P' \rangle$ where $q' = \delta_A(q, a)$ and $P' = \delta_B(P, a) \cup \{s_B\}$ if $q' \in F_A$, $P' = \delta_B(P, a)$ otherwise;
and $F_C = \{\langle q, P \rangle \mid P \cap F_B \neq \emptyset\}$.
Now we show that at most mn states of Q_C are reachable from s_C.

First we assume that in A there is a final state f in the loop of A's transition diagram. Then $\delta_A(s_A, a^t) = f$ and $\delta_A(f, a^l) = f$ for some non-negative integers $t < m$ and $l \leq m$. Let $j_1, \ldots, j_r, 0 < j_1 < \ldots < j_r < l$, be all the integers such that $\delta_A(f, a^{j_i}) \in F_A$ for each $1 \leq i \leq r$. Denote
$P_0 = \{s_B\}$,
$P_1 = \{\delta_B(s_B, a^l), \delta_B(s_B, a^{l-j_1}), \ldots, \delta_B(s_B, a^{l-j_r})\}$,
and for $i \geq 2$ we define
$P_i = \delta_B(P_{i-1}, a^l)$.
Let $\delta_C(s_C, a^t) = \langle f, S \rangle$. Denote $S_0 = S - \{s_B\}$ and $S_i = \delta_B(S_{i-1}, a^l)$ for each $i \geq 1$. Then we have the following state transition sequence of C:

$$s_C \vdash_C^t \langle f, P_0 \cup S_0 \rangle$$
$$\vdash_C^l \langle f, P_0 \cup P_1 \cup S_1 \rangle$$
$$\cdots\cdots$$
$$\vdash_C^l \langle f, P_0 \cup P_1 \cup \ldots \cup P_{n-1} \cup S_{n-1} \rangle$$
$$\vdash_C^l \langle f, P_0 \cup P_1 \cup \ldots \cup P_n \cup S_n \rangle$$

Here $p \vdash_C^k q$ stands for $\delta_C(p, a^k) = q$. Denote $P_0 \cup \ldots \cup P_i$ by $\mathcal{P}_i$, $i \geq 0$. Let i be the smallest integer such that $\mathcal{P}_{i-1} = \mathcal{P}_i$. It is clear that $i \leq n$ since B has n states. If $i = n$, then $\mathcal{P}_{n-1} = Q_B$ and

$$\langle f, \mathcal{P}_{n-1} \cup S_{n-1} \rangle = \langle f, \mathcal{P}_n \cup S_n \rangle = \langle f, Q_B \rangle.$$

Therefore, C needs at most $m + l(n-1) \leq m + m(n-1) = mn$ states. If $i < n$, consider the set $S'_{i-1} = S_{i-1} - \mathcal{P}_{i-1}$. Note that every state in S'_{i-1} is in the loop of the transition diagram of B. If for each element r of S'_{i-1}, there exists j, $0 \leq j \leq n - i$, such that $\delta_B(r, a^{jl}) \in \mathcal{P}_{i-1}$ (i.e., $\mathcal{P}_{n-1}$), then the proof is concluded as above. Otherwise, there is an element r_0 of S'_{i-1} and a transition sequence

$$r_0 \vdash_B^l r_1 \vdash_B^l \ldots \vdash_B^l r_{n-i}$$

such that, for some $j, k \leq n - i$ and $j < k$, $r_j = r_k$. (There are at most $n - i$ states not in $\mathcal{P}_{i-1}$.) Then it is easy to verify that $S_{i-1+j} = S_{i-1+k}$. Therefore, $\langle f, \mathcal{P}_{i-1+j} \cup S_{i-1+j} \rangle = \langle f, \mathcal{P}_{i-1+k} \cup S_{i-1+k} \rangle$. Thus, the number of states that are reachable from s_C is at most $t + 1 + l(n-1) \leq (m-1) + 1 + m(n-1) = mn$.

Finally we consider the case when no final states of A are in the loop. Let $Q_A = \{0, \ldots, m-1\}$ where $s_A = 0$ and $\delta_A(0, a^i) = i$ for $0 \leq i \leq m-1$. We can assume that $m-2$ is a final state and $m-1$ loops to itself. Otherwise, $L(A)$ can be accepted by a complete DFA with less than m states. Consider the following $m + n - 1$ transition steps of C

$$s_C \vdash_C^{m-2} \langle m-2, T \rangle \vdash_C \langle m-1, T_0 \rangle \vdash_C \langle m-1, T_1 \rangle$$

$$\vdash_C \ldots \vdash_C \langle m-1, T_n \rangle.$$

Let the state $\delta_B(s_B, a^{i+1})$ be t_i, for each $i \geq 0$. Note that $s_B \in T$ and t_i is in T_i. It is clear that there exist j, k such that $0 \leq j < k \leq n$ and $t_j = t_k$. Then it is not difficult to see that $\langle m-1, T_j \rangle = \langle m-1, T_k \rangle$. Therefore, at most $m+n$ states are necessary for C. (We have $m+n < mn$ for $m, n \geq 2$.) □

For the union, intersection, and catenation operations, we have considered only the cases when $(m, n) = 1$. For $(m, n) = t > 1$, we have not obtained exact formulas for those cases. Note that neither $mn/(m, n)$ nor $lcm(m, n)$ (the least common multiple of m and n) is the solution. For example, $a(a^5)^*$ and $(a^9)^*$ are accepted by a 6-state and a 9-state DFA, respectively, but the union of them needs at least 45 states rather than $6 \times 9/3 = 18$ states.

We now consider the last operation in the table, i.e., the *star* operation on infinite regular languages. We find that the proofs for both directions are interesting.

Theorem 3.2 *The number of states which is sufficient and necessary in the worst case for a DFA to accept the star of an n-state DFA language, $n > 1$, over a one-letter alphabet is $(n-1)^2 + 1$.*

Proof. For $n = 2$, the necessity is shown by a 2-state DFA which accepts $(aa)^*$. For each $n > 2$, the necessary condition can be shown by the DFA $A = (\{0, \ldots, n-1\}, \{a\}, \delta, 0, \{n-1\})$ where $\delta(i, a) = i + 1 \mod n$ for each i, $0 \le i \le n - 1$. The star of $L(A)$ is the language $\{a^i \mid i = c(n-1) + dn$, for some integers $c > 0$ and $d \ge 0$, or $i = 0\}$. By (ii) of Lemma 2.2, the largest i such that $a^i \notin L(A)^*$ is $(n-2)n$. So, the minimal DFA that accepts $(L(A))^*$ has $(n-2)n + 2$, i.e., $(n-1)^2 + 1$, states.

The proof for showing that $(n-1)^2 + 1$ states are sufficient is more interesting. Let $A = (Q, \{a\}, \delta, s, F)$ be an arbitrary n-state DFA, $n > 1$, and $R = L(A)$. If s is the only final state of A, then $R^* = R$. So, we assume that there is at least one final state f such that $f \ne s$. Clearly, R^* (excluding ε if $s \notin F$) is accepted by the NFA $A' = (Q, \{a\}, \delta', s, F)$ where $\delta' = \delta \cup \{(q, \varepsilon, s) \mid q \in F\}$. For any $X \subseteq Q$, denote by $closure(X)$ the set $X \cup \{q \in Q \mid (p, \varepsilon, q) \in \delta'$ for some $p \in X\}$. Now we follow the subset construction approach to build a DFA $B = (P, \{a\}, \eta, \{s\}, F_P)$ from A' to accept R^* such that $P \subseteq 2^Q$, $\eta(X, a) = closure(\{q \in Q \mid$ there exists $p \in X$ such that $(p, a, q) \in \delta'\})$, and $F_P = \{X \in P \mid X \cap F \ne \emptyset$ or $X = \{s\}\}$. Let f be the first final state from s in A and a^t is the shortest word such that $\delta(s, a^t) = f$. Then $\eta(\{s\}, a^t) = \{s, f\}$. Denote by p_{k_i} the state $\eta(\{s\}, a^{it})$ in P, $i \ge 0$, which is a subset of Q.

We claim that $p_{k_i} \supseteq p_{k_{i-1}}$ for all $i \ge 1$. It is true for $i = 1$, because $\eta(\{s\}, a^t) = \{s, f\}$, and also true for $i > 1$ since

$$p_{k_i} = \eta(\{s\}, a^{it}) = \eta(\{s, f\}, a^{(i-1)t}) = \eta(\{s\}, a^{(i-1)t}) \cup \eta(\{f\}, a^{(i-1)t})$$

$$= p_{k_{i-1}} \cup \eta(\{f\}, a^{(i-1)t}).$$

Then one of the following must be true:

(1) $p_{k_i} = p_{k_{i-1}}$ for some $i \le n - 1$;
(2) $p_{k_{n-1}} = Q$.

This is because if (1) is false, then $p_{k_{n-1}}$ contains at least n states and, therefore, (2) is true. Note that if (2) is true, then $\eta(p_{k_{n-1}}, a) = p_{k_{n-1}}$. In any of the cases, the number of states of B is no more than $t(n-1)+1$ which is at most $(n-1)^2+1$.   □

For the transformation from an n-state NFA to a DFA, it is clear in the case of finite languages over a one-letter alphabet that at most n states are needed. However, in the case of an infinite regular language over a one-letter alphabet, the problem is still open.

4   Finite Versus Regular Languages Over an Arbitrary Alphabet

For the one-letter alphabet case which we have discussed in the previous section, the state complexities for most operations on finite languages are of a lower order than their counterpart for regular languages. However, this is no longer true in the case when the size of the alphabet is arbitrary. Although none of the operations on finite languages, except the complementation, can reach the exact bound for regular languages, most of them have a complexity that is of the same order as the corresponding operation on regular languages.

We list the state complexity of the basic operations for both finite and regular languages over an arbitrary alphabet below. All the results for regular languages are given as exact numbers. However, we use the big "O" notation for most of the results for finite languages due to the fact that either the formulas we have obtained are acutely nonintuitive or we do not have an exact result, yet. More detailed explanations follow the table.

We assume that L_1 and L_2 are accepted by an m-state DFA $A_1 = (Q_1, \Sigma, \delta_1, s_1, F_1)$ and an n-state DFA $A_2 = (Q_2, \Sigma, \delta_2, s_2, F_2)$, respectively, and $m, n > 1$. We use t to denote the number of final states in A_1.

	Finite	Regular
$L_1 \cup L_2$	$O(mn)$	mn
$L_1 \cap L_2$	$O(mn)$	mn
$\Sigma^* - L_1$	m	m
$L_1 L_2$	$O(mn^{t-1} + n^t)$	$(2m-1)2^{n-1}$
L_1^R	$O(2^{m/2})$, for $|\Sigma| = 2$	2^m
L_1^*	$2^{m-3} + 2^{m-4}$, for $m \geq 4$	$2^{m-1} + 2^{m-2}$

For the *union* and the *intersection* of finite languages, it was expected that their state complexities would be linear, more specifically $O(m+n)$, but it turns out that both of them are of the order of mn although neither of them can reach the exact bound mn.

It is easy to show that mn states are sufficient for both union and intersection by the following simple argument. We can construct a DFA $A = (Q, \Sigma, \delta, s, F)$ which is the cross-product of A_1 and A_2, i.e., $Q = Q_1 \times Q_2$, $\delta = \delta_1 \times \delta_2$ that is $\delta((q_1, q_2), a) = (\delta(q_1, a), \delta_2(q_2, a))$, $s = (s_1, s_2)$. For the union operation, $F = \{(q_1, q_2) \mid q_1 \in F_1 \text{ or } q_2 \in F_2\}$ and for the intersection operation, $F = \{(q_1, q_2) \mid q_1 \in F_1 \text{ and } q_2 \in F_2\}$.

Note that the pairs of the form (s_1, q_2) where $q_2 \neq s_2$ and (q_1, s_2) where $q_1 \neq s_1$ are never reached from (s_1, s_2), and therefore, are useless. So, $mn - (m + n - 2)$ states are sufficient for both the union and intersection of two finite languages accepted by an m-state and an n-state DFA, respectively. However, this is a very rough upper bound. Much tighter upper bounds for the union and intersection of finite languages are given in [2], which unfortunately are in a very complicated and highly incomprehensible form. Thus, we will not quote them in this paper.

It is more interesting to show that the state complexities of those two operations are indeed of the order of mn but not lower. The following examples were originally given by Shallit [16]. Automaton-based examples are given in [2], which give better lower bounds than the examples below. We choose to present the following examples due to their clarity and intuitiveness.

For the intersection of two finite languages, consider the following example. Let $\Sigma = \{a, b\}$ and

$$L_1 = \{w \in \Sigma^* \mid \#_a(w) + \#_b(w) = 2n\},$$
$$L_2 = \{w \in \Sigma^* \mid \#_a(w) + 2\#_b(w) = 3n\}.$$

Clearly, L_1 is accepted by a DFA with $2n + 2$ states and L_2 by a DFA with $3n + 2$ states. The intersection $L = L_1 \cap L_2$ is

$$\{w \in \Sigma^* \mid \#_a(w) = \#_b(w) = n\}$$

One can prove that any DFA accepting L needs at least n^2 states by the Myhill-Nerode Theorem [7].

For the union of two finite languages, the example is slightly more complicated. Let $\Sigma = \{a, b\}$ and

$$L_1 = \{w \in \Sigma^* \mid \#(w) \leq 3t \text{ and } \#_a(w) + \#_b(w) \neq 2t\},$$
$$L_2 = \{w \in \Sigma^* \mid \#_a(w) + 2\#_b(w) < 3t\}.$$

It is clear that $L_1 \cup L_2$ includes all words in Σ^* of length less than or equal to $3t$ except those words w such that $\#(w) = 2t$ and $\#_b(w) \geq t$. One can prove that any DFA accepting $L_1 \cup L_2$ needs more than t^2 states by checking the number of the equivalence classes of $\equiv_{L_1 \cup L_2}$.

We now consider the *catenation* operation. Notice that for the finite language case, if the number of final states in A_1 is a constant, then the state complexity of catenation is a polynomial in terms of m and n. In particular, if A_1 has only one final state, then the state complexity is linear, i.e., $m+n$. In contrast, for infinite regular languages, there are examples in which A_1 has only one final state but any DFA accepting the catenation of the two languages needs at least $(2m - 1)2^{n-1}$ states [19], [18]. This is one of a few cases in which the state complexities for finite and infinite regular languages, respectively, are in different orders.

We now give the proof for the finite language case. For the general case for the catenation of regular languages, the reader may refer to [19] or [18].

Without loss of generality, we assume that all the DFA we are considering are reduced and ordered. A DFA $A = (Q, \Sigma, \delta, 0, F)$ with $Q = \{0, 1, \ldots, n\}$ is called an *ordered* DFA if, for any $p, q \in Q$, the condition $\delta(p, a) = q$ implies that $p \leq q$.

For convenience, we introduce the following notation:

$$\binom{n}{\leq i} = \sum_{j=0}^{i} \binom{n}{j}.$$

Theorem 4.1 *Let* $A_i = (Q_i, \Sigma, \delta_i, 0, F_i)$, $i = 1, 2$, *be two DFA accepting finite languages* L_i, $i = 1, 2$, *respectively, and* $\#Q_1 = m$, $\#Q_2 = n$, $\#\Sigma = k$, *and* $\#F_1 = t$. *There exists a DFA* $A = (Q, \Sigma, \delta, s, F)$ *such that* $L(A) = L(A_1)L(A_2)$ *and*

$$\#Q \leq \sum_{i=0}^{m-2} \min\left\{ k^i, \binom{n-2}{\leq i}, \binom{n-2}{\leq t-1} \right\}$$

$$+ \min\left\{ k^{m-1}, \binom{n-2}{\leq t} \right\}. \tag{$*$}$$

Proof. The DFA A is constructed in two steps. First, an NFA A' is constructed from A_1 and A_2 by adding a λ-transition from each final state in F_1 to the starting state 0 of A_2. Then, we construct a DFA A from the NFA A' by the standard subset construction. Again, we assume that A is reduced and ordered.

It is clear that we can view each $q \in Q$ as a pair (q_1, P_2), where $q_1 \in Q_1$ and $P_2 \subseteq Q_2$. The starting state of A is $s = (0, \emptyset)$ if $0 \notin F_1$ and $s = (0, \{0\})$ if $0 \in F_1$. Let us consider all states $q \in Q$ such that $q = (i, P)$ for a particular state $i \in Q_1 - \{m - 1\}$ and some set $P \subseteq Q_2$. Since A_1 is ordered and acyclic, the number of such states in Q is restricted by the following three bounds: (1) k^i, (2) $\binom{n-2}{\leq i}$, and (3) $\binom{n-2}{\leq t-1}$. We explain these bounds below informally.

We have (1) as a bound since all states of the form $q = (i, P)$ are at a level $\leq i$, which have at most k^{i-1} predecessors. By saying that a state p is at level i we mean that the length of the longest path from the starting state to q is i.

We now consider (2). Notice that if $q, q' \in Q$ such that $\delta(q, a) = q'$, $q = (q_1, P_2)$ and $q' = (q_1', P_2')$, then $\delta_1(q_1, a) = q_1'$ and $P_2' = \{\delta_2(p, a) \mid p \in P_2\}$ if $q_1' \notin F_1$ and $P_2' = \{0\} \cup \{\delta_2(p, a) \mid p \in P_2\}$ if $q_1' \in F_1$. So, $\#P_2' > \#P_2$ is possible only when $q_1' \in F_1$. Therefore, for $q = (i, P)$, $\#P \leq i$ if $i \notin F_1$ and $\#P \leq i + 1$ if $i \in F_1$. In both cases, the maximum number of distinct sets P is $\binom{n-2}{\leq i}$. The number $n - 2$ comes from the exclusion of the sink state $n - 1$ and starting state 0 of A_2. Note that, for a fixed i, either $0 \in P$ for all $(i, P) \in Q$ or 0 is not in any set P such that $(i, P) \in Q$.

(3) is a bound since for each state $i \in Q_1 - \{m - 1\}$, there are at most $t - 1$ final states on the path from the starting state to i (not including i).

For the second term of $(*)$, it suffices to explain that for each $(m-1, P)$, $P \subseteq Q_2$, $\#P$ is bounded by the total number of final states in F_1. $\square$

Corollary 4.1 *Let $A_i = (Q_i, \Sigma, \delta_i, 0, F_i)$, $i = 1, 2$, be two DFA accepting finite languages L_i, $i = 1, 2$, respectively, and $\#Q_1 = m$, $\#Q_2 = n$, and $\#F_1 = t$, where $t > 0$ is a constant. Then there exists a DFA $A = (Q, \Sigma, \delta, s, F)$ of $O(mn^{t-1} + n^t)$ states such that $L(A) = L(A_1)L(A_2)$.*

It has been shown in [1] that the bound given in Theorem 4.1 can be reached in the case $|\Sigma| = 2$.

About the state complexity of the *reversal* of an m-state DFA language, one may easily have a misconception. Many thought, without any hesitation, that it should be linear (in terms of m), especially in the case of finite languages. In fact, it is not even polynomial for both finite and infinite regular languages. We break the misconception by giving two examples in the following: one for finite languages and the other for infinite regular languages. Note that a nontrivial proof for a tight upper bound on the state complexity of the reversal of finite languages can be found in [1].

Example 4.1 Let $m = 2n + 2$ and $L = \{a,b\}^n a\{a,b\}^{\leq n}$, where $\{a,b\}^{\leq n}$ denotes
$$\lambda \cup \{a,b\} \cup \{a,b\}^2 \cup \cdots \cup \{a,b\}^n.$$
It is clear that L is a finite language accepted by an m-state DFA. One can prove that any DFA accepting L^R needs at least 2^n states.

Example 4.2 An n-state DFA that accepts an infinite regular language is shown in Figure 1. A proof showing that any DFA accepting the reversal of this language requires at least 2^n states can found in [19].

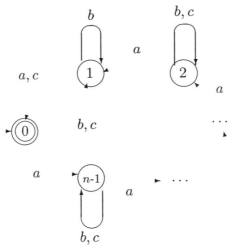

Figure 1: An n-state DFA such that $L(A)^R$ requires 2^n states

For the star operation, the difference between the state complexity for finite and infinite regular languages is that the latter is 4 times the former.

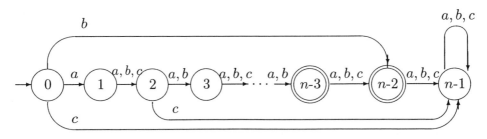

Figure 2: An n-state DFA such that $L(A)^*$ needs $2^{n-3} + 2^{n-4}$ states

Both bounds have been shown to be tight [1], [19]. Here we only give two examples to demonstrate that the bound given on the table can be reached.

Example 4.3 The n-state DFA A shown in Figure 2 accepts a finite language. It is shown in [1] that any DFA accepting $L(A)^*$ needs at least $2^{n-3} + 2^{n-4}$ states (assuming that n is even).

Example 4.4 Let L be the language accepted by the DFA shown in Figure 3. It is shown in [18] that any DFA accepting L^* requires at least $2^{n-1}+2^{n-2}$ states.

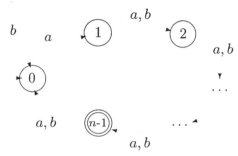

Figure 3: An n-state DFA such that $L(A)^*$ requires $2^{n-1} + 2^{n-2}$ states

5   Cover-Automata for Finite Languages

Let L be a finite language. The number of states of a deterministic finite automaton accepting L can, in the worst case, be exponential in the maximal length of words appearing in L, maxlen(L). The number of states of the minimal deterministic automaton for L can be significantly reduced if we require only that the automaton gives the "correct answer" for words of length at most maxlen(L), and the automaton may also accept words having length greater than maxlen(L). The more relaxed definition is natural in many applications since the system can keep track of the input word length separately. In a high-level programming language environment the length-control function can be readily implemented by an integer variable.

The above idea leads to the definition of *cover-automata* [3]. A similar notion called automaticity is used in [17] as a descriptive complexity measure for arbitrary languages.

Let L be a finite language over Σ and maxlen(L) = l. A DFA $A = (Q, \Sigma, \delta, q_0, F)$ is said to be a *deterministic finite cover-automaton* (DFCA)

of L if $L(A) \cap \Sigma^{\leq l} = L$. We say that A is a *minimal* DFCA of L if A has the smallest number of states among all the cover-automata of L.

It is clear that a minimal cover-automaton of a finite language L may require a smaller number of states than the minimal DFA accepting L. For instance, the language $\{a, a^3, a^5\}$ has a minimal cover-automaton with two states whereas any DFA accepting it requires seven states.

It is well known that the minimal DFA accepting a regular language L is unique up to isomorphism [7], [15], [18], the number of states is given by the index of the Nerode congruence $\equiv_L$. The cover-automata do not enjoy this uniqueness property. For instance let $\Sigma = \{a\}$ and consider the finite language $L_1 = \{\varepsilon, a, a^3\}$. Since $a^2 \notin L_1$ it is easy to see that any DFCA of L_1 needs at least three states. Thus

$$A = (\{0, 1, 2\}, \Sigma, \delta, 0, \{0, 1\}),$$

where $\delta(i, a) = (i + 1 \mod 3)$ is a minimal DFCA of L_1. On the other hand, also $A' = (\{0, 1, 2\}, \Sigma, \delta', 0, \{0, 1\})$ is a minimal DFCA of L_1 where $\delta'(i, a) = \delta(i, a)$, $i = 0, 1$, and $\delta'(2, a) = 1$.

However, it turns out that for a finite language L the number of states of any minimal DFCA of L is unique [3]. In order to show this we consider an L-similarity relation on Σ^* and a corresponding similarity relation defined on the set of states of a cover-automaton. The notion of L-similarity generalizes the Nerode congruence $\equiv_L$ and it has first been introduced in [8].

Let $L \subseteq \Sigma^*$ be a finite language and denote $l = \text{maxlen}(L)$. The L-similarity relation $\sim_L \subseteq \Sigma^* \times \Sigma^*$ is defined by setting $x \sim_L y$ iff the following condition holds:

for all $z \in \Sigma^*$ such that $\#(xz) \leq l$ and $\#(yz) \leq l$, $xz \in L$ iff $yz \in L$.

The relation $\sim_L$ is reflexive and symmetric, but not transitive. For instance if $L = \{a, b, ab\}$ then $a \sim_L ab$, $ab \sim_L b$ and $a \not\sim_L b$. However, the L-similarity relation has some "transitive-like" properties when we impose additional conditions on the lengths of the words. The properties are listed in the lemma below, the proof of which is straightforward.

Lemma 5.1 *Let $L \subseteq \Sigma^*$ be a finite language and $x, y, z \in \Sigma^*$, $\#(x) \leq \#(y) \leq \#(z)$. The following statements hold:*

(i) *If $x \sim_L y$, $x \sim_L z$, then $y \sim_L z$.*

(ii) *If $x \sim_L y$, $y \sim_L z$, then $x \sim_L z$.*

(iii) *If $x \sim_L y$, $y \not\sim_L z$, then $x \not\sim_L z$.*

The relation $\sim_L$ is not an equivalence relation. However we can consider L-similarity sets where all elements are pairwise similar. Then following [3] we can introduce canonical dissimilar sequences of L where the elements are pairwise dissimilar and the sequence contains an element from a maximal number of L-similarity sets. The length of canonical dissimilar sequences of L turns out to give the number of states of a minimal DFCA of L. These notions are formalized in the definition below.

Definition 5.1 *Let $L \subseteq \Sigma^*$ be a finite language.*

(i) *A set $S \subseteq \Sigma^*$ is called an L-similarity set if $x \sim_L y$ for any $x, y \in S$.*

(ii) *A sequence of words $[x_1, \ldots, x_n]$ over Σ is called a dissimilar sequence of L if $x_i \not\sim_L x_j$ for each pair i, j, $1 \leq i < j \leq n$.*

(iii) *A dissimilar sequence $[x_1, \ldots, x_n]$ is called a canonical dissimilar sequence of L if there exists a partition $\pi = \{S_1, \ldots, S_n\}$ of Σ^* such that for each i, $1 \leq i \leq n$, $x_i \in S_i$, and S_i is an L-similarity set.*

(iv) *A dissimilar sequence $[x_1, \ldots, x_n]$ of L is called a maximal dissimilar sequence of L if for any dissimilar sequence $[y_1, \ldots, y_m]$ of L, $m \leq n$.*

Note that maximal dissimilar sequences always exist since all words $w \in \Sigma^*$ such that $\#(w) > \text{maxlen}(L)$ are pairwise L-similar.

Theorem 5.1 *Let L be a finite language. A dissimilar sequence of L is a canonical dissimilar sequence of L if and only if it is a maximal dissimilar sequence of L.*

Proof. Let $[x_1, \ldots, x_n]$ be a canonical dissimilar sequence of L and $\pi = \{S_1, \ldots, S_n\}$ the corresponding partition of Σ^*. Let $[y_1, \ldots, y_m]$ be an arbitrary dissimilar sequence of L. Assume that $m > n$. Then there are y_i and y_j, $i \neq j$, such that $y_i, y_j \in S_k$ for some k, $1 \leq k \leq n$. Since S_k is an L-similarity set, $y_i \sim_L y_j$. This is impossible.

Conversely, let $[x_1, \ldots, x_n]$ be a maximal dissimilar sequence of L. Without loss of generality we can suppose that $\#(x_1) \leq \ldots \leq \#(x_n)$. For $i = 1, \ldots, n$, define

$$X_i = \{y \in \Sigma^* \mid y \sim_L x_i \text{ and } y \notin X_j \text{ for } j < i\}.$$

Note that for each $y \in \Sigma^*$, $y \sim_L x_i$ for at least one i, $1 \le i \le n$, since $[x_1, \ldots, x_n]$ is a *maximal* dissimilar sequence. Thus, $\pi = \{X_1, \ldots, X_n\}$ is a partition of Σ^*. The remaining task is to show that each X_i, $1 \le i \le n$, is a similarity set.

For the sake of contradiction assume that for some i, $1 \le i \le n$, there exist $y, z \in X_i$ such that $y \nsim_L z$. We know that $x_i \sim_L y$ and $x_i \sim_L z$ by the definition of X_i. We have the following three cases: (1) $\#(x_i) < \#(y), \#(z)$, (2) $\#(y) \le \#(x_i) \le \#(z)$ (or $\#(z) \le \#(x_i) \le \#(y)$), and (3) $\#(x_i) > \#(y), \#(z)$. If (1) or (2), then $y \sim_L z$ by Lemma 5.1 which is impossible. In case (3) it is easy to prove that $y \nsim_L x_j$ and $z \nsim_L x_j$, for all $j \ne i$, using Lemma 5.1 and the definition of X_i. Then we can replace x_i by both y and z to obtain a longer dissimilar sequence $[x_1, \ldots, x_{i-1}, y, z, x_{i+1}, \ldots, x_n]$. This contradicts the fact that $[x_1, \ldots, x_{i-1}, x_i, x_{i+1}, \ldots, x_n]$ is a maximal dissimilar sequence of L. Hence, $y \sim_L z$ and X_i is a similarity set. $\square$

The above theorem implies that the number of elements in any canonical dissimilar sequence of L is fixed. We denote this number by $N(L)$.

It turns out that any minimal DFCA of a finite language L will have exactly $N(L)$ states. In the following, if $A = (Q, \Sigma, \delta, q_0, F)$ is a DFA, in order to simplify the notation, we always assume that $Q = \{0, 1, \ldots, |Q|\}$ and $q_0 = 0$. First we define the level of states and a similarity relation on the states of a DFCA.

Definition 5.2 *Let $A = (Q, \Sigma, \delta, 0, F)$ be a DFA. We define, for each state $q \in Q$,*

$$level(q) = min\{\#(w) \mid \delta(0, w) = q\},$$

i.e., $level(q)$ is the length of the shortest path from the initial state to q.

Definition 5.3 *Let $A = (Q, \Sigma, \delta, 0, F)$ be a DFCA of a finite language L and $l = \text{maxlen}(L)$. Let $level(p) = i$ and $level(q) = j$, $m = \max\{i, j\}$. We say that $p \sim_A q$ (state p is L-similar to q in A) if for every $w \in \Sigma^{\le l-m}$, $\delta(p, w) \in F$ iff $\delta(q, w) \in F$.*

Again $\sim_A$ is a weakened version of the Nerode equivalence relation defined on states of a DFA. Note that, strictly speaking, the relation $\sim_A$ depends also on L (and not just on A). It follows from the definition that A is a minimal DFCA of L if and only if no distinct pair of states of A is in the L-similarity relation.

If $A = (Q, \Sigma, \delta, 0, F)$ is a DFA, for each $q \in Q$, we denote $x_A(q) = min\{w \mid \delta(0, w) = q\}$, where the minimum is taken according to the quasi-lexicographical order on Σ^*. When the automaton A is understood, we write x_q instead of $x_A(q)$.

Lemma 5.2 *Let $A = (Q, \Sigma, \delta, 0, F)$ be a DFCA of a finite language L. Let $x, y \in \Sigma^*$ be such that $\delta(0, x) = p$ and $\delta(0, y) = q$. If $p \sim_A q$, then $x \sim_L y$.*

Proof. Let $level(p) = i$ and $level(q) = j$, $m = \max\{i, j\}$, and $p \sim_A q$. Choose an arbitrary $w \in \Sigma^*$ such that $\#(xw) \leq l$ and $\#(yw) \leq l$. Because $i \leq \#(x)$ and $j \leq \#(y)$ it follows that $\#(w) \leq l - m$. Since $p \sim_A q$ we have that $\delta(p, w) \in F$ iff $\delta(q, w) \in F$, that is, $\delta(0, xw) \in F$ iff $\delta(0, yw) \in F$. This implies that $xw \in L(A)$ iff $yw \in L(A)$. Hence $x \sim_L y$. □

Lemma 5.3 *Let $A = (Q, \Sigma, \delta, 0, F)$ be DFCA of a finite language L. Let $level(p) = i$ and $level(q) = j$, $m = \max\{i, j\}$, and $x \in \Sigma^i$, $y \in \Sigma^j$ be such that $\delta(0, x) = p$ and $\delta(0, y) = q$. If $x \sim_L y$, then $p \sim_A q$.*

Proof. Let $x \sim_L y$ and $w \in \Sigma^{\leq l-m}$. If $\delta(p, w) \in F$, then $\delta(0, xw) \in F$. Because $x \sim_L y$, it follows that $\delta(0, yw) \in F$, so $\delta(q, w) \in F$. Using symmetry we get $p \sim_A q$. □

Now we can prove the main result of this section.

Theorem 5.2 *Let $L \subseteq \Sigma^*$ be finite and $A = (Q, \Sigma, \delta, 0, F)$ be a minimal DFCA of L. Then A has exactly $N(L)$ states.*

Proof. First assume that $|Q| < N(L)$. Let $[y_1, \ldots, y_{N(L)}]$ be a canonical dissimilar sequence of L. Then there exist i, j, $1 \leq i < j \leq N(L)$, such that $\delta(0, y_i) = \delta(0, y_j) = q$ for some $q \in Q$. Using Lemma 5.2 and reflexivity of the relation $\sim_A$ we get $y_i \sim_L y_j$. This is a contradiction.

Secondly, consider the possibility that $N(L) < |Q|$. By Theorem 5.1 there exist $p, q \in Q$, $p \neq q$, such that $x_p \sim_L x_q$. By the definition of the words x_p, x_q we have $\#(x_p) = level(p)$ and $\#(x_q) = level(q)$. Now Lemma 5.3 gives that $p \sim_A q$. Without loss of generality we can assume that $level(p) \leq level(q)$. We define a DFA $A' = (Q', \Sigma, \delta', 0, F')$ where $Q' = Q - \{q\}$, $F' = F - \{q\}$, and

$$\delta'(s, a) = \begin{cases} \delta(s, a) & \text{if } \delta(s, a) \neq q, \\ p & \delta(s, a) = q \end{cases}$$

for each $s \in Q'$ and $a \in \Sigma$. It can be verified that $L(A') \cap \Sigma^{\leq(\mathrm{maxlen}(L))} = L$ and thus A' is a DFCA of L. This contradicts the minimality of A. $\Box$

In general, a minimal DFCA of a finite language L is smaller than the minimal DFA accepting L. In [3] it is shown that if A is a minimal DFCA of L then A is minimal also as a DFA, i.e., it is the smallest DFA accepting the language $L(A)$. (Note that $L(A)$ and L may differ on words of length greater than $\mathrm{maxlen}(L)$.) The paper [3] also gives an algorithm for constructing the L-similarity relation of a DFCA A, and using it a minimization algorithm for cover-automata.

References

[1] C. Câmpeanu, K. Culik II, K. Salomaa, S. Yu, State complexity of basic operations on finite languages, *Proc. Fourth International Workshop on Implementing Automata, WIA'99*, to appear.

[2] C. Câmpeanu, K. Culik II, S. Yu, Finite languages and cover automata, in preparation.

[3] C. Câmpeanu, N. Sântean, S. Yu, Minimal cover automata for finite languages, *Proc. Third International Workshop on Implementing Automata, WIA'98, Lecture Notes in Computer Science*, **1660**, Springer-Verlag, Berlin, 1999, 33–42.

[4] A. K. Chandra, D. C. Kozen, L. J. Stockmeyer, Alternation, *Journal of the ACM*, **28** (1981), 114–133.

[5] A. Fellah, H. Jürgensen, S. Yu, Constructions for alternating finite automata, *Intern. J. Computer Math.*, **35** (1990) 117–132.

[6] J. E. Hopcroft, An $n \log n$ algorithm for minimizing the states in a finite automaton, *The Theory of Machines and Computations* (Z. Kohavi, ed.), Academic Press, New York, 189–196.

[7] J. E. Hopcroft, J. D. Ullman, *Introduction to Automata Theory, Languages, and Computation*, Addison Wesley, Reading, Mass., 1979.

[8] J. Kaneps, R. Freivalds, Minimal nontrivial space complexity of probabilistic one-way Turing machines, *Proc. Mathematical Foundations of Computer Science*, Banská Bystryca, Czechoslovakia, August 1990, *Lecture Notes in Computer Science*, **452**, Springer-Verlag, Berlin, 1990, 355–361.

[9] E. Leiss, Succinct representation of regular languages by boolean automata, *Theoretical Computer Science*, **13** (1981), 323–330.

[10] A. R. Meyer, M. J. Fischer. Economy of description by automata, grammars, and formal systems, *FOCS*, **12** (1971), 188–191.

[11] B. Ravikumar, Some applications of a technique of Sakoda and Sipser, *SIGACT News*, **21** (1990), 73–77.

[12] B. Ravikumar, O. H. Ibarra, Relating the type of ambiguity of finite automata to the succinctness of their representation, *SIAM J. Comput.*, **18**, 6 (1989), 1263–1282.

[13] D. Revuz, Minimisation of acyclic deterministic automata in linear time, *Theoretical Computer Science*, **92** (1992), 181–189.

[14] T. Jiang, B. Ravikumar, Minimal NFA problems are hard, *Proc. 18th ICALP*, *Lecture Notes in Computer Science*, **510**, Springer-Verlag, Berlin, 1991, 629–640.

[15] A. Salomaa, *Theory of Automata*, Pergamon Press, Oxford, 1969.

[16] J. Shallit, Private communication, Nov. 1996.

[17] J. Shallit and Y. Breitbart, Automaticity I: Properties of a Measure of Descriptional Complexity, *Journal of Computer and System Sciences*, **53** (1996), 10–25.

[18] S. Yu, Regular languages, in vol. 2 of *Handbook of Formal Languages* (G. Rozenberg, A. Salomaa, eds.), Springer-Verlag, Berlin, 1997.

[19] S. Yu, Q. Zhuang, K. Salomaa, The state complexity of some basic operations on regular languages, *Theoretical Computer Science*, **125** (1994), 315–328.

A Century of Controversy over the Foundations of Mathematics[1]

Gregory J. Chaitin

IBM T. J. Watson Research Center
30 Saw Mill River Road
Hawthorne, NY 10532, USA
E-mail: chaitin@watson.ibm.com

1   Introduction

I would like to talk about some crazy stuff. The general idea is that sometimes ideas are very powerful. I'd like to talk about theory, about the computer as a concept, a philosophical concept.

We all know that the computer is a very practical thing out there in the real world! It pays for a lot of our salaries, right? But what people don't remember as much is that really—I'm going to exaggerate, but I'll say it—the computer was invented in order to help to clarify a question about the foundations of mathematics, a philosophical question about the foundations of mathematics.

Now that sounds absurd, but there's some truth in it. There are actually lots of threads that led to the computer, to computer technology, which come from mathematical logic and from philosophical questions about the limits and the power of mathematics.

The computer pioneer Turing was inspired by these questions. Turing was trying to settle a question of Hilbert's having to do with the philosophy of mathematics, when he invented a thing called the Turing machine, which is a mathematical model of a toy computer. Turing did this before there were any real computers, and then he went on to actually build computers. The first computers in England were built by Turing.

And von Neumann, who was instrumental in encouraging the creation of computers as a technology in the United States (unfortunately as part

---

[1]Lecture given Friday 30 April 1999 at UMass-Lowell. The lecture was videotaped; this is an edited transcript.

of the war effort, as part of the effort to build the atom bomb), he knew Turing's work very well. I learned of Turing by reading von Neumann talking about the importance of Turing's work.

So what I said about the origin of the computer isn't a complete lie, but it is a forgotten piece of intellectual history. In fact, let me start off with the final conclusion of this talk... In a way, a lot of this came from work of Hilbert. Hilbert, who was a very well-known German mathematician around the beginning of this century, had proposed formalizing completely all of mathematics, all of mathematical reasoning—deduction. And this proposal of his is a tremendous, glorious failure!

In a way, it's a spectacular failure. Because it turned out that you couldn't formalize mathematical reasoning. That's a famous result of Gödel's that I'll tell you about, done in 1931.

But in another way, Hilbert was really right, because formalism has been the biggest success of this century. Not for reasoning, not for deduction, but for programming, for calculating, for computing, that's where formalism has been a tremendous success. If you look at work by logicians at the beginning of this century, they were talking about formal languages for reasoning and deduction, for doing mathematics and symbolic logic, but they also invented some early versions of programming languages. And **these** are the formalisms that we all live with and work with now all the time! They're a tremendously important technology.

So formalism for reasoning did not work. Mathematicians don't reason in formal languages. But formalism for computing, programming languages, are, in a way, what was right in the formalistic vision that goes back to Hilbert at the beginning of this century, which was intended to clarify epistemological, philosophical questions about mathematics.

So I'm going to tell you this story, which has a very surprising outcome. I'm going to tell you this surprising piece of intellectual history.

2   The Crisis in Set Theory

So let me start roughly a hundred years ago, with Georg Cantor.

The point is this. Normally you think that pure mathematics is static, unchanging, perfect, absolutely correct, absolute truth... Right? Physics may be tentative, but math, things are certain there! Well, it turns out that's not exactly the case.

In this century, in this past century there was a lot of controversy over the foundations of mathematics, and how you should do math, and what's right and what isn't right, and what's a valid proof. Blood was almost shed over this... People had terrible fights and ended up in insane asylums over this. It was a fairly serious controversy. This isn't well known, but I think it's an interesting piece of intellectual history.

More people are aware of the controversy over relativity theory. Einstein was very controversial at first. And then of the controversy over quantum mechanics... These were the two revolutions in the physics of this century. But what's less well known is that there were tremendous revolutions and controversies in pure mathematics too. I'd like to tell you about this. It really all starts in a way from Cantor.

What Cantor did was to invent a theory of infinite sets.

He did it about a hundred years ago; it's really a little more than a hundred years ago. And it was a tremendously revolutionary theory, it was *extremely* adventurous. Let me tell you why.

Cantor said, let's take 1, 2, 3, ... We've all seen these numbers, right?! And he said, well, let's add an infinite number after this.

$$1, 2, 3, \ldots \omega$$

He called it ω, lowercase Greek omega. And then he said, well, why stop here? Let's go on and keep extending the number series.

$$1, 2, 3, \ldots \omega, \omega + 1, \omega + 2, \ldots$$

Omega plus one, omega plus two, then you go on for an infinite amount of time. And what do you put afterwards? Well, two omega? (Actually, it's omega times two for technical reasons.)

$$1, 2, 3, \ldots \omega \ldots 2\omega$$

Then two omega plus one, two omega plus two, two omega plus three, two omega plus four...

$$1, 2, 3, \ldots 2\omega, 2\omega + 1, 2\omega + 2, 2\omega + 3, 2\omega + 4, \ldots$$

Then you have what? Three omega, four omega, five omega, six omega,...

$$1, 2, 3, \ldots, 3\omega, \ldots, 4\omega, \ldots, 5\omega, \ldots, 6\omega, \ldots$$

Well, what will come after all of these? Omega squared! Then you keep going, omega squared plus one, omega squared plus six omega squared plus eight... Okay, you keep going for a long time, and the next interesting thing after omega squared will be? Omega cubed! And then you have omega to the fourth, omega to the fifth, and much later?

$$1, 2, 3, \ldots, \omega, \ldots, \omega^2, \ldots, \omega^3, \ldots, \omega^4, \ldots, \omega^5$$

Omega to the omega!

$$1, 2, 3, \ldots, \omega, \ldots, \omega^2, \ldots, \omega^\omega$$

And then much later it's omega to the omega to the omega an infinite number of times!

$$1, 2, 3, \ldots, \omega, \ldots, \omega^2, \ldots, \omega^\omega, \ldots, \omega^{\omega^{\omega^{\omega^{\cdots}}}}$$

I think this is usually called epsilon nought.

$$\varepsilon_0 = \omega^{\omega^{\omega^{\omega^{\cdots}}}}$$

It's a pretty mind-boggling number! After this point things get a little complicated...

And this was just one little thing that Cantor did as a warm-up exercise for his main stuff, which was measuring the size of infinite sets! It was spectacularly imaginative, and the reactions were extreme. Some people loved what Cantor was doing, and some people thought that he should be put in an insane asylum! In fact he had a nervous breakdown as a result of those criticisms. Cantor's work was very influential, leading to point-set topology and other abstract fields in the mathematics of the twentieth century. But it was also very controversial. Some people said, it's theology, it's not real, it's a fantasy world, it has nothing to do with serious math! And Cantor never got a good position and he spent his entire life at a second-rate institution.

3   Bertrand Russell's Logical Paradoxes

Then things got even worse, due mainly, I think, to Bertrand Russell, one of my childhood heroes.

Bertrand Russell was a British philosopher who wrote beautiful essays, very individualistic essays, and I think he got the Nobel prize in literature for his wonderful essays. Bertrand Russell started off as a mathematician and then degenerated into a philosopher and finally into a humanist; he went downhill rapidly! Anyway, Bertrand Russell discovered a whole bunch of disturbing paradoxes, first in Cantor's theory, then in logic itself. He found cases where reasoning that seemed to be okay led to contradictions.

And I think that Bertrand Russell was tremendously influential in spreading the idea that there was a serious crisis and that these contradictions had to be resolved somehow. The paradoxes that Russell discovered attracted a great deal of attention, but strangely enough only one of them ended up with Russell's name on it! For example, one of these paradoxes is called the Burali-Forti paradox, because when Russell published it he stated in a footnote that it had been suggested to him by reading a paper by Burali-Forti. But if you look at the paper by Burali-Forti, you don't see the paradox!

But I think that the realization that something was seriously wrong, that something was rotten in the state of Denmark, that reasoning was bankrupt and something had to be done about it pronto, is due principally to Russell. Alejandro Garciadiego, a Mexican historian of math, has written a book which suggests that Bertrand Russell really played a much bigger role in this than is usually realized: Russell played a key role in formulating not only the Russell paradox, which bears his name, but also the Burali-Forti paradox and the Berry paradox, which don't. Russell was instrumental in discovering them and in realizing their significance. He told everyone that they were important, that they were not just childish word-play.

Anyway, the best known of these paradoxes is called the Russell paradox nowadays. You consider the set of all sets that are not members of themselves. And then you ask, "Is this set a member of itself or not?" If it is a member of itself, then it shouldn't be, and vice versa! It's like the barber in a small, remote town who shaves all the men in the town who don't shave themselves. That seems pretty reasonable, until you ask "Does the barber shave himself?" He shaves himself if and only if he doesn't shave himself, so he can't apply that rule to himself!

Now you may say, "Who cares about this barber!" It was a silly rule anyway, and there are always exceptions to the rule! But when you're dealing with a *set*, with a mathematical concept, it's not so easy to dismiss the problem. Then it's not so easy to shrug when reasoning that seems to be okay gets you into trouble!

By the way, the Russell paradox is a set-theoretic echo of an earlier paradox, one that was known to the ancient Greeks and is called the Epimenides paradox by some philosophers. That's the paradox of the liar: "This statement is false!" "What I'm now saying is false, it's a lie." Well, is it false? If it's false, if something is false, then it doesn't correspond with reality. So if I'm saying this statement is false, that means that it's not false—which means that it must be true. But if it's true, and I'm saying it's false, then it must be false! So whatever you do you're in trouble!

So you can't get a definite logical truth value, everything flip flops, it's neither true nor false. And you might dismiss this and say that these are just meaningless word games, that it's not serious. But Kurt Gödel later built his work on these paradoxes, and he had a very different opinion.

He said that Bertrand Russell made the amazing discovery that our logical intuitions, our mathematical intuitions, are self-contradictory, they're inconsistent! So Gödel took Russell very seriously, he didn't think that it was all a big joke.

Now I'd like to move on and tell you about David Hilbert's rescue plan for dealing with the crisis provoked by Cantor's set theory and by Russell's paradoxes.

4   David Hilbert to the Rescue with Formal Axiomatic Theories

One of the reactions to the crisis provoked by Cantor's theory of infinite sets was the suggestion to escape into *formalism*. If we get into trouble with reasoning that seems okay, then one solution is to use symbolic logic, to create an artificial language where we're going to be very careful and say what the rules of the game are, and make sure that we don't get the contradictions. Because here's a piece of reasoning that looks okay but it leads to a contradiction, we'd like to get rid of that. Natural language is ambiguous—you never know what a pronoun refers to. So let's create an artificial language and make things very, very precise and make sure that we get rid of all the contradictions! So this was the notion of formalism.

Now I don't think that Hilbert actually intended that mathematicians should work in such a perfect artificial language. It would sort of be like a programming language, but for reasoning, for doing mathematics, for deduction, not for computing, that was Hilbert's idea. But he never expressed it that way, because there were no programming languages back then.

So what are the ideas here? First of all, Hilbert stressed the importance of the axiomatic method.

The notion of doing mathematics that way goes back to the ancient Greeks and particularly to Euclidean geometry, which is a beautifully clear mathematical system. But that's not enough; Hilbert was also saying that we should use symbolic logic.

And symbolic logic also has a long history: Leibniz, Boole, Frege, Peano...These mathematicians wanted to make reasoning like algebra. Here's how Leibniz put it: He talked about avoiding disputes—and he was probably thinking of political disputes and religious disputes—by calculating who was right instead of arguing about it! Instead of fighting, you should be able to sit down at a table and say, "Gentleman, let us compute!" What a beautiful fantasy!...

So the idea was that mathematical logic should be like arithmetic and you should be able to just grind out a conclusion, no uncertainty, no questions of interpretation. By using an artificial math language with a symbolic logic you should be able to achieve *perfect rigor*. The idea is that an argument is either completely correct or else it's total nonsense, with nothing in between. And a proof that is formulated in a formal axiomatic system should be absolutely clear, it should be completely sharp!

In other words, Hilbert's idea was that we should be completely precise about what the rules of the game are, and about the definitions, the elementary concepts, and the grammar and the language—all the rules of the game—so that we can all agree on how mathematics should be done. In practice it would be too much work to use such a formal axiomatic system, but it would be philosophically significant because it would settle once and for all the question of whether a piece of mathematical reasoning is correct or incorrect.

Hilbert's idea seemed fairly straightforward. He was just following the axiomatic and the formal traditions in mathematics. Formal as in formalism, as in using formulas, as in calculating! He wanted to go all the way, to the very end, and formalize all of mathematics, but it seemed like a fairly reasonable plan. Hilbert wasn't a revolutionary, he was a conservative...The amazing thing, as I said before, was that it turned out that Hilbert's rescue plan *could not work*, that it couldn't be done, that it was *impossible* to make it work!

Hilbert was just following the whole mathematics tradition up to that point: the axiomatic method, symbolic logic, formalism...He wanted to avoid the paradoxes by being absolutely precise, by creating a completely formal axiomatic system, an artificial language, that avoided the paradoxes, that made them impossible, that *outlawed* them! And most mathematicians probably thought that Hilbert was right, that *of course* you could do this— it's just the notion that in mathematics things are absolutely clear, black or white, true or false.

So Hilbert's idea was just an extreme, an exaggerated version of the normal notion of what mathematics is all about: the idea that we can decide and agree on the rules of the game, all of them, once and for all. The big surprise is that it turned out that this *could not* be done. Hilbert turned out to be wrong, but wrong in a tremendously fruitful way, because he had asked a very good question. In fact, by asking this question he actually created an entirely new field of mathematics called **meta**mathematics.

Metamathematics is mathematics turned inward, it's an introspective field of math in which you study what mathematics can achieve or can't achieve.

5   What is Metamathematics?

That's my field—metamathematics! In it you look at mathematics from above, and you use mathematical reasoning to discuss what mathematical reasoning can or cannot achieve. The basic idea is this: Once you entomb mathematics in an artificial language *à la* Hilbert, once you set up a completely formal axiomatic system, then you can forget that it has any meaning and just look at it as a game that you play with marks on paper that enables you to deduce theorems from axioms. You can forget about the meaning of this game, the game of mathematical reasoning, it's just combinatorial play with symbols! There are certain rules, and you can study these rules and forget that they have any meaning!

What things do you look at when you study a formal axiomatic system from above, from the outside? What kind of questions do you ask?

Well, one question you can ask is if you can prove that "0 equals 1"?

Hopefully you can't, but how can you be sure? It's hard to be sure!

And for any question A, for any affirmation A, you can ask if it's possible to settle the matter by either proving A or the opposite of A, not A. That's called *completeness*.

A formal axiomatic system is complete if you can settle any question A, either by proving it (A), or by proving that it's false $(\neg A)$. That would be nice! Another interesting question is if you can prove an assertion (A) and you can also prove the contrary assertion $(\neg A)$. That's called *inconsistency*.

So what Hilbert did was to have the remarkable idea of creating a new field of mathematics whose subject would be mathematics itself. But you can't do this until you have a completely formal axiomatic system. Because as long as any "meaning" is involved in mathematical reasoning, it's all subjective. Of course, the reason we do mathematics is because it has meaning, right?. But if you want to be able to study mathematics, the power of mathematics, using mathematical methods, you have to "desiccate" it to "crystallize out" the meaning and just be left with an artificial language with completely precise rules, in fact, with one that has a *mechanical proof-checking algorithm.*

The key idea that Hilbert had was to envision this perfectly desiccated or crystallized axiomatic system for all of mathematics, in which the rules would be so precise that if someone had a proof there would be a referee, there would be a mechanical procedure, which would either say "This proof obeys the rules" or "This proof is wrong; it's breaking the rules". That's how you get the criterion for mathematical truth to be completely objective and not to depend on meaning or subjective understanding: by reducing it all to calculation. Somebody says "This is a proof", and instead of having to submit it to a human referee who takes two years to decide if the paper is correct, instead you just give it to a machine. And the machine eventually says "This obeys the rules" or "On line 4 there's a misspelling" or "This thing on line 4 that supposedly follows from line 3, actually doesn't". And that would be the end, no appeal!

The idea was not that mathematics should actually be done this way. I think that that's calumny, that's a false accusation. I don't think that Hilbert really wanted to turn mathematicians into machines. But the idea was that if you could take mathematics and do it this way, then you could use mathematics to study the power of mathematics. And *that* is the important new thing that Hilbert came up with. Hilbert wanted to do this in order to reaffirm the traditional view of mathematics, in order to justify himself. . .

He proposed having one set of axioms and this formal language, this formal system, which would include all of mathematical reasoning, that we could all agree on, and that would be *perfect!* We'd then know all the rules of the game. And he just wanted to use metamathematics to show that this formal axiomatic system was good—that it was consistent and that it was complete—in order to convince people to accept it. This would have settled once and for all the philosophical questions "When is a proof correct?" and "What is mathematical truth?" Like this everyone could agree on whether a mathematical proof is correct or not. And in fact we used to think that this was an objective thing.

In other words, Hilbert's just saying, if it's really objective, if there's no subjective element, and a mathematical proof is either true or false, well, then there should be certain rules for deciding that and it shouldn't depend, if you fill in all the details, it shouldn't depend on interpretation. It's important to fill in all the details—that's the idea of mathematical logic, to "atomize" mathematical reasoning into such tiny steps that *nothing* is left to the imagination, nothing is left out! And if nothing is left out, then a proof can be checked automatically, that was Hilbert's point, that's really what symbolic logic is all about.

And Hilbert thought that he was actually going to be able to do this. He was going to formalize all of mathematics, and we were all going to agree that these were in fact the rules of the game. Then there'd be just one version of mathematical truth, not many variations. We don't want to have a German mathematics and a French mathematics and a Swedish mathematics and an American mathematics, no, we want a *universal* mathematics, one universal criterion for mathematical truth! Then a paper that is done by a mathematician in one country can be understood by a mathematician in another country. Doesn't that sound reasonable?! So you can imagine just how very, very shocking it was in 1931 when Kurt Gödel showed that it wasn't at all reasonable, that it could *never* be done!

6 Kurt Gödel Discovers Incompleteness

Gödel did this is Vienna, but he was from the Czech Republic, from the city of Brünn or Brno. And later he was at the Institute for Advanced Study in Princeton,

Gödel sort of exploded this whole view of what mathematics is all about. He came up with a famous incompleteness result, "Gödel's incompleteness theorem".

There's a lovely book explaining the way Gödel originally did it. It's by Nagel and Newman, and it's called *Gödel's Proof*. I read it when I was a child, and forty years later it's still in print!

What is this amazing result of Gödel's? Gödel's amazing discovery is that Hilbert was *wrong*, that it cannot be done, that there's *no way* to take all of mathematical truth and to agree on a set of rules and to have a formal axiomatic system for all of mathematics in which it is crystal clear whether something is correct or not!

More precisely, what Gödel discovered was that if you just try to deal with elementary arithmetic, with 0, 1, 2, 3, 4,... and with addition and multiplication

$$+ \times 0, 1, 2, 3, 4, \ldots$$

(this is "elementary number theory" or "arithmetic") and you just try to have a set of axioms for this—the usual axioms are called Peano arithmetic—even this can't be done! Any set of axioms that tries to have **the whole truth** and *nothing but the truth* about addition, multiplication, and 0, 1, 2, 3, 4, 5, 6, 7, 8, 9, 10,... will have to be incomplete. More precisely, it'll either be inconsistent or it'll be incomplete. So if you assume that it only tells the truth, then it won't tell the whole truth. There's no way to capture all the truth about addition, multiplication, and 0, 1, 2, 3, 4,...! In particular, if you assume that the axioms don't allow you to prove false theorems, then it'll be incomplete, there'll be true theorems that you cannot prove from these axioms!

This is an absolutely devastating result, and all of traditional mathematical philosophy ends up in a heap on the floor! At the time this was considered to be absolutely devastating. However you may notice that in 1931 there were also a few other problems to worry about. The situation in Europe was bad. There was a major depression, and a war was brewing. I agree, not all problems are mathematical! There's more to life than epistemology! But you begin to wonder, well, if the traditional view of mathematics isn't correct, then *what is correct?* Gödel's incompleteness theorem was very surprising and a terrible shock.

How did Gödel do it? Well, Gödel's proof is very clever. It almost looks crazy, it's very paradoxical. Gödel starts with the paradox of the liar, "I'm false!", which is neither true nor false.

"This statement is false!"

And what Gödel does is to construct a statement that says of itself "I'm unprovable!"

"This statement is unprovable!"

Now if you can construct such a statement in elementary number theory, in arithmetic, a mathematical statement—I don't know how you make a mathematical statement say it's unprovable, you've got to be very clever—but if you can do it, it's easy to see that you're in trouble. Just think about it a little bit. It's easy to see that you're in trouble. Because if it's provable, it's false, right? So you're in trouble, you're proving false results. And if it's unprovable and it says that it's unprovable, then it's true, and mathematics is incomplete. So either way, you're in trouble! Big trouble!

And Gödel's original proof is very, very clever and hard to understand. There are a lot of complicated technical details. But if you look at his original paper, it seems to me that there's a lot of LISP programming in it, or at least something that looks a lot like LISP programming. Anyway, now we'd call it LISP programming. Gödel's proof involves defining a great many functions recursively, and these are functions dealing with lists, which is precisely what LISP is all about. So even though there were no programming languages in 1931, with the benefit of hindsight you can clearly see a programming language in Gödel's original paper. And the programming language I know that's closest to it is LISP, pure LISP, LISP without side-effects, interestingly enough—that's the heart of LISP.

So this was a very, very shocking result, and people didn't really know what to make of it.

Now the next major step forward comes only five years later, in 1936, and it's by Alan Turing.

7   Alan Turing Discovers Uncomputability

Turing's approach to all these questions is completely different from Gödel's, and much deeper. Because Turing brings it out of the closet! What he brings out of the closet is the computer! The computer was implicit in Gödel's paper, but this was really not visible to any ordinary mortal, not at that time, only with hindsight. And Turing really brings it out in the open.

Hilbert had said that there should be a "mechanical procedure" to decide if a proof obeys the rules or not. And Hilbert never clarified what he meant by a mechanical procedure, it was all words. But, Turing said, what you really mean is a machine, and a machine of a kind that we now call a Turing machine—but it wasn't called that in Turing's original paper. In fact, Turing's original paper contains a programming language, just like Gödel's paper does, what we would now call a programming language. But the two programming languages are very different. Turing's programming language isn't a high-level language like LISP, it's more like a machine language. In fact, it's a horrible machine language, one that nobody would want to use today, because it's too simple.

But Turing makes the point that even though Turing machines are very simple, even though their machine language is rather primitive, they're very flexible, very general-purpose machines. In fact, he claims, any computation that a human being can perform, should be *possible* to do using such a machine. Turing's train of thought now takes a very dramatic turn. What, he asks, is *impossible* for such a machine? What can't it do? And he immediately finds a question that no Turing machine can settle, a problem that no Turing machine can solve. That's *the halting problem,* the problem of deciding in advance if a Turing machine or a computer program will eventually halt.

So the shocking thing about this 1936 paper is that first of all he comes up with the notion of a general-purpose or universal computer, with a machine that's flexible, that can do what any machine can do. One calculating machine that can do *any* calculation, which is, we now say, a general-purpose computer. And then he immediately shows that there are limits to what such a machine can do. And how does he find something that cannot be done by any such machine? Well, it's very simple! It's the question of whether a computer program will eventually halt, with no time limit.

If you put a time limit, it's very easy. If you want to know if a program halts in a year, you just run it for a year, and either it halted or doesn't. What Turing showed is that you get in terrible trouble if there's no time limit. Now you may say, "What good is a computer program that takes more than a year, that takes more than a thousand years?! There's always a time limit!" I agree, this is pure math, this is not the real world. You only get in trouble with *infinity*! But Turing shows that if you put no time limit, then you're in real difficulties.

So this is called *the halting problem.* And what Turing showed is that there's no way to decide in advance if a program will eventually halt.

If it *does* halt, by running it you can eventually discover that, if you're just patient. The problem is you don't know when to give up. And Turing was able to show with a very simple argument which is just Cantor's diagonal argument—coming from Cantor's theory of infinite sets, by the way—I don't have time to explain all this—with a very simple argument Turing was able to show that this problem cannot be solved.

No computer program can tell you in advance if another computer program will eventually halt or not. And the problem is the ones that don't halt, that's really the problem. The problem is knowing when to give up.

So now the interesting thing about this is that Turing immediately deduces as a corollary that if there's no way to decide in advance *by a calculation* if a program will halt or not, then there cannot be any way to *deduce* it in advance using reasoning either. No formal axiomatic system can enable you to deduce in advance whether a program will halt or not.

Because if you can use a formal axiomatic system to always deduce whether a program will halt or not, that will give you a way to calculate in advance whether a program will halt or not. You simply run through all possible deductions—you can't do this in practice—but in principle you can run through all possible proofs in size order, checking which ones are correct, until either you find a proof that the program will halt eventually or you find a proof that it's never going to halt.

This is using the idea of a completely formal axiomatic system where you don't need a mathematician—you just run through this calculation on a computer—it's mechanical to check if a proof is correct or not. So if there were a formal axiomatic system which always would enable you to prove, to deduce, whether a program will halt or not, that would give you a way to calculate in advance whether a program will halt or not. And that's impossible, because you get into a paradox like "This statement is false!" You get a program that halts if and only if it doesn't halt, that's basically the problem. You use an argument having the same flavor as the Russell paradox.

So Turing went more deeply into these questions than Gödel. As a student I read Gödel's proof, and I could follow it step by step. But I couldn't really feel that I was coming to grips with Gödel's proof, that I could really understand it. The whole thing seemed too delicate, it seemed too fragile, it seemed too superficial... And there's this business in the closet about computing, that's there in Gödel, but it's hidden, it's not in the open, we're not really coming to terms with it.

Now Turing is really going, I think, much deeper into this whole matter. And he's showing, by the way, that it's not just one particular axiomatic system, the one that Gödel studied, that can't work, but that *no* formal axiomatic system can work. But it's in a slightly different context. Gödel was really looking at 0, 1, 2, 3, 4... and addition and multiplication, and Turing is looking at a rather strange mathematical question, which is does a program halt or not. It's a mathematical question *that did not exist* at the time of Gödel's original paper. So you see, Turing worked with completely new concepts...

But Gödel's paper is not only tremendously clever, he had to have the courage to imagine that Hilbert might be wrong. There's another famous mathematician of that time, von Neumann. Von Neumann was probably as clever as Gödel or anyone else, but it never occurred to him that Hilbert could be wrong. And the moment that he heard Gödel explain his result, von Neumann immediately appreciated it and immediately started deducing consequences. But von Neumann said, "I missed it, I missed the boat, I didn't get it right!" And Gödel did, so he was much more profound...

Now Turing's paper is also full of technical details, like Gödel's paper, because there is a programming language in Turing's paper, and Turing also gives a rather large program, which of course has bugs, because he wasn't able to run it and debug it—it's the program for a universal Turing machine. But the basic thing is the ideas, and the new ideas in Turing's work are just breathtaking! So I think that Turing went beyond Gödel, but you have to recognize that Gödel took the first step, and the first step is historically the most difficult one and takes the most courage. To imagine that Hilbert could be wrong, which never occurred to von Neumann, that was something!

8   I Discover Randomness in Pure Mathematics

Okay, so then what happened? Then World War II begins. Turing starts working on cryptography, von Neumann starts working on how to calculate atom bomb detonations, and people forget about incompleteness for a while.

This is where I show up on the scene. The generation of mathematicians who were concerned with these questions basically passes from the scene with World War II. And I'm a kid in the 1950s in the United States reading the original article by Nagel and Newman in *Scientific American* in 1956 that became their book.

And I didn't realize that mathematicians really preferred to forget about Gödel and go on working on their favorite problems. I'm fascinated by incompleteness and I want to understand it. Gödel's incompleteness result fascinates me, but I can't really understand it, I think there's something fishy... As for Turing's approach, I think it goes much deeper, but I'm still not satisfied, I want to understand it better.

And I get a funny idea about randomness... I was reading a lot of discussions of another famous intellectual issue when I was a kid—not the question of the foundations of mathematics, the question of the foundations of *physics*! These were discussions about relativity theory and cosmology and even more often about quantum mechanics, about what happens in the atom. It seems that when things are very small the physical world behaves in a completely crazy way that is totally unlike how objects behave in real life. In fact things are **random**—intrinsically unpredictable—in the atom.

Einstein hated this. Einstein said that "God doesn't play dice!" By the way, Einstein and Gödel were friends at Princeton.

I was reading about all of this, and I began to wonder—in the back of my head I began to ask myself—could it be that there was also randomness in pure mathematics?

The idea in quantum mechanics is that randomness is fundamental, it's a basic part of the universe. In normal, everyday life we know that things are unpredictable, but in theory, in Newtonian physics and even in Einstein's relativity theory—that's all called *classical* as opposed to *quantum* physics—in theory in classical physics you can predict the future. The equations are *deterministic*, not *probabilistic*. If you know the initial conditions exactly, with infinite precision, you apply the equations and you can predict with infinite precision any future time and even in the past, because the equations work either way, in either direction. The equations don't care about the direction of time...

This is that wonderful thing sometimes referred to as *Laplacian determinism.* I think that it's called that because of Laplace's *Essai Philosophique sur les Probabilités,* a book that was published almost two centuries ago. At the beginning of this book Laplace explains that by applying Newton's laws, in principle a demon could predict the future arbitrarily far, or the past arbitrarily far, if it knew the exact conditions at the current moment. This is not the type of world where you talk about free will and moral responsibility, but if you're doing physics calculations it's a great world, because you can calculate everything!

But in the 1920s with quantum mechanics it began to look like God plays dice in the atom, because the basic equation of quantum mechanics is the Schrödinger equation, and the Schrödinger equation is an equation that talks about the *probability* that an electron will do something. The basic quantity is a probability and it's a wave equation saying how a probability wave interferes with itself. So it's a completely different kind of equation, because in Newtonian physics you can calculate the precise trajectory of a particle and know exactly how it's going to behave. But in quantum mechanics the fundamental equation is an equation dealing with probabilities!

You *can't know* exactly where an electron is and what its velocity vector is—exactly what direction and how fast it's going. It doesn't have a specific state that's known with infinite precision the way it is in classical physics. If you know very accurately where an electron is, then its velocity—its momentum—turns out to be wildly uncertain. And if you know exactly in which direction and at what speed it's going, then its position becomes infinitely uncertain. That's the infamous *Heisenberg uncertainty principle,* there's a trade-off, that seems to be the way the physical universe works...

It's an interesting historical fact that before people used to hate this— Einstein hated it—but now people think that they can **use** it! There's a crazy new field called *quantum computing* where the idea is that maybe you can make a brand new technology using something called *quantum parallelism*. If a quantum computer is uncertain, maybe you can have it uncertainly do many computations at the same time! So instead of fighting it, the idea is to use it, which is a great idea.

But when I was a kid people were still arguing over this. Even though he had helped to create quantum mechanics, Einstein was still fighting it, and people were saying, "Poor guy, he's obviously past his prime!"

I began to think that maybe there's also randomness in pure mathematics. I began to suspect that maybe that's *the real reason for incompleteness*. A case in point is elementary number theory, where there are some very difficult questions. Take a look at the prime numbers. Individual prime numbers behave in a very unpredictable way, if you're interested in their detailed structure. It's true that there are *statistical* patterns. There's a thing called *the prime number theorem* that predicts fairly accurately the over-all average distribution of the primes. But as for the detailed distribution of individual prime numbers, that looks pretty random!

So I began to think about *randomness*...I began to think that maybe that's what's really going on, maybe that's a deeper reason for all this incompleteness. So in the 1960s I, and independently some other people, came up with some new ideas. And I like to call this new set of ideas *algorithmic information theory*.

That name makes it sound very impressive, but the basic idea is just to look at the size of computer programs. You see, it's just a *complexity measure*, it's just a kind of *computational complexity*...

I think that one of the first places that I heard about the idea of computational complexity was from von Neumann. Turing came up with the idea of a computer as a mathematical concept—it's a perfect computer, one that never makes mistakes, one that has as much time and space as it needs to work—it's always finite, but the calculation can go on as long as it has to. After Turing comes up with this idea, the next logical step for a mathematician is to study the time, the work needed to do a calculation—its complexity. And in fact I think that around 1950 von Neumann suggested somewhere that there should be a new field which looks at the *time* complexity of computations, and that's now a very well-developed field. So of course if most people are doing that, then I'm going to try something else!

My idea was not to look at the **time**, even though from a practical point of view time is very important. My idea was to look at the *size* of computer programs, at the amount of information that you have to give a computer to get it to perform a given task. From a practical point of view, the amount of information required isn't as interesting as the running time, because of course it's very important for computers to do things as fast as possible... But it turns out that from a conceptual point of view, it's not that way at all. I believe that from a fundamental philosophical point of view, the right question is to look at the *size* of computer programs, not at the *time*. The reason is because program-size complexity connects with a lot of fundamental stuff in physics.

You see, in physics there's a notion called *entropy,* which is how disordered a system is. Entropy played a particularly crucial role in the work of the famous 19th century physicist Boltzmann, and it comes up in the field of statistical mechanics and in thermodynamics. Entropy measures how disordered, how chaotic, a physical system is. A crystal has low entropy, and a gas at high temperature has high entropy. It's the amount of chaos or disorder, and it's a notion of randomness that physicists like.

And entropy is connected with some fundamental philosophical questions—it's connected with the question of the *arrow of time,* which is another famous controversy. When Boltzmann invented this wonderful thing called statistical mechanics—his theory is now considered to be one of the masterpieces of 19th century physics, and all physics is now statistical physics—he ended up by committing suicide, because people said that his theory was *obviously* wrong! Why was it obviously wrong? Because in Boltzmann's theory entropy has got to increase and so there's an arrow of time. But if you look at the equations of Newtonian physics, they're time reversible. There's no difference between predicting the future and predicting the past. If you know at one instant exactly how everything is, you can go in either direction, the equations don't care, there's no direction of time, backward is the same as forward.

But in everyday life and in Boltzmann statistical mechanics, there is a difference between going backward and forward. Glasses break, but they don't reassemble spontaneously! And in Boltzmann's theory entropy has got to increase, the system has to get more and more disordered. But people said, "You can't deduce that from Newtonian physics!" Boltzmann was pretending to. He was looking at a gas. The atoms of a gas bounce around like billiard balls, it's a billiard ball model of how a gas works. And each interaction is reversible. If you run the movie backwards, it looks the same. If you look at a small portion of a gas for a small amount of time, you can't tell whether you're seeing the movie in the right direction or the wrong direction.

But Boltzmann gas theory says that there is an arrow of time—a system will start off in an ordered state and will end up in a very mixed up disordered state. There's even a scary expression in German, *heat death.* People said that according to Boltzmann's theory the universe is going to end up in a horrible ugly state of maximum entropy or heat death! This was the dire prediction! So there was a lot of controversy about his theory, and maybe that was one of the reasons that Boltzmann killed himself.

And there is a connection between my ideas and Boltzmann's, because looking at the size of computer programs is very similar to this notion of the degree of disorder of a physical system. A gas takes a large program to say where all its atoms are, but a crystal doesn't take as big a program, because of its regular structure. Entropy and program-size complexity are closely related...

This idea of program-size complexity is also connected with the philosophy of the scientific method. You've heard of *Occam's razor,* of the idea that the simplest theory is best? Well, what's a theory? It's a computer program for predicting observations. And the idea that the simplest theory is best translates into saying that a *concise* computer program is the best theory. What if there is no concise theory, what if the most concise program or the best theory for reproducing a given set of experimental data is *the same size* as the data? Then the theory is no good, it's cooked up, and the data is incomprehensible, it's random. In that case the theory isn't doing a useful job. A theory is good to the extent that it compresses the data into a much smaller set of theoretical assumptions. The greater the compression, the better!—That's the idea...

So this idea of program size has a lot of philosophical resonances, and you can define randomness or maximum entropy as something that cannot be compressed at all. It's an object with the property that basically the only way you can describe it to someone is to say "this is it" and show it to them. Because it has no structure or pattern, there is no concise description, and the thing has to be understood as "a thing in itself", it's irreducible.

Randomness = Incompressibility

The other extreme is an object that has a very regular pattern so you can just say that it's "a million 0s" or "half a million repetitions of 01", pairs 01, 01, 01 repeated half a million times. These are very long objects with a very concise description. Another long object with a concise description is an *ephemeris,* I think it's called that, it's a table giving the positions of the planets as seen in sky, daily, for a year. You can compress all this astronomical information into a small FORTRAN program that uses New-tonian physics to calculate where the planets will be seen in the sky every night.

But if you look at how a roulette wheel behaves, then there is no pattern, the series of outcomes cannot be compressed. Because if there were a pattern, then people could use it to win, and having a casino wouldn't be such a good business! The fact that casinos make lots of money shows that there is no way to predict what a roulette wheel will do, there is no pattern—the casinos make it their job to ensure that!

So I had this new idea, which was to use program-size complexity to define randomness. And when you start looking at the size of computer programs—when you begin to think about this notion of program-size or information complexity instead of run-time complexity—then the inter-esting thing that happens is that everywhere you turn you immediately find incompleteness! You immediately find things that escape the power of mathematical reasoning, things that escape the power of any computer program. It turns out that they're everywhere!

It's very dramatic! In only three steps we went from Gödel, where it's very surprising that there are limits to reasoning, to Turing, where it looks much more natural, and then when you start looking at program size, well, incompleteness, the limits of mathematics, it just hits you in the face! Why?! Well, *the very first question* that you ask in my theory gets you into trouble. What's that? Well, in my theory I measure the complexity of something by the size of the smallest computer program for calculating it. But how can I be sure that I have the smallest computer program?

Let's say that I have a particular calculation, a particular output, that I'm interested in, and that I have this nice, small computer program that calculates it, and I think that it's the smallest possible program, the most concise one that produces this output. Maybe a few friends of mine and I were trying to do it, and this was the best program that we came up with; nobody did any better. But how can you be **sure**? Well, the answer is that *you can't be sure.* It turns out you *can never be sure!* You can *never* be sure that a computer program is what I like to call *elegant,* namely that it's the most concise one that produces the output that it produces. *Never ever!* This escapes the power of mathematical reasoning, amazingly enough.

But for any computational task, once you fix the computer program-ming language, once you decide on the computer programming language, and if you have in mind a particular output, there's got to be at least one program that is the smallest possible. There may be a tie, there may be several, right?, but there's got to be at least *one* that's smaller than all the others. But you can never be sure that you've found it!

And the precise result, which is one of my favorite incompleteness results, is that if you have N bits of axioms, you can never prove that a program is elegant—smallest possible—if the program is more than N bits long. That's basically how it works. So any given set of mathematical axioms, any formal axiomatic system in Hilbert's style, can only prove that *finitely many* programs are elegant, are the most concise possible for their output.

To be more precise, you get into trouble with an elegant program if it's larger than a computerized version of the axioms—It's really the size of the proof-checking program for your axioms. In fact, it's the size of the program that runs through all possible proofs producing all possible theorems. If you have in mind a particular programming language, and you need a program of a certain size to implement a formal axiomatic system, that is to say, to write the proof-checking algorithm and to write the program that runs through all possible proofs filtering out all the theorems, if that program is a certain size in a language, and if you look at programs in that same language that are larger, then you can never be sure that such a program is elegant, you can never prove that such a program is elegant using the axioms that are implemented in the same language by a smaller program. That's basically how it works.

So there are an infinity of elegant programs out there. For any computational task there's got to be at least one elegant program, and there may be several, but you can never be sure except in a finite number of cases.

So it turns out that you can't calculate the program-size complexity, you can never be sure what the program-size complexity of anything is. Because to determine the program-size complexity of something is to know the size of the most concise program that calculates it—but that means—it's essentially the same problem—then I would know that this program is the most concise possible, I would know that it's an elegant program, and you can't do that if the program is larger than the axioms. So if it's N bits of axioms, you can never determine the program-size complexity of anything that has more than N bits of complexity, which means almost everything, because almost everything has more than N bits of complexity. Almost everything has more complexity than the axioms that you're using.

Why do I say that? The reason for using axioms is because they're simple and believable. So the sets of axioms that mathematicians normally use are fairly concise, otherwise no one would believe in them! Which means that in practice there's this vast world of mathematical truth out there, which is an infinite amount of information, but any given set of axioms only captures a tiny finite amount of this information! And that's why we're in trouble, that's my bottom line, that's my final conclusion, that's the real dilemma.

So in summary, I have two ways to explain why I think Gödel incompleteness is natural and inevitable rather than mysterious and surprising. The two ways are—that the idea of randomness in physics, that some things make no sense, also happens in pure mathematics, is one way to say it. But a better way to say it, is that mathematical truth is an infinite amount of information, but any particular set of axioms just has a finite amount of information, because there are only going to be a finite number of principles that you've agreed on as the rules of the game. And whenever any statement, any mathematical assertion, involves more information than the amount in those axioms, then it's very natural that it will escape the ability of those axioms.

So you see, the way that mathematics progresses is you trivialize everything! The way it progresses is that you take a result that originally required an immense effort, and you reduce it to a trivial corollary of a more general theory!

Let me give an example involving Fermat's "last theorem", namely the assertion that

$$x^n + y^n = z^n$$

has no solutions in positive integers x, y, z, and n with n greater than 2. Andrew Wiles's recent proof of this is hundreds of pages long, but, probably, a century or two from now there will be a one-page proof! But that one-page proof will require a whole book inventing a theory with concepts that are the natural concepts for thinking about Fermat's last theorem. And when you work with those concepts it'll appear immediately obvious—Wiles's proof will be a trivial afterthought—because you'll have imbedded it in the appropriate theoretical context.

And the same thing is happening with incompleteness.

Gödel's result, like any very fundamental basic result, starts off by being very mysterious and complicated, with a long impenetrable proof. People said about Gödel's original paper the same thing that they said about Einstein's theory of relativity, which is that there are less than five people on this entire planet who understand it.

So in 1931 Gödel's proof was like that. If you look at his original paper, it's very complicated. The details are programming details we would say now—really it's a kind of complication that we all know how to handle now—but at the time it looked very mysterious. This was a 1931 mathematics paper, and all of a sudden you're doing what amounts to LISP programming, thirty years before LISP was invented! And there weren't even any computers then!

But when you get to Turing, he makes Gödel's result seem much more natural. And I think that my idea of program-size complexity and information—really, algorithmic information content—makes Gödel's result seem more than natural, it makes it seem, I'd say, obvious, inevitable. But of course that's the way it works, that's how we progress.

9 Where Do We Go from Here?!

I should say, though, that if this were really true, if it were *that* simple, then that would be the end of the field of metamathematics. It would be a sad thing, because it would mean that this whole subject is dead. But I don't think that it is!

I'we been giving versions of this talk for many years. In these talks I like to give examples of things that might escape the power of normal mathematical reasoning. And my favorite examples were Fermat's last theorem, the Riemann hypothesis, and the four-color conjecture. When I was a kid these were the three most outstanding open questions in all of mathematics.

But a funny thing happened. First the four-color conjecture was settled by a computer proof, and recently the proof has been greatly improved. The latest version has more ideas and less computation, so that's a big step forward. And then Wiles settled Fermat's last theorem. There was a misstep, but now everyone's convinced that the new proof is correct.

Fortunately the Riemann hypothesis is still open at this point, as far as I know!

But I was using Fermat's last theorem as a possible example of incompleteness, as an example of something that might be beyond the power of the normal mathematical methods. I needed a good example, because people used to say to me, "Well, this is all very well and good, Algorithmic Information Theory is a nice theory, but give me an example of a specific mathematical result that you think escapes the power of the usual axioms." And I would say, well, maybe Fermat's last theorem!

So there's a problem. Algorithmic information theory is very nice and shows that there are lots of things that you can't prove, but what about individual mathematical questions? How about a natural mathematical question? Can these methods be applied? Well, the answer is no, my methods are not as general as they sound. There are technical limitations. I can't analyze Fermat's last theorem with these methods. Fortunately! Because if I had announced that my methods show that Fermat's last theorem can't be settled, then it's very embarrassing when someone settles it!

So now the question is, how come in spite of these negative results, mathematicians are making so much progress? How come mathematics works so well in spite of incompleteness? You know, I'm not a pessimist, but my results have the wrong kind of feeling about them, they're much too pessimistic!

So I think that a very interesting question now is to look for positive results...There are already too many negative results! If you take them at face value, it would seem that there's no way to do mathematics, that mathematics is impossible. Fortunately for those of us who do mathematics, that doesn't seem to be the case. So I think that now we should look for positive results...The fundamental questions, like the questions of philosophy, they're great, because you never exhaust them. Every generation takes a few steps forward...So I think there's a lot more interesting work to be done in this area.

And here's another very interesting question: program size is a complexity measure, and we know that it works great in metamathematics, but does it have anything to do with complexity in the real world? For example, what about the complexity of biological organisms? What about a theory of evolution?

Von Neumann talked about a general theory of the evolution of life. He said that the first step was to define complexity. Well, here's a definition of complexity, but it doesn't seem to be the correct one to use in theoretical biology. And there is no such thing as theoretical biology, not yet!

As a mathematician, I would love it if somebody would prove a general result saying that under very general circumstances life has to evolve. But I don't know how you define life in a general mathematical setting. We know it when we see it, right? But as a mathematician I don't know how to tell the difference between a beautiful deer running across the road and the pile of garbage that my neighbor left out in the street! Well, actually that garbage is connected with life, it's the debris produced by life...

So let's compare a deer with a rock instead. Well, the rock is harder, but that doesn't seem to go to the essential difference that the deer is alive and the rock is a pretty passive object. It's certainly very easy for us to tell the difference in practice, but what is the fundamental difference? Can one grasp that mathematically?

So what von Neumann was asking for was a general mathematical theory. Von Neumann used to like to invent new mathematical theories. He'd invent one before breakfast every day: the theory of games, the theory of self-reproducing automata, the Hilbert space formulation of quantum mechanics... Von Neumann wrote a book on quantum mechanics using Hilbert spaces—that was done by von Neumann, who had studied under Hilbert, and who said that this was the right mathematical framework for doing quantum mechanics.

Von Neumann was always inventing new fields of mathematics, and since he was a childhood hero of mine, and since he talked about Gödel and Turing, well, I said to myself, if von Neumann could do it, I think I'll give it a try. Von Neumann even suggested that there should be a theory of the complexity of computations. He never took any steps in that direction, but I think that you can find someplace where he said that this has got to be an interesting new area to develop, and he was certainly right.

Von Neumann also said that we ought to have a general mathematical theory of the evolution of life... But we want it to be a very general theory, we don't want to get involved in low-level questions like biochemistry or geology... He insisted that we should do things in a more general way, because von Neumann believed, and I guess I do too, that if Darwin is right, then it's probably a very general thing.

For example, there is the idea of *genetic programming*, that's a computer version of this. Instead of writing a program to do something, you sort of evolve it by trial and error. And it seems to work remarkably well, but can you prove that this has got to be the case? Or take a look at Tom Ray's *Tierra*... Some of these computer models of biology almost seem to work too well—the problem is that there's no theoretical understanding why they work so well. If you run Ray's model on the computer you get these parasites and hyperparasites, you get a whole ecology. That's just terrific, but as a pure mathematician I'm looking for theoretical understanding, I'm looking for a general theory that starts by defining what an organism is and how you measure its complexity, and that *proves* that organisms have to evolve and increase in complexity. That's what I want, wouldn't that be nice?

And if you could do that, it might shed some light on how general the phenomenon of evolution is, and whether there's likely to be life elsewhere in the universe. Of course, even if mathematicians never come up with such a theory, we'll probably find out by visiting other places and seeing if there's life there... But anyway, von Neumann had proposed this as an interesting question, and at one point in my deluded youth I thought that maybe program-size complexity had something to do with evolution... But I don't think so anymore, because I was never able to get anywhere with this idea...

So I think that there's a lot of interesting work to be done! And I think that we live in exciting times. In fact, sometimes I think that maybe they're even a little bit too exciting!... And I hope that if this talk were being given a century from now, in 2099, there would be *another* century of exciting controversy about the foundations of mathematics to summarize, one with different concerns and preoccupations... It would be interesting to hear what that talk would be like a hundred years from now!

Further Reading

[1] G. J. Chaitin, *The Unknowable,* Springer-Verlag, 1999.
[2] G. J. Chaitin, *The Limits of Mathematics,* Springer-Verlag, 1998.

Finite Versus Infinite in Singularity Dynamics

Florin Diacu[1]

Department of Mathematics and Statistics
University of Victoria, P.O. Box 3045
Victoria, B.C., Canada, V8W 3P4
E-mail: diacu@math.uvic.ca

Abstract. We study the dynamics of particle systems near solution singularities for different types of potentials. Solution singularities appear in finite time, but they can be blown up and removed to infinity with the help of certain transformations, thus creating the framework of a dynamical system. In this paper we present several examples of solution singularities and various ways of removing them by the blow-up technique. These will put into the evidence the benefits of blow-up transformations towards understanding the motion near singularities.

1 Introduction

The beginnings of the study of singularities in the dynamics of particles and systems are difficult to trace in the scattered mathematical and astronomical literature. It is clear, however, that this problem was first considered in connection with the Newtonian n-body problem of celestial mechanics, characterized by Whittaker [16] as "the most celebrated of all dynamical problems." The first significant study appeared in the last chapter of [15], a treatise based on Painlevé's lectures delivered in 1895 at the University of Stockholm upon the invitation of King Oscar II of Sweden and Norway. Details on the origin, content, and history of these lectures can be found in [5] and [9].

The understanding of the qualitative behaviour of solutions near singularities is an important but difficult task. It is important because motion near singularities is unpredictable and it can drastically change the "fate"

---

[1]Supported in part by Grant OGP0122045 of the NSERC of Canada.

not only of an orbit but of the whole phase-space picture and consequently of the global dynamical behaviour. It is difficult because most methods fail to provide significant results. Exact techniques are powerless and numerical ones prove far from reliable for anything but short time intervals. A qualitative method that offers some insight into the matter is that of blow up, introduced in celestial mechanics in the context of regularization techniques by the Italian mathematician Tullio Levi-Cività (see [12] and [13]). In 1974 Richard McGehee of the University of Minnesota (see [14]) improved this method and applied it to understand the triple collision in the rectilinear 3-body problem. The advancement of the field would have been hard to imagine without McGehee's seminal contribution. The blow-up techniques we present below are based on McGehee's idea.

The goal of this paper is to discuss the qualitative blow-up technique for solution singularities and see how it applies to different situations. We will describe it, emphasize its advantages for understanding the behaviour of the solutions near a singularity, and make the connection between the finite and the infinite aspects of it.

In Section 2 we define the notion of *solution singularity*. We will show that a singularity is either due to a physical *collision* or to a so-called *pseudocollision*, which may occur if the motion reaches no limit position while becoming unbounded in finite time.

In Section 3 we will discuss collisions for the Manev potential, which is an approximation of general relativity in the framework of classical mechanics, as proved in [10]. More precisely, the Manev gravitational law is the analogue of the Schwarzschild solution to Einstein's field equations, assuming that the study is restricted to almost circular orbits, for example to planetary motion.

In Section 4 we consider the anisotropic case of the Manev problem, proposed by the present author in 1995. Unlike the classical case, the anisotropic Manev problem is not integrable. This problem stays at the intersection between classical mechanics, quantum mechanics, and general relativity, and can be also seen as an approximation of special relativity. It therefore presents interest for understanding the possible connections and for laying bridges between these fields.

In Section 5 we discuss the blow-up technique for 2-body problems with drag or thrust. Such problems appear in astronomy and astrophysics when studying the motion of microcosmic particles near the gravitational field of a celestial body. The drag or the thrust can be due to the friction with the atmosphere, to magnetic interactions, to radiation pressure, to the solar wind, or to other nongravitational forces that act in the neighbourhood of planets or stars.

In Section 6 we consider the Mücket-Treder gravitational law, which unlike the previous models is given by a potential that is neither homogeneous nor quasihomogeneous. This model was proposed in 1977 by the German astronomers J.P. Mücket and H.-J. Treder (see [7]). We will show that the blow-up technique can also be adapted to such situations and we will apply it to the 3-body problem.

In Section 7 we draw conclusions regarding the importance of blow-up techniques in particle-systems dynamics. We mention the interplay between finite and infinite for understanding the qualitative behaviour of motion near singularities and the constructive and generative role of this dichotomy in mathematics. This paper illustrates one of the various aspects of this role.

2 Solution Singularities

In this paper we will study systems of the form

$$\begin{cases} \dot{\mathbf{q}} = \mathbf{M}^{-1}\mathbf{p} \\ \dot{\mathbf{p}} = \nabla U(\mathbf{q}), \end{cases} \tag{1}$$

where $\mathbf{q} = (\mathbf{q}_1, \mathbf{q}_2, \ldots, \mathbf{q}_n)$ is the *configuration* of an n-particle systems, $\mathbf{q}_i = (q_i^1, q_i^2, q_i^3)$, $i = 1, 2, \ldots, n$, are the position vectors, $\mathbf{p} = \mathbf{M}\dot{\mathbf{q}}$ is the *momentum*, $\mathbf{M}$ is the $3n$-dimensional matrix of the masses, having 0 everywhere except on the main diagonal which is $m_1, m_1, m_1, m_2, m_2, m_2, \ldots, m_n, m_n, m_n$, where $m_i > 0$ for each $i = 1, 2, \ldots, n$. U is a real function of $\mathbf{q}$ called *potential function*, which signifies the negative of the potential energy, ∇ is the gradient operator, and the upper dot denotes differentiation with respect to the independent time-variable t. From the physical point of view such a system describes the motion of n point masses under a force law defined by the potential function U. For example, the potential

$$U(\mathbf{q}) = G \sum_{1 \leq i < j \leq n} \frac{m_i m_j}{|\mathbf{q}_i - \mathbf{q}_j|}, \tag{2}$$

where G is the gravitational constant, defines the Newtonian n-body problem; for

$$U(\mathbf{q}) = G \sum_{1 \leq i < j \leq n} \frac{m_i m_j}{|\mathbf{q}_i - \mathbf{q}_j|} + \frac{3G^2}{2c^2} \sum_{1 \leq i < j \leq n} \frac{m_i m_j (m_i + m_j)}{|\mathbf{q}_i - \mathbf{q}_j|^2}, \tag{3}$$

equations (1) are those of the Manev n-body problem of celestial mechanics (see [2], [4], [6], or [10]). The Lennard-Jones potential (see e.g. [11])

$$U(\mathbf{q}) = - \sum_{2 \leq i < j \leq n} \frac{1}{|\mathbf{q}_i - \mathbf{q}_j|^6} + \sum_{1 \leq i < j \leq n} \frac{1}{|\mathbf{q}_i - \mathbf{q}_j|^{12}} \qquad (4)$$

and the equations (1) with $m_1 = m_2 = \ldots = m_n = 1$ are used in physical chemistry to explain crystal formation. The Liboff potential

$$U(\mathbf{q}) = \sum_{1 \leq i < j \leq n} \frac{1}{|\mathbf{q}_i - \mathbf{q}_j|^3} + \sum_{1 \leq i < j \leq n} \frac{1}{|\mathbf{q}_i - \mathbf{q}_j|^4} \qquad (5)$$

appears in electromagnetics.

Notice that all the above potentials are *quasihomogeneous*, i.e., sums of homogeneous functions. (In fact the potential (2) is homogeneous but (3), (4), and (5) are not.) For U quasihomogeneous, standard results of the theory of differential equations ensure the existence and the uniqueness of an analytic solution for the initial value problem given by system (1) with initial conditions $(\mathbf{q}, \mathbf{p})(0) \in (\mathbb{R}^{3n} \setminus \boldsymbol{\Delta}) \times \mathbb{R}^{3n}$, where

$$\boldsymbol{\Delta} = \bigcup_{1 \leq i < j \leq n} \{\mathbf{q} \in \mathbb{R}^{3n} | \mathbf{q}_i = \mathbf{q}_j\} \qquad (6)$$

is the *collision set*. This solution can be extended analytically to its maximal domain $[0, t^*)$. If $t^* = \infty$, the solution is called *regular*. If $t^* < \infty$, the solution is called *singular*.

In his 1895 Stockholm lectures, Painlevé proved a result concerning singularities for the Newtonian potential (see [3]). Its extension to quasihomogeneous potentials is straightforward. If $(\mathbf{q}, \mathbf{p})$ is an analytic solution of system (1) given by a quasihomogeneous potential U and if this solution is defined on a maximal interval $[0, t^*)$, then t^* is a singularity if and only if $\lim_{t \to t^*} \min(\mathbf{q}(t)) = 0$, where $\min(\mathbf{q}(t))$ is the minimum of the mutual distances between particles at the given time t.

This result allows us to differentiate between two types of solution singularities: those that have a limit and those that don't. We can formulate this as follows. A singularity t^* is due to a collision if $\mathbf{q}$ has a definite limit when $t \to t^*$. The time instant t^* is called a *collision singularity*. If $\mathbf{q}$ has no limit or becomes unbounded when $t \to t^*$, we call t^* a *noncollision singularity* or a *pseudocollision*.

This definition has the following interpretation. If a solution reaches the set $\boldsymbol{\Delta}$, the denominator of $\nabla U(\mathbf{q})$ cancels and system (1) loses its meaning.

The set Δ, however, can be approached either with asymptotic phase (i.e., the solution tends to a certain element of Δ) or without asymptotic phase (i.e., the solution tends to Δ but oscillates among its elements without reaching any of them). In the former case we have a collision singularity and in the latter a noncollision singularity.

The dynamics near Δ is very different for each of the above potentials. For the Newtonian one, the set of initial conditions leading to collisions has measure zero and is of the first Baire category. For $n = 4$ the set of initial conditions leading to pseudocollisions has measure zero and is also of the first Baire category. But the size of this set is unknown in the general case. For the Manev potential the set of initial data leading to collisions has positive measure. The same is true for the Liboff potential. In the case of the Lennard-Jones potential this set is empty because when particles get close, the repelling force overcomes the attractive one, so collisions cannot occur. For more details regarding this type of result see [3].

In his Stockholm lectures Painlevé proved that pseudocollisions do not exist in the Newtonian 2- and 3-body problems and conjectured that they can appear in the Newtonian n-body problem for $n \geq 4$. This problem is notoriously difficult and it took a lot of energy and many attempts to answer it. In 1992, almost a century after the conjecture was stated, Zihong Xia from Northwestern University produced in his Ph.D. thesis the first example of a pseudocollision in the spatial 5-body problem (see [17] and [5]). Shortly after, Joe Gerver from Rutgers University came up with the first example of a pseudocollision for particles moving in a plane (see [3]) for an n-body problem with n large and which exhibits certain symmetries. In principle, Xia's example can be generalized to any $n \geq 5$ but not to $n = 4$. So the existence of pseudocollisions in the 4-body problem is still an open problem. It is important to mention that Xia's example makes extensive use of blow-up techniques, which we will discuss below, first in the context of the Manev problem.

3 The Manev Potential

Let us now consider system (1) with potential (3) for $n = 2$, called the Manev problem after the Bulgarian physicist G. Manev, who proposed it in the 1920s as a classical alternative to general relativity. It is interesting that although Manev's physical derivation of this law was initially based on shaky physical grounds (like ether, for example), his law explains the perihelion advance of the inner planets with the same accuracy as Einstein's theory. In fact, as we proved in [10], there is a strong similarity between the Schwarzschild solution of Einstein's field equations and the Manev potential.

The Manev problem can be written in Hamiltonian form with the help of the Hamiltonian function

$$H(\mathbf{p}, \mathbf{q}) = \frac{1}{2}(|\mathbf{p_1}|^2 + |\mathbf{p_2}|^2) - \frac{1}{|\mathbf{q_1} - \mathbf{q_2}|} - \frac{k}{|\mathbf{q_1} - \mathbf{q_2}|^2}, \tag{7}$$

where $k = \frac{3G^2}{2c^2}$ (see (3)). In this setting, the McGehee transformations for removing the singularity of the equations occurring when $\mathbf{q_1} = \mathbf{q_2}$, reduce to first writing the Hamiltonian in polar coordinates $r > 0$, $\theta \in S^1$, where S^1 is the segment $[0, 2\pi]$ with the end points identified. Thus the Hamiltonian takes the form

$$H(p_r, p_\theta, r) = (1/2)(p_r^2 + p_\theta^2/r^2) - 1/r - k/r^2, \tag{8}$$

where p_r, p_θ are the new polar variables of the momenta.

The next step consists of blowing up the collision singularity that now occurs at $r = 0$. For this we can formally multiply the energy integral by r^2, which takes the form

$$(1/2)r^2 p_r^2 + p_\theta^2 - r - k = hr^2. \tag{9}$$

Further introducing the transformations $v = rp_r$ and $u = p_\theta$ and scaling the time variable by using $dt = r^2 d\tau$, we complete the sequence of blow-up transformations. The equations given by the Hamiltonian (8) become

$$\begin{cases} r' = rv \\ v' = r(1 + 2hr) \\ \theta' = u \\ u' = 0, \end{cases} \tag{10}$$

where the prime denotes differentiation with respect to the fictitious time variable τ. The energy relation (9) gets transformed into

$$v^2 + u^2 - 2r - 2hr^2 = 2k. \tag{11}$$

We define the *collision manifold* as the set of solutions given by relation (11) when $r = 0$. Notice that, geometrically, the collision manifold is a cylinder in the 3-dimensional space of coordinates (u, θ, v), and since $\theta \in [0, 2\pi]$, it follows that this cylinder can be identified with a torus. In fact, the 2-dimensional torus representing the collision manifold is embedded in the 4-dimensional phase space of the coordinates (r, u, θ, v). The equations (10) show that the flow on the collision manifold is formed almost exclusively by periodic orbits, except the upper and lower circles of the torus given by $r = 0$, $u = 0$, $v = \pm\sqrt{2k}$, which consist of equilibrium points (see Figure 1a).

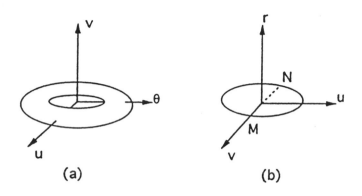

(a)   (b)

Figure 1: (a) The collision manifold embedded in the 4-dimensional phase space, (b) the collision manifold in the reduced phase space

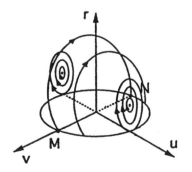

Figure 2: The flow in the reduced phase space for negative energy

Since θ does not appear explicitly in the equations (10) or in the energy relation (11), we can further reduce the 4-dimensional phase space to a

3-dimensional one by factorizing the flow to S^1. Exploiting this symmetry
we can obtain clear pictures of the global flow in phase space. In fact the
global description can be reduced to different energy levels by regarding
the energy constant h as a parameter. Factoring the collision manifold to
S^1, the torus becomes a circle (see Figure 1b). The points M and N on
this circle correspond to the circles of equilibria on the torus, while all the
other points correspond to the periodic orbits on the torus. In the case of
negative energy $h < 0$, for example, the flow in the reduced phase space
looks as in Figure 2.

Through the above blow-up technique we have removed the collision singularity by shifting it to infinity in time, so we can now study the properties
of solutions passing close to this type of collision. The complete qualitative
analysis of these solutions for the Manev potential in the particular case
$n = 2$ is done in [2] and generalized in [10].

4   The Anisotropic Manev Potential

The Manev problem is integrable but its relative, the anisotropic Manev
problem, is not. Its flow is also far from fully understood. The first problem of this type, the anisotropic Kepler problem, was proposed by Martin
Gutzwiller in the late 1960s in an attempt to find a bridge between classical and quantum mechanics (for references see [1]). The anisotropic Manev
problem, suggested by the present author in 1995, attempts even more. It
is a bridge between classical mechanics, quantum mechanics, and general
relativity. As we have shown in [1], this problem exhibits properties from
all three fields.

The anisotropic Manev problem is given by system (1) with potential

$$U(\mathbf{q}) = \frac{1}{\sqrt{\mu q_1^2 + q_2^2}} + \frac{b}{\mu q_1^2 + q_2^2},$$

where $\mathbf{q} = (q_1, q_2)$, and $\mu > 0$ and $b > 0$ are constants. These equations
define the motion of two particles of unit mass in an anisotropic space, i.e.,
one in which the attraction forces act differently in every direction. The
above potential defines the anisotropy of the space as a function of the
parameter μ. If $\mu < 1$, the attraction is the weakest in the direction of the
q_1-axis and the strongest in that of the q_2-axis, the situation being reversed
if $\mu > 1$. If $\mu = 1$, the space is isotropic and we are in the case of the
classical Manev problem discussed above. Since both remaining cases have

a weakest-force and a strongest-force direction, we can assume, without loss of generality, that $\mu > 1$.

The Hamiltonian function of the system is given by

$$H(\mathbf{p}(t), \mathbf{q}(t)) = \frac{1}{2}|\mathbf{p}(t)|^2 - U(\mathbf{q}(t)),$$

the sum of kinetic and potential energies, which yields the energy integral

$$H(\mathbf{p}(t), \mathbf{q}(t)) = h.$$

However, since the force ∇U is not central, the angular momentum $L(t) = \mathbf{p}(t) \times \mathbf{q}(t)$ is *not* an integral of the system, as it is in the classical Manev problem.

To perform the McGehee transformations of the second kind we proceed as follows. We first change the dependent variables using the formulas

$$\begin{cases} r = |\mathbf{q}| \\ \theta = \arctan(q_2/q_1) \\ y = \dot{r} = (q_1 p_1 + q_2 p_2)/|\mathbf{q}| \\ x = r\dot{\theta} = (q_1 p_2 - q_2 p_1)/|\mathbf{q}|, \end{cases}$$

and

$$\begin{cases} v = ry \\ u = rx, \end{cases}$$

and then change the independent variable using

$$d\tau = r^{-2}dt.$$

Composing these transformations, which are analytic diffeomorphisms in their respective domains, the energy relation becomes

$$u^2 + v^2 - 2r\delta^{-1/2} - 2b\delta^{-1} = 2r^2 h,$$

and the equations of motion take the form

$$\begin{cases} r' = rv \\ v' = 2r^2 h + r\delta^{-1/2} \\ \theta' = u \\ u' = (1/2)(\mu - 1)(r\delta^{-3/2} + 2b\delta^{-2})\sin 2\theta, \end{cases}$$

where $\delta = \mu \cos^2 \theta + \sin^2 \theta$. The new variables $(r, v, \theta, u) \in (0, \infty) \times \mathbb{R} \times S^1 \times \mathbb{R}$ depend on the fictitious time τ, so the prime denotes here differentiation with respect to the new independent variable τ. Note that the new equations extend analytically to $r = 0$, an invariant manifold which physically corresponds to binary collisions.

Notice that the sets $\{(r, v, \theta, u) \mid r = 0\}$ and $\{(r, v, \theta, u) \mid r > 0\}$ are invariant manifolds for the new system. The set

$$C = \{(r, v, \theta, u) \mid r = 0 \text{ and the energy relation holds}\}$$

is the *collision manifold*. It replaces the set of singularities $\{(\mathbf{q}, \mathbf{p}) \mid \mathbf{q} = \mathbf{0}\}$ of the original system with a 2-dimensional manifold in the space of the new variables. This 2-dimensional manifold is embedded in $\mathbb{R}^3 \times S^1$ and is given by the equations

$$r = 0 \quad \text{and} \quad u^2 + v^2 = 2b\delta^{-1}.$$

This shows that C is homeomorphic to a torus (see Figure 3).

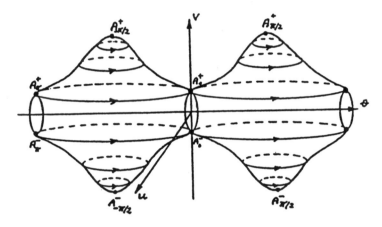

Figure 3: The flow on the collision manifold in the anisotropic Manev problem

A qualitative analysis of the flow near the collision manifold and the physical interpretation of the results are given in [1]. As in the case of the Manev problem, we have transformed a finite-in-time phenomenon to an infinite-in-time one in order to study the behaviour of solutions near the singularity.

5   Potentials with Drag or Thrust

The blow-up technique can be used in a larger context. For example, it also applies to systems in which additional forces appear. The 2-body problem with drag or thrust is such an example, rooted in the astronomical literature of the beginning of the 20th century. It models the motion of 2 point masses, one of negligible mass, subject to Newtonian gravitation and a drag force that involves the velocity (due to atmospheric friction, radiation pressure, solar wind, magnetic interaction, etc.).

Assume that the motion of two point masses (a particle and a primary) is subject to a perturbation force of the type $F = -r^{-2}(\alpha v_r + \beta v_t)$, where r is the distance between bodies, v_r and v_t are the radial and the tangential velocity, respectively, and α, β are real, nonzero constants. With respect to a frame whose origin is in the primary, the equations of motion in polar coordinates (r, θ) are given by the system

$$\begin{cases} \ddot{r} - r\dot{\theta}^2 = -\mu r^{-2} - \alpha \dot{r} r^{-2} \\ \frac{d}{dt}(r^2\dot{\theta}) = -\beta\dot{\theta}, \end{cases} \tag{12}$$

where μ combines several physical constants (mass, gravitational constant, charge, etc.). Unlike in the absence of drag or thrust, the total energy and the angular momentum are not conserved quantities.

All possible choices of signs for α, β, and μ present practical interest, from the one modeling the motion of small satellites near the Earth to the one involving microcosmic particles under radiation pressure. Therefore we will further discuss the qualitative aspects of each case. Notice that $\mu > 0$ corresponds to an *attractive* force, $\mu < 0$ represents a *repelling* one, while $\alpha > 0$ implies the presence of a radial *drag* and $\alpha < 0$ that of a radial *thrust*.

Using the substitutions $\dot{r} = u$ and $\dot{\theta} = \varphi$, we transform equations (12) into the first-order system

$$\begin{cases} \dot{r} = u \\ \dot{\theta} = \varphi \\ \dot{u} = r\varphi^2 - \mu r^{-2} - \alpha u r^{-2} \\ \dot{\varphi} = -\beta\varphi r^{-2} - 2u\varphi r^{-1}. \end{cases}$$

Since the vector field of this system is independent of θ, we can drop the second equation and thus obtain the simpler form

$$\begin{cases} \dot{r} = u \\ \dot{u} = r\varphi^2 - \mu r^{-2} - \alpha u r^{-2} \\ \dot{\varphi} = -\beta\varphi r^{-2} - 2u\varphi r^{-1}, \end{cases}$$

which under the time-rescaling analytic diffeomorphism $d\tau = r^{-2}dt$ is converted into the system

$$\begin{cases} r' = r^2 u \\ u' = -\alpha u + r^3 \varphi^2 - \mu \\ \varphi' = -\beta \varphi - 2u\varphi r. \end{cases} \tag{13}$$

Equations (13) are defined for $(r, u, \varphi) \in [0, \infty) \times \mathbb{R}^2$ and $'$ represents the derivative with respect to the new fictitious time-variable τ. Notice that though equations (12) were undefined at $r = 0$, system (13) extends to this set, i.e., the plane $r = 0$ is pasted to the 3-dimensional phase space of the variables (r, u, φ). The *collision plane* $r = 0$ is thus an invariant manifold for the equations (13). (Recall that an *invariant manifold* is a union of orbits, i.e., initial data in the set constrain the whole orbit to the set.)

Since system (13) has only one equilibrium solution at $(0, -\frac{\mu}{\alpha}, 0)$, we will shift the origin of the frame into it with the help of the transformation $v = u + \frac{\mu}{\alpha}$. Thus system (13) becomes

$$\begin{cases} r' = r^2(v - \frac{\mu}{\alpha}) \\ v' = -\alpha v + r^3 \varphi^2 \\ \varphi' = -\beta \varphi - 2(v - \frac{\mu}{\alpha})r\varphi, \end{cases} \tag{14}$$

and will be the main object of our further qualitative investigations. System (14) has the unique equilibrium solution $(r, v, \varphi) = (0, 0, 0)$. Due to the embedding of the collision manifold into the phase space, system (14) is well suited for understanding the behaviour of orbits near collision.

We will also study an alternative system, suitable for understanding the behaviour of orbits at infinity. To obtain it, we transform system (14) via the change of variable

$$x = \frac{1}{r},$$

which is an analytic diffeomorphism that brings infinity into the phase space. The new equations are

$$\begin{cases} x' = -(v - \frac{\mu}{\alpha}) \\ v' = -\alpha v + x^{-3} \varphi^2 \\ \varphi' = -\beta \varphi - 2(v - \frac{\mu}{\alpha})x^{-1}\varphi. \end{cases} \tag{15}$$

The time-rescaling transformation

$$ds = x^{-3} d\tau,$$

which is an analytic diffeomorphism, changes system (15) into the regularized system

$$\begin{cases} x' = -x^3(v - \frac{\mu}{\alpha}) \\ v' = x^3 - \alpha x^3 v + \varphi^2 \\ \varphi' = -\beta x^3 \varphi - 2(v - \frac{\mu}{\alpha})x^2\varphi, \end{cases} \quad (16)$$

where ' denotes differentiation with respect to the new fictitious time variable s. The invariant plane $x = 0$, called the *infinity manifold*, represents the physical points at infinity. In other words, a solution tending to this plane reaches infinity in physical space. On the other hand a solution of system (16) that tends to infinity in phase space reaches a collision in physical space. Equations (16) form the alternative system to be studied.

Unlike in previous cases, we can now use the blow-up technique not only for the study of motion near the singularity but also at infinity. We pasted an *infinity manifold* to the phase space, which we can use to draw conclusions about the qualitative behaviour of solutions at infinity. The analysis of the above aspects of the motion is done in [8].

6 Mücket-Treder Potentials

All the above cases involve quasihomogeneous potentials. Blow-up techniques, however, go beyond the class of quasihomogeneous functions. In 1977 the German astronomers J. P. Mücket and H.-J. Treder proposed a gravitational law, which for the 3-body problem is given by a potential of the form

$$U = V + W, \quad (17)$$

where the functions V and W are

$$V(\mathbf{q}) = G \sum_{1 \leq i < j \leq 3} \frac{m_i m_j}{|\mathbf{q}_i - \mathbf{q}_j|}$$

and

$$W(\mathbf{q}) = \sum_{1 \leq i < j \leq 3} \frac{\alpha m_i m_j (1 + \ln|\mathbf{q}_i - \mathbf{q}_j|)}{|\mathbf{q}_i - \mathbf{q}_j|}.$$

Here G is the gravitational constant and α is a small negative constant. If $\alpha > 0$, singularities do not occur and the equations (1) are globally defined (see [7]). In our case, $\alpha < 0$, noncollision singularities are excluded but collisions can take place. In what follows we will blow up the singularity that appears due to triple collisions. For this we first consider the transformation

$$\begin{cases} r = (\mathbf{q}^T M \mathbf{q})^{\frac{1}{2}} \\ \mathbf{s} = r^{-1}\mathbf{q} \\ y = \mathbf{p}^T \mathbf{s} \\ \mathbf{x} = \mathbf{s} - yM\mathbf{s}. \end{cases} \qquad (18)$$

Notice that the following relations take place:

$$\mathbf{s}^T M \mathbf{s} = 1, \quad \mathbf{s}^T \mathbf{x} = 0$$
$$V(\mathbf{q}) = r^{-1}V(\mathbf{s}), \quad \nabla V(\mathbf{q}) = r^{-2}\nabla V(\mathbf{s})$$
$$W(\mathbf{q} = r^{-1}W\mathbf{s}) + \alpha r^{-1}(\ln r)V(\mathbf{s})$$
$$\nabla W(\mathbf{q}) = r^{-2}\nabla W(\mathbf{s}) + \alpha r^{-2}(\ln r)\nabla V(\mathbf{s}).$$

Since V is a homogeneous function of degree -1, Euler's relation yields

$$\mathbf{q}^T \nabla V(\mathbf{q}) = -V(\mathbf{q}).$$

Using the transformation (18), equations (1) given by the potential (17) take the form

$$\begin{cases} \dot{r} = y \\ \dot{y} = r^{-1}\mathbf{x}^T M^{-1}\mathbf{x} + r^{-2}\mathbf{s}^T \nabla W(\mathbf{s}) - (1 + \alpha \ln r)r^{-2}V(\mathbf{s}) \\ \dot{\mathbf{s}} = r^{-1}M^{-1}\mathbf{x} \\ \dot{\mathbf{x}} = -r^{-1}y\mathbf{x} - r^{-1}(\mathbf{x}M^{-1}\mathbf{x} + \nabla V(\mathbf{s})) + (1 + \alpha \ln r)^{-2}[V(\mathbf{s})M\mathbf{s} + \\ \qquad r^{-2}\nabla W(\mathbf{s})] - r^{-2}[\mathbf{s}^T \nabla W(\mathbf{s})]M\mathbf{s}. \end{cases} \qquad (19)$$

For r positive and small, which implies that the 3 particles are close together, we further consider the change of variables

$$\begin{cases} v = r^{\frac{1}{2}}(-\ln r)^{-\frac{1}{2}}y \\ \mathbf{u} = r^{\frac{1}{2}}(-\ln r)^{-\frac{1}{2}}\mathbf{x}, \end{cases} \qquad (20)$$

which transforms equations (19) into

$$\begin{cases} \dot{r} = r^{-\frac{1}{2}}(-\ln r)^{\frac{1}{2}}v \\ \dot{v} = r^{-\frac{3}{2}}(-\ln r)^{\frac{1}{2}}[\frac{v^2}{2}(1 - \frac{1}{\ln r}) + \mathbf{u}^T M^{-1}\mathbf{u} - \frac{\mathbf{s}^T \nabla W(\mathbf{s})}{\ln r} + \\ \qquad (\alpha + \frac{1}{\ln r})V(\mathbf{s})] \\ \dot{\mathbf{s}} = r^{-\frac{3}{2}}(-\ln r)^{\frac{1}{2}}M^{-1}\mathbf{u} \\ \dot{\mathbf{u}} = -r^{-\frac{3}{2}}(-\ln r)^{\frac{1}{2}}\Big[(1 + \frac{1}{\ln r})\frac{\mathbf{u}v}{2} + (\mathbf{u}^T M^{-1}\mathbf{u}M\mathbf{s} + \\ \qquad (\alpha + \frac{1}{\ln r})[V(\mathbf{s})M\mathbf{s} + \nabla V(\mathbf{s})] + \frac{1}{\ln r}[\nabla W(\mathbf{s}) + \\ \qquad (\mathbf{s}^T \nabla W(\mathbf{s}))M\mathbf{s}]\Big]. \end{cases} \qquad (21)$$

To eliminate the negative powers of r and $-\ln r$ in the above equations, we consider the time rescaling transformation

$$d\tau = r^{-\frac{3}{2}}(-\ln r)^{\frac{1}{2}} dt,$$

which introduces the fictitious time variable τ. With this transformation system (21) becomes

$$
\begin{cases}
r' = rv \\
v' = \frac{v^2}{2}(1 - \frac{1}{\ln r}) + u^T M^{-1} u - \frac{s^T \nabla W(s)}{\ln r} + (\alpha + \frac{1}{\ln r})V(s) \\
s' = M^{-1} u \\
u' = -\left[(1 + \frac{1}{\ln r})\frac{uv}{2} + (u^T M^{-1} u Ms + (\alpha + \frac{1}{\ln r})[V(s)Ms+ \right. \\
\qquad \left. \nabla V(s)] + \frac{1}{\ln r}[\nabla W(s) + (s^T \nabla W(s))Ms]\right],
\end{cases}
\tag{22}
$$

where the prime denotes differentiation with respect to the independent variable τ. With the above transformations the energy relation takes the form

$$\frac{1}{2}(uM^{-1}u + v^2) + \frac{V(s) + W(s)}{\ln r} + \alpha V(s) = -\frac{hr}{\ln r}.$$

With the transformation

$$\rho = -\frac{1}{\ln r}$$

system (22) becomes

$$
\begin{cases}
\rho' = \rho^2 v \\
v' = (1 + \rho)\frac{v^2}{2} + u^T M^{-1} u + \rho\frac{s^T \nabla W(s)}{\ln r} + (\alpha - \rho)V(s) \\
s' = M^{-1} u \\
u' = (\rho - 1)\frac{uv}{2} + u^T M^{-1} u Ms - (\rho - \alpha)[V(s)Ms + \nabla V(s)]+ \\
\qquad \rho[\nabla W(s) + (s^T \nabla W(s))Ms],
\end{cases}
\tag{23}
$$

and the energy relation changes to

$$\frac{1}{2}(uM^{-1}u + v^2) - \rho[V(s) + W(s] + \alpha V(s) = h\rho e^{-\frac{1}{\rho}}.$$

Since the triple collision takes place at $\rho = 0$, system (23) is now defined to include this singularity. However, notice that the energy relation is still undefined at $\rho = 0$. To avoid this difficulty, we extend the energy relation to

$$\frac{1}{2}(uM^{-1}u + v^2) - \rho[V(s) + W(s] + \alpha V(s) = hg(\rho),$$

where

$$g(\rho) = \begin{cases} \rho e^{-\frac{1}{\rho}}, & \text{if } \rho \neq 0 \\ 0, & \text{if } \rho = 0. \end{cases}$$

The function g is analytic for $\rho \neq 0$ and of class C^∞ at $\rho = 0$. Notice that this is the only way to extend g to 0 in the class C^∞. Now both the equations of motion and the energy relation make sense at $\rho = 0$. Therefore we can define the *triple collision set* to be

$$\mathbf{C} = \{(\rho, y, \mathbf{s}, \mathbf{u})|\rho = 0 \ \text{ and } \ \frac{1}{2}(\mathbf{u}M^{-1}\mathbf{u} + v^2) + \alpha V(\mathbf{s}) = 0\},$$

which is a C^∞-manifold. The flow of this manifold provides information on the dynamics near triple collision. The study of this flow is done in [7].

7   Conclusions

As we have seen above, the blow-up techniques connect finite and infinite time phenomena in the dynamics of particles. Some finite-time phenomena are better understood through infinite-time techniques, whereas some asymptotic properties at infinity can be better grasped with the help of finite-space investigations. This can be done for various classes of potentials. We can thus see that the strong finite-versus-infinite relationship is not restricted to discrete mathematics. It leaves its imprint in all branches of the field, including the qualitative theory of differential equations and dynamical systems, and it can be successfully applied to many practical problems.

References

[1] S. Craig, F. Diacu, E.A. Lacomba, E. Pérez, The anisotropic Manev problem, *Journ. Math. Phys.*, **40**, 3 (1999), 1359–1375.

[2] J. Delgado, F. Diacu, E.A. Lacomba, A. Mingarelli, V. Mioc, E. Pérez, C. Stoica, The global flow of the Manev problem, *Journ. Math. Phys.*, **37**, 6 (1996), 2748–2761.

[3] F. Diacu, *Singularities of the N-Body Problem–An Introduction to Celestial Mechanics*, Les Publications CRM, Montréal, 1992.

[4] F. Diacu, The planar isosceles problem for Maneff's gravitational law, *Journ. Math. Phys.*, **34**, 12 (1993), 5671–5690.

[5] F. Diacu, Painlevé's conjecture, *Math. Intelligencer*, **15** (1993), 6–12.

[6] F. Diacu, Near-collision dynamics for particle systems with quasihomogeneous potentials, *Journ. Differential Equations*, **128** (1996), 58–77.

[7] F. Diacu, On the Mücket-Treder gravitational law, in *New Trends for Hamiltonian Systems and Celestial Mechanics* (E. A. Lacomba, J. Llibre, eds.), World Scientific, London, 1996, 127–139.

[8] F. Diacu, Two-body problems with drag or thrust: qualitative results, *Celestial Mechanics* (to appear).

[9] F. Diacu, P. Holmes, *Celestial Encounters–The Origins of Chaos and Stability*, Princeton Univ. Press, Princeton, N.J., 1996.

[10] F. Diacu, V. Mioc, C. Stoica, Phase-space structure and regularization of Manev-type problems, *Nonlinear Analysis* (to appear).

[11] M. R. Hoare, P. Pal, Statics and stability of small cluster nuclei, *Nature* (*Physical Science*), **230** (1971), 5–8.

[12] T. Levi-Cività, Sulla regolarizzazione del problema piano dei tre corpi, *Rendiconti dell'Accademia Nazionale dei Lincei*, **24**, 5 (1915), 61–75.

[13] T. Levi-Cività, Sur la régularisation du problème des trois corps, *Acta Mathematica*, **42** (1918), 92–144.

[14] R. McGehee, Triple collision in the collinear three-body problem, *Invent. Math.*, **27** (1974), 191–227.

[15] P. Painlevé, *Leçons sur la théorie analytique des équations differentielles*, Hermann, Paris, 1897.

[16] E.T. Whittaker, *A Treatise on the Analytical Dynamics of Particles and Rigid Bodies*, 4th edition, Cambridge Univ. Press, 1937.

[17] Z. Xia, The existence of noncollision singularities in the N-body problem, *Annals of Math.*, **135** (1992), 411–468.

An Interesting Serendipitous Real Number

John Ewing, Ciprian Foias

Mathematics Department
Indiana University
Bloomington, IN 47405, USA
E-mail: foias@indiana.edu, ewing@ams.org

1 Introduction

This is the story of a remarkable real number, the discovery of which was due to a misprint. Namely, in the midseventies, while Ciprian was at the University of Bucharest, one of his former students[1] approached him with the following question:

(Q) If $x_1 > 0$ and $x_{n+1} = (1 + \frac{1}{x_n})^n$ $(n = 1, 2, \ldots)$, can $x_n \to \infty$?

This was listed in a fall issue of the "Gazeta Matematică" as one of the problems given at the previous summer admission examination for prospective freshmen in the Department of Mathematics at the University of Bucharest. Ciprian found the answer in about one day, but considered that the problem was even above the sophomore level. He also found that (Q) is a misprinted version of the following question (given by Professor N. Boboc):

(Q') If $x_1 > 0$ and $x_{n+1} = (1 + \frac{1}{x_n})^{x_n}$ $(n = 1, 2, \ldots)$, can $x_n \to \infty$?

This was an appropriate exam question since the answer is clearly "No." Years later, in the 1980s, Ciprian told the story to Professor P. Halmos, who in turn told the story to John, but mischieveously did not mention at all Ciprian's answer to (Q), so a day later John also found the answer. This answer is given by the following

---

[1] If by chance that gentleman reads this article, we would appreciate if he will let us know his name.

Theorem 1.1 *There exists exactly one real number[2] $a \sim 1.187$ such that if $x_1 = a$ then $x_n \to \infty$. Moreover in this case*

$$x_n \frac{\ln n}{n} \to 1 \ for \ n \to \infty. \tag{1}$$

Relation (1) can be rewritten as

$$\lim_{n \to \infty} \frac{x_n}{\pi(n)} = 1 \tag{2}$$

where $\pi(n)$ is the number of primes less than n. However after many attempts to establish a deeper connection with the Prime Number Theorem, we came to believe that relation (2) is fortuitous. A strong argument for this opinion is provided by Theorem 3.1 below in which we show that the estimate for the error

$$x_n \frac{\ln n}{n} - 1$$

differs from its analog in the Prime Number Theorem.

We dedicate this short note to Professor Solomon Marcus, for his 75th birthday. Among many things Ciprian learned from him was also the interest in the anecdotical aspects of mathematics which explains the little story above.

2   Proof of Theorem 1.1

To start we recall that on $(0, \infty)$, $(1 + 1/\beta)^\beta$ increases from 1 to e and $(1 + 1/\beta)^{\beta+1}$ decreases from ∞ to e. It follows

$$x_{n+1} < e^{n/x_n} \ \text{and} \ x_{n+1} > e^{n/(x_n+1)}, \tag{3}$$

for all $n \geq 1$.

Lemma 2.1 *Let n satisfy*

$$\frac{\ln(n + 3)}{n + 2} \leq \frac{1}{1 + e} \tag{4}$$

and

$$x_n \leq \frac{n}{\ln(n + 1)} - 1. \tag{5}$$

Then

$$x_{n+2k} \to 1, \ x_{n+2k+1} \to 0 \ \ for \ \ k \to \infty. \tag{6}$$

---

[2] As computed by Professor Clay C. Ross $a = 1.18745235112650105459548\ 0158396\ldots$

Proof. We have, using (3) with n replaced by $n + 1$,

$$x_{n+2} = \left(1 + \frac{1}{x_{n+1}}\right)^{n+1} < e^{\frac{n+1}{x_{n+1}}}$$

whence, due to (3) and (5),

$$\frac{n+1}{x_{n+1}} < (n+1)e^{-n/(x_n+1)} < 1.$$

It follows $x_{n+2} < e$. Relation (4) now yields

$$x_{n+2} \leq \frac{n+2}{\ln(n+3)} - 1,$$

and therefore $x_{n+4} \leq 2$. Now note that if (4) holds, so does

$$\frac{1}{1+e} \geq \frac{\ln(n+k+1)}{n+k} \quad \text{for all } k \geq 2.$$

Thus $x_{n+2k} < e$ for all $k = 1, 2, \ldots$ This in turn implies

$$x_{n+2k+1} > e^{n/(x_{n+2k}+1)} > e^{n/(e+1)} \to \infty \text{ for } k \to \infty$$

and consequently

$$x_{n+2k+2} < e^{n/x_{n+2k+1}} \to 1 \text{ for } k \to \infty.$$

$\square$

Lemma 2.2 *If n satisfies (4) and*

$$x_n \geq \frac{n}{\ln y_n}, \quad \text{where } y_n = \frac{n+1}{\ln(n+2)} - 1, \tag{7}$$

then

$$x_{n+2k} \to \infty, \; x_{n+2k+1} \to 1 \quad \text{for } k \to \infty. \tag{8}$$

Proof. If (4) holds, so does

$$\frac{\ln(n+4)}{n+3} \leq \frac{1}{1+e}. \tag{9}$$

Moreover

$$x_{n+1} < e^{\frac{n}{x_n}} \leq \frac{n+1}{\ln(n+2)} - 1$$

so that we apply Lemma 2.1 with $n + 1$ replacing n, and obtain (8).   □

For convenience, we denote

$$f_n(x) = \left(1 + \frac{1}{x}\right)^n, \quad x \in (0, \infty), \ n = 1, 2, \ldots \qquad (10)$$

Thus if $x_1 = x$, the sequence $(x_n)_{n=1}^\infty$ is given by

$$x_{n+1} = (f_n \circ f_{n-1} \circ \cdots \circ f_1)(x) \text{ for } n = 1, 2, \ldots \qquad (11)$$

Clearly f_n is decreasing, $f((0, \infty)) = (1, \infty)$ and $f_n((1, \infty)) = (1, 2^n), n = 2, 3, \ldots$. For $n = 1, 2, \ldots$ we define the disjoint subsets A_n and B_n of $(0, \infty)$ by

$$A_n = \{x : x_{n+m+1} = (f_{m+n} \circ \cdots \circ f_n)(x)(m = 1, 2, \ldots) \text{ satisfies } (6)\}. \qquad (12)$$

$$B_n = \{x : x_{n+m+1} = (f_{m+n} \circ \cdots \circ f_n)(x)(m = 1, 2, \ldots) \text{ satisfies } (8)\}. \qquad (13)$$

Lemma 2.3 *Let $x \in A_n$ and $y \in B_n$. Then*

$$f_n(x) \in B_{n+1} \text{ and } (0, x] \subset A_n, \qquad (14)$$

$$f_n(y) \in A_{n+1} \text{ and } [y, \infty) \subset B_n, \qquad (15)$$

and

$$x < y. \qquad (16)$$

Proof. The first relations in (14), (15) are direct consequences of the definitions (12), (13); these together with the monotonicity property of the functions f_n also directly imply the second properties in (14), (15). These properties then yield (16).   □

As a corollary, if we set

$$a_n = \sup A_n, \ b_n = \inf B_n \ (n = 1, 2, \ldots), \qquad (17)$$

then

$$f_n(a_n) = b_{n+1}, f_n(b_n) = a_{n+1} \text{ and } a_n \le b_n. \qquad (18)$$

Moreover the proofs of Lemmas 2.1 and 2.2 also show that if n satisfies (4) then

$$\frac{n}{\ln(n+1)} - 1 \le a_n, \ b_n \le \frac{n}{\ln y_n}. \qquad (19)$$

Thus

$$0 \le \varepsilon_n = (b_n - a_n)\frac{\ln(n+1)}{n} \to 0 \text{ for } n \to \infty. \qquad (20)$$

Lemma 2.4 *For all $n = 1, 2, \ldots$*

$$\min\{|f'_n(x)| : a_n \le x \le b_n\} = |f'_n(b_n)|. \tag{21}$$

Moreover

$$|f'_n(b_n)| \to \infty \text{ for } n \to \infty. \tag{22}$$

Proof. Relation (21) follows from

$$\frac{d}{dx}|f'_n(x)| = n\frac{d}{dx}\left[\left(1 + \frac{1}{x}\right)^{n-1}\frac{1}{x^2}\right] < 0.$$

For (22), we observe first that for n large enough

$$
\begin{aligned}
\ln|f'(b_n)| &= \ln\left[n\left(1 + \frac{1}{b_n}\right)^{n-1}\frac{1}{b_n^2}\right] \ge \ln\left[\frac{(\ln y_n)^2}{n}e^{\frac{n-1}{b_n+1}}\right] \\
&\ge 2\ln\ln y_n - \ln n + \left(1 - \frac{1}{n}\right)(\ln y_n)\left(1 + \frac{1}{n}\ln y_n\right)^{-1} \\
&= 2\ln\ln y_n + \ln(y_n/n) + \alpha_n = \ln(y_n(\ln y_n)^2/n) + \alpha_n \\
&= \ln\left(\frac{(\ln y_n)^2}{\ln(n+2)} + \beta_n\right) + \alpha_n,
\end{aligned}
$$

where $\alpha_n, \beta_n \to 0$ for $n \to \infty$. Since

$$\ln y_n = \ln(n + 1) - \ln\ln(n + 2) + \gamma_n,$$

with $\gamma_n \to 0$, the lower bound of $\ln|f'(b_n)|$ established above goes to ∞ as $\ln\ln(n + 1)$ for $n \to \infty$. This establishes relation (22). $\qquad\square$

We can now conclude the proof of Theorem 1.1. Indeed we have, for some n_0 large enough,

$$|f'_n(b_n)| \ge 2, \text{ for all } n \ge n_0$$

and consequently (using (21))

$$|f(b_n) - f(a_n)| \ge 2|b_n - a_n|, \text{ for } n \ge n_0.$$

The first two equalities in (18) imply

$$|b_{n+1} - a_{n+1}| \ge 2|b_n - a_n|, \text{ for } n \ge n_0. \tag{23}$$

It is obvious that (20) and (23) are compatible only if $b_{n_0} = a_{n_0}$ and consequently only if

$$a_n = b_n \text{ for } n = 1, 2, \ldots$$

Let $a = a_1 = b_1$. Then, due to (18), the sequence

$$x_1 = a, \ x_{n+1} = (f_n \circ \cdots \circ f_1)(a) \text{ for } n = 1, 2, \ldots$$

satisfies

$$x_n = a_n = b_n \text{ for all } n = 1, 2, \ldots. \tag{24}$$

By virtue of (19),

$$\frac{x}{\ln(n+1)} - 1 \le x_n \le \frac{n}{\ln y_n}, \tag{25}$$

which obviously implies (1). Moreover, since

$$(0, \infty) \backslash \{a\} = A_1 \cup B_1$$

it follows that for any $x \ne a$ the sequence

$$x_1 = x, \ x_{n+1} = \left(1 + \frac{1}{x_n}\right)^n \text{ for } n = 1, 2, \ldots$$

will satisfy

$$\liminf_{n \to \infty} x_n = 1, \ \limsup_{n \to \infty} x_n = \infty.$$

This concludes the proof of Theorem 1.1.

3 The "Disconnection" from the Prime Number Theorem

For our purpose referred in the Introduction, it suffices to recall la Vallée Poussin's version of the Prime Number Theorem[3] namely,

$$\pi(n) = Li(n)(1 + \varepsilon(n)) \tag{26}$$

where for n large enough

$$|\varepsilon(n)| \le Ce^{-\sqrt{c \ln n}} \tag{27}$$

---

[3]See H. M. Edwards, *Riemann's Zeta Function*, Academic Press, 1974, Ch. 5.

and c and C are adequate positive constants. The function $Li(x)$ is defined for $x > 1$ by

$$Li(x) = (\text{p.v. at } 1) \int_0^x \frac{dt}{\ln t} = \lim_{\varepsilon \downarrow 0} \left(\int_0^{1-\varepsilon} \frac{dt}{\ln t} + \int_{1+\varepsilon}^x \frac{dt}{\ln t} \right).$$

Clearly

$$Li(n) = \frac{n}{\ln n} + \frac{n}{(\ln n)^2} + \cdots + (k-1)! \frac{n}{(\ln n)^k} + O\left(\frac{n}{(\ln n)^{k+1}} \right).$$

In particular,

$$\pi(n) \frac{\ln n}{n} - 1 = \frac{1}{\ln n} + O\left(\frac{1}{(\ln n)^2} \right),$$

so that

$$\left(\pi(n) \frac{\ln n}{n} - 1 \right) \ln n \to 1 \text{ for } n \to \infty. \tag{28}$$

This is in dire contrast with the behaviour of our sequence in Theorem 1.1 as displayed in the next theorem.

Theorem 3.1 *For the sequence* $(x_n)_{n=1}^\infty$ *considered in Theorem 1.1, we have*

$$\left(x_n \frac{\ln n}{n} - 1 \right) \ln n - \ln \ln n \to 0, \text{ for } n \to \infty. \tag{29}$$

Proof. We start by noticing that (see (3))

$$\frac{n}{\ln x_{n+1}} - 1 < x_n < \frac{n}{\ln x_{n+1}}. \tag{30}$$

Writing (30) with $n + 1$ in place of n, and introducing the result back into (30), we obtain

$$\frac{n}{\ln(n+1) - \ln \ln x_{n+2}} - 1 < x_n < \frac{n}{\ln\left(\frac{n+1}{\ln x_{n+2}} - 1 \right)}$$

$$= \frac{n}{\ln(n+1) - \ln \ln x_{n+2} + \ln(1 - (\ln x_{n+2})/(n+1))}$$

Using (1), it follows

$$\frac{x_n \ln(n+1)}{n} = \frac{1}{1 - \frac{\ln \ln x_{n+2}}{\ln(n+1)}} + O\left(\frac{\ln n}{n} \right). \tag{31}$$

Writing (31) with $n + 2$ in place of n, one obtains

$$\ln \ln x_{n+2} = \ln \ln n - \frac{\ln \ln n}{\ln n} + O\left(\frac{\ln \ln n}{(\ln n)^2}\right).$$

Introducing this into (31) yields

$$\frac{x_n \ln n}{n} = \frac{1}{1 - \frac{\ln \ln n}{\ln n}} + O\left(\frac{\ln \ln n}{(\ln n)^2}\right),$$

whence

$$\left(x_n \frac{\ln n}{n} - 1\right) \ln n - \ln \ln n =$$
$$\frac{\ln \ln n}{\ln n - \ln \ln n} + O\left(\frac{\ln \ln n}{(\ln n)^2}\right) = O\left(\frac{\ln \ln n}{\ln n}\right),$$

which establishes (29).   □

Algebraic Representations
of Regular Array Languages

Rudolf Freund

Institut für Computersprachen, Technische Universität Wien
Karlsplatz 13, A-1040 Wien, Austria
E-mail: rudi@logic.at

Alexandru Mateescu, Arto Salomaa

Turku Center for Computer Science
Lemminkäisenkatu 14 A, 20520 Turku, Finland
E-mail: mateescu, asalomaa@utu.fi

Abstract. We consider formal sequences of d-dimensional vectors and symbols representing d-dimensional arrays and introduce the operations of catenation as well as iterated catenation of these formal sequences (d-dimensional formal arrays). Together with the usual set union, these operations allow us to define d-dimensional regular array expressions and thus to develop an algebraic representation of regular array languages generated by specific d-dimensional array grammars. In that way, specific infinite regular array languages allow for a finite representation as regular array expressions. Whereas, in general, it is undecidable whether the array language generated by a given regular array grammar is empty, finite or infinite, for these specific regular array grammars, these questions are decidable.

1   Introduction

The generalization of one-dimensional strings to d-dimensional arrays has been studied during the last two decades; many interesting results could be obtained not only in the theoretical modelling of generating and accepting devices, but also with respect to applications in pattern recognition (e.g., see [1], [5], [9], [12]). Yet so far no algebraic approach to characterize specific d-dimensional array languages is known. In this paper we start the investigation of these topics by introducing specific operations on d-dimensional arrays which can be seen as the d-dimensional generalization of string catenation and its iterated variant. Moreover, using these d-dimensional catenation operations we define specific d-dimensional regular array expressions and show that they characterize specific subclasses of the family of d-dimensional regular array languages.

After recalling some well-known notions on string grammars and d-dimensional array grammars, in the third section we introduce the notion of d-dimensional formal arrays and the operations of catenation of d-dimensional formal arrays and of iterated catenation of d-dimensional formal arrays. Using these operations and the set union, we then define d-dimensional regular array expressions. In the fifth section we focus on the algebraic representation of specific d-dimensional regular array languages; in particular, we exhibit the correspondence between d-dimensional array languages generated by specific d-dimensional regular array grammars and the corresponding d-dimensional regular array expressions; we also show how a d-dimensional regular array expression describing the array language generated by a specific regular array grammar can be obtained as the fixpoint solution of a linear system of equations directly constructed from the given regular array grammar. Furthermore, both fixed and general membership for d-dimensional regular array grammars, even in the unary case, is NP-complete, while it is undecidable whether the array language generated by a given d-dimensional regular array grammar is empty, finite or infinite, if $d \geq 2$ (see [3], [8]). These results correspond with similar results observed in other picture describing formalisms (e.g., see [6], [11]). Yet for specific classes of d-dimensional regular array languages allowing for a characterization by d-dimensional regular array expressions, these questions can be shown to be decidable as for the case of regular string grammars. In the last section, we provide a list of further research topics.

2 Preliminaries

First, we recall some basic notions from the theory of formal languages (for more details, the reader is referred to [10]).

For an alphabet V, by V^* we denote the free monoid generated by V under the operation of catenation; the *empty string* is denoted by λ, and

$V^* - \{\lambda\}$ is denoted by V^+. Any subset of V^* is called a *(string) language*.

A *(string) grammar* is a quadruple $G = (V_N, V_T, P, S)$, where V_N and V_T are finite sets of *non-terminal* and *terminal symbols*, respectively, with $V_N \cap V_T = \emptyset$, P is a finite set of *productions* $\alpha \to \beta$ with $\alpha \in V^+$ and $\beta \in V^*$, where $V = V_N \cup V_T$, and $S \in V_N$ is the *start symbol*. For $x, y \in V^*$ we say that y *is directly derivable from x in G*, denoted by $x \Longrightarrow_G y$, if and only if for some $\alpha \to \beta$ in P and $u, v \in V^*$ we get $x = u\alpha v$ *and* $y = u\beta v$. Denoting the reflexive and transitive closure of the derivation relation $\Longrightarrow_G$ by $\Longrightarrow_G^*$, the *(string) language generated by G* is $L(G) = \{w \in V_T^* \mid S \Longrightarrow_G^* w\}$.

A grammar is called *regular,* if every production in P is of the form $\alpha \to \beta$ with $\alpha \in V_N$ and $\beta \in V_T \cup V_T V_N$ or of the form $S \to \lambda$; yet if $S \to \lambda \in P$, then S must not appear on the right-hand side β of a production $\alpha \to \beta \in P$. The families of (string) languages generated by arbitrary and regular grammars with the terminal alphabet V_T are denoted by $L(enum(V_T))$ and $L(reg(V_T))$, respectively.

In the second part of this section, we recall the definitions and notations for d-dimensional arrays and d-dimensional array grammars (e.g., see [1], [5], [9], [12]).

Let Z denote the set of integers, let N denote the set of positive integers, $N = \{1, 2, \ldots\}$, and let $d \in N$. Then a *d-dimensional array* $\mathcal{A}$ over an alphabet V is a function $\mathcal{A} : Z^d \to V \cup \{\#t\}$, where $shape(\mathcal{A}) = \{v \in Z^d \mid \mathcal{A}(v) \neq \#\}$ is finite and $\# \notin V$ is called the *background* or *blank symbol*. We usually shall write $\mathcal{A} = \{(v, \mathcal{A}(v)) \mid v \in shape(\mathcal{A})\}$.

The set of all d-dimensional arrays over V is denoted by V^{*d}. The *empty array* in V^{*d} with empty shape is denoted by Λ_d. Moreover, we define $V^{+d} = V^{*d} - \{\Lambda_d\}$. Any subset of V^{*d} is called a *d-dimensional array language.*

Let $v \in Z^d$. Then the *translation* $\tau_v : Z^d \to Z^d$ is defined by $\tau_v(w) = w + v$ for all $w \in Z^d$, and for any array $\mathcal{A} \in V^{*d}$ we define $\tau_v(\mathcal{A})$, the corresponding d-dimensional array translated by v, by $(\tau_v(\mathcal{A}))(w) = \mathcal{A}(w - v)$ for all $w \in Z^d$. The vector $(0, \ldots, 0) \in Z^d$ is denoted by Ω_d, and $Z_0^d = Z^d - \{\Omega_d\}$.

The equivalence class with respect to linear translations $[\mathcal{A}]$ of an array $\mathcal{A} \in V^{*d}$ is defined by $[\mathcal{A}] = \{\mathcal{B} \in V^{*d} \mid \mathcal{B} = \tau_v(\mathcal{A}) \text{ for some } v \in Z^d\}$. The set of all equivalence classes of d-dimensional arrays over V with respect to linear translations is denoted by $[V^{*d}]$ etc.

A *d-dimensional array production p* over V is a triple $(W, \mathcal{A}_1, \mathcal{A}_2)$, where $W \subseteq Z^d$ is a finite set and $\mathcal{A}_1$ and $\mathcal{A}_2$ are mappings from W to $V \cup \{\#\}$; p is called *Λ-free* if $shape(\mathcal{A}_2) \neq \emptyset$, where we define $shape(\mathcal{A}_i) = \{v \in W \mid$

$\mathcal{A}_i(v) \neq \#\}$, $1 \leq i \leq 2$. We say that the array $\mathcal{C}_2 \in V^{*d}$ is *directly derivable* from the array $\mathcal{C}_1 \in V^{*d}$ by the d-dimensional array production $(W, \mathcal{A}_1, \mathcal{A}_2)$ if and only if there exists a vector $v \in Z^d$ such that $\mathcal{C}_1(w) = \mathcal{C}_2(w)$ for all $w \in Z^d - \tau_v(W)$ as well as $\mathcal{C}_1(w) = \mathcal{A}_1(\tau_{-v}(w))$ and $\mathcal{C}_2(w) = \mathcal{A}_2(\tau_{-v}(w))$ for all $w \in \tau_v(W)$, i.e., the subarray of $\mathcal{C}_1$ corresponding to $\mathcal{A}_1$ is replaced by $\mathcal{A}_2$, thus yielding $\mathcal{C}_2$; we also write $\mathcal{C}_1 \Longrightarrow_p \mathcal{C}_2$. In the following we will represent the d-dimensional array production $(W, \mathcal{A}_1, \mathcal{A}_2)$ also by writing $\mathcal{A}_1 \to \mathcal{A}_2$, i.e., $\{(v, \mathcal{A}_1(v)) \mid v \in W\} \to \{(v, \mathcal{A}_2(v)) \mid v \in W\}$.

A d-*dimensional array grammar* is a septuple

$$G = (d, V_N, V_T, \#, P, v_0, S),$$

where V_N is the alphabet of *non-terminal symbols*, V_T is the alphabet of *terminal symbols*, $V_N \cap V_T = \emptyset$, $\# \notin V_N \cup V_T$; P is a finite set of d-dimensional array productions over $V_N \cup V_T$, v_0 is the start vector, and S is the *start symbol*; $\{(v_0, S)\}$ is called the *start array (axiom)*.

We say that the array $\mathcal{B}_2 \in V^{*d}$ is *directly derivable* from the array $\mathcal{B}_1 \in V^{*d}$ in G, denoted $\mathcal{B}_1 \Longrightarrow_G \mathcal{B}_2$, if and only if there exists a d-dimensional array production $p = (W, \mathcal{A}_1, \mathcal{A}_2)$ in P such that $\mathcal{B}_1 \Longrightarrow_p \mathcal{B}_2$. Let $\Longrightarrow_G^*$ be the reflexive transitive closure of $\Longrightarrow_G$. Then the $(d$-*dimensional) array language generated by* G, $L(G)$, is defined by $L(G) = \{\mathcal{A} \mid \mathcal{A} \in V_T^{*d}, \{(v_0, S)\} \Longrightarrow_G^* \mathcal{A}\}$.

A d-dimensional array grammar G, $G = (d, V_N, V_T, \#, P, v_0, S)$, is said to be $\#$-*context-free*, if for every d-dimensional array production $p = (W, \mathcal{A}_1, \mathcal{A}_2)$ in P we have $card(shape(\mathcal{A}_1)) = 1$. G is called *regular*, if:

1. $W = \{\Omega_d, v\}$ for some $v \in U_d$, where $U_d = \{(i_1, \ldots, i_d) \mid \sum_{k=1}^d |i_k| = 1\}$, and $\mathcal{A}_1 = \{(\Omega_d, A), (v, \#t)\}$, $\mathcal{A}_2 = \{(\Omega_d, a), (v, B)\}$, with $A, B \in V_N$ and $a \in V_T$ – we also write $A \to avB$ –, or

2. $W = \{\Omega_d\}$, $\mathcal{A}_1 = \{(\Omega_d, A)\}$, $\mathcal{A}_2 = \{(\Omega_d, a)\}$, with $A \in V_N$ and $a \in V_T$ – we also write $A \to a$ –, or

3. $W = \{\Omega_d\}$, $\mathcal{A}_1 = \{(\Omega_d, S)\}$, $\mathcal{A}_2 = \{(\Omega_d, \#)\}$ – we also write $S \to \#$; yet this array production $S \to \#$ is only allowed if for any other array production $A \to avB$ in P we have $B \neq S$.

The families of d-dimensional array languages generated by arbitrary, $\#$-context-free, and regular d-dimensional array grammars with the terminal alphabet V_T are denoted by $L(X(V_T))$, $X = d$-*enum*, d-$\#$-*cf*, d-*reg*.

An interesting feature of d-dimensional array grammars is the fact that even regular array productions make use of some special context, namely the context of blank symbols #. This blank-sensing ability induces a relatively high generating power for even regular two-dimensional array grammars and yields some rather astonishing results, e.g., the set of all solid squares can be generated by a regular two-dimensional array grammar [13].

Moreover, both fixed and general membership for regular d-dimensional array grammars, even in the unary case, is NP-complete, while the emptiness problem is undecidable for d-dimensional regular array grammars, if $d \geq 2$ (see [3], [8]). Moreover, every recursively enumerable string language can be represented by an at least two-dimensional #-context-free array grammar (see [3]), so that the membership problem for this grammar and language type is undecidable.

3 Catenation of Arrays and Iterated Catenation

The formal concepts we use to define the catenation operations on arrays are elaborated in this section.

Definition 3.1 *Let V be a finite alphabet, $k \in N$, $a_i \in V$, and $v_i \in Z^d$ for $1 \leq i \leq k$; then the sequence $\langle v_1 a_1 \ldots v_k a_k \rangle$ (we also write $\langle v_i a_i \rangle_{i=1}^{k}$) is called a (well-defined) d-dimensional formal array, if for arbitrary i and m with $i \neq m$ and $1 \leq i, m \leq k$ we have $\sum_{j=1}^{i} v_j \neq \sum_{j=1}^{m} v_j$. The formal array $\langle v_i a_i \rangle_{i=1}^{k}$ represents the d-dimensional array $\mathcal{A}$ with $shape(\mathcal{A}) = \{\sum_{j=1}^{i} v_j \mid 1 \leq i \leq k\}$ and $\mathcal{A}(\sum_{j=1}^{i} v_j) = a_i$ for $1 \leq i \leq k$; we also write $array(\langle v_i a_i \rangle_{i=1}^{k}) = \mathcal{A}$. The corresponding formal sequence $\left[\langle v_i a_i \rangle_{i=1}^{k} \right]$ represents the equivalence class of d-dimensional arrays $[\mathcal{A}]$. The empty formal arrays $\langle \emptyset_d \rangle$ and $[\langle \emptyset_d \rangle]$ represent the empty array (we also write $\langle v_i a_i \rangle_{i=1}^{0}$ and $[\langle v_i a_i \rangle_{i=1}^{0}]$).*

The set of all well-defined formal arrays of the form $\langle v_i a_i \rangle_{i=1}^{k}$ with $k \in N$, $a_i \in V$, and $v_i \in Z^d$ for $1 \leq i \leq k$ such that for arbitrary i and m with $i \neq m$ and $1 \leq i, m \leq k$ we have $\sum_{j=1}^{i} v_j \neq \sum_{j=1}^{m} v_j$, is denoted by $\langle V^{+d} \rangle$, the corresponding set of formal arrays $\left[\langle v_i a_i \rangle_{i=1}^{k} \right]$ is denoted by $\left[\langle V^{+d} \rangle \right]$. Moreover, we define $\langle V^{*d} \rangle = \langle V^{+d} \rangle \cup \{\langle \emptyset_d \rangle\}$ and $\left[\langle V^{*d} \rangle \right] = \left[\langle V^{+d} \rangle \right] \cup \{[\langle \emptyset_d \rangle]\}$.

Any subset L of $\langle V^{*d} \rangle$ ($\left[\langle V^{*d} \rangle \right]$) is called a language of d-dimensional formal arrays; if L does not contain the empty array, then L is called Λ-free.

Obviously, the d-dimensional array assigned to a d-dimensional formal array is uniquely defined. On the other hand, for a d-dimensional array there may be a finite number of d-dimensional formal arrays representing this d-dimensional array. The mapping $array$ defines a function from $\langle V^{*d} \rangle$ to V^{*d} as defined above with $array(\langle \emptyset_d \rangle) = \Lambda_d$.

The concept of formal arrays now allows us to introduce catenation operations $\odot_v$ with $v \in Z_0^d$ to define the catenation of two such formal arrays:

Definition 3.2 *Let V be a finite alphabet, $k, l \in N$, $a_i, b_j \in V$, and $v_i, w_j \in Z^d$ for $1 \leq i \leq k$ and $1 \leq j \leq l$; then for every $w \in Z_0^d$ the formal array*

$$\langle v_1 a_1 \ldots v_k a_k w b_1 \ldots w_l b_l \rangle,$$

if well-defined, denotes the $\odot_w$-catenation

$$\langle v_1 a_1 \ldots v_k a_k \rangle \odot_w \langle w_1 b_1 \ldots w_l b_l \rangle$$

of the two well-defined formal arrays $\langle v_1 a_1 \ldots v_k a_k \rangle$ and $\langle w_1 b_1 \ldots w_l b_l \rangle$.
The formal array

$$\left[\langle v_i a_i \rangle_{i=1}^k \odot_w \langle w_j b_j \rangle_{j=1}^l \right]$$

represents the $\odot_w$-catenation

$$\left[\langle v_i a_i \rangle_{i=1}^k \right] \odot_w \left[\langle w_j b_j \rangle_{j=1}^l \right]$$

of the two formal arrays $\left[\langle v_i a_i \rangle_{i=1}^k \right]$ and $\left[\langle w_j b_j \rangle_{j=1}^l \right]$.
In addition, for the empty (formal) arrays $\langle \emptyset_d \rangle$ and $[\langle \emptyset_d \rangle]$, respectively, we define

$$\langle v_i a_i \rangle_{i=1}^k \odot_w \langle \emptyset_d \rangle = \langle v_i a_i \rangle_{i=1}^k \text{ and } \langle \emptyset_d \rangle \odot_w \langle v_i a_i \rangle_{i=1}^k = \langle v_i a_i \rangle_{i=1}^k$$

as well as

$$\left[\langle v_i a_i \rangle_{i=1}^k \right] \odot_w [\langle \emptyset_d \rangle] = \left[\langle v_i a_i \rangle_{i=1}^k \right] \text{ and } [\langle \emptyset_d \rangle] \odot_w \left[\langle v_i a_i \rangle_{i=1}^k \right] = \left[\langle v_i a_i \rangle_{i=1}^k \right],$$

i.e., the empty formal arrays $\langle \emptyset_d \rangle$ and $[\langle \emptyset_d \rangle]$, respectively, are unit elements with respect to each catenation operation $\odot_w$.

The associativity of the catenation operations is a direct consequence of the definitions:

Lemma 3.1 *Let* $x, y, z \in \langle V^{*d} \rangle$ *or* $x, y, z \in \left[\langle V^{*d} \rangle \right]$, *respectively, and* $v, w \in Z_0^d$. *Then, if all catenations are well-defined,*

$$(x \odot_v y) \odot_w z = x \odot_v (y \odot_w z).$$

For example, any (well-defined) formal array $\langle v_i a_i \rangle_{i=1}^k \in \langle V^{+d} \rangle$ can also be written as $\langle v_1 a_1 \rangle \odot_{v_2} \langle \Omega_d a_2 \rangle \ldots \odot_{v_k} \langle \Omega_d a_k \rangle$.

Obviously, the definitions concerning the catenation operations $\odot_w$ can be extended to sets of formal arrays and non-empty sets $C \subseteq Z_0^d$ in the usual natural way, e.g., for two sets of formal arrays L_1 and L_2 we have

$$L_1 \odot_C L_2 = \{ z \mid z = x \odot_w y \text{ is well-defined}, x \in L_1, y \in L_2, \text{ and } w \in C \}.$$

We now define the iterated versions of the catenation operations $\odot_w$:

Definition 3.3 *Let* V *be a finite alphabet,* C *be a (non-empty) subset of* Z_0^d, *and* $L \subseteq \langle V^{*d} \rangle$. *Then the iterations of* L *with respect to* C *are defined in the following way:*

- $L^{\odot_C^0} = \{ \langle \emptyset_d \rangle \}$,

- $L^{\odot_C^{m+1}} = \bigcup_{w \in C} (L \odot_w L^{\odot_C^m})$ *for all* $m \geq 0$,

- $L^{\odot_C^+} = \bigcup_{m \geq 1} (L^{\odot_C^m})$,

- $L^{\odot_C^*} = L^{\odot_C^+} \cup \{ \langle \emptyset_d \rangle \}$.

Moreover, by obvious extensions of these definitions, we obtain $[L]^{\odot_C^m} = \left[L^{\odot_C^m} \right]$ *for* $m \in N \cup \{0, *, +\}$.

Example 3.1 *For example,* $\langle V^{*d} \rangle = \{ v_0 a_0 \mid v_0 \in Z^d, a_0 \in V \}^{\odot_{Z_0^d}^*}$; *observe that there is no completely finite representation of* $\langle V^{*d} \rangle$, *i.e., if* M *is a finite subset of* $\langle V^{*d} \rangle$ *and* C *is a finite subset of* Z_0^d, *then* $M^{\odot_C^*} \subset \langle V^{*d} \rangle$.

As many results for d-dimensional arrays for a special d can be carried over immediately for higher dimensions, we introduce the following notion.

Let $n, m \in N$ with $n \leq m$. For $n < m$, the natural embedding $i_{n,m} : Z^n \to Z^m$ is defined by $i_{n,m}(v) = (v, \Omega_{m-n})$ for all $v \in Z^n$; for $n = m$ we define $i_{n,n} : Z^n \to Z^n$ by $i_{n,n}(v) = v$ for all $v \in Z^n$. To an n-dimensional array $\mathcal{A} \in V^{*n}$ with $\mathcal{A} = \{ (v, \mathcal{A}(v)) \mid v \in shape(\mathcal{A}) \}$ we assign the m-dimensional array $i_{n,m}(\mathcal{A}) = \{ (i_{n,m}(v), \mathcal{A}(v)) \mid v \in shape(\mathcal{A}) \}$. In a similar way, the natural embedding of an n-dimensional formal array $\langle v_i a_i \rangle_{i=1}^k$ is defined by $\langle i_{n,m}(v_i) a_i \rangle_{i=1}^k$.

4 Regular Array Sets and Regular Array Grammars

In this section we consider specific regular array grammars and the corresponding specific regular array sets. The special representation of the (formal) arrays in these specific array sets is based on the following definitions:

Definition 4.1 *Let $C \subseteq Z_0^d$ and $u \in Z^d$. A (C, u)-representation of a d-dimensional array $\mathcal{A}$ from V^{+d} is a formal array $\langle v_0 a_0 \ldots v_k a_k \rangle$ such that $array(\langle v_0 a_0 \ldots v_k a_k \rangle) = \mathcal{A}$ and $v_0 = u$ as well as $v_i \in C$, $1 \le i \le k$; moreover, $\langle \emptyset_d \rangle$ is a (C, u)-representation of Λ_d. A formal array $\left[\langle v_i a_i \rangle_{i=1}^k \right]$ is called a C-representation of the array $[\mathcal{A}]$ if and only if $\langle \Omega_d a_0 \ldots v_k a_k \rangle$ is a (C, Ω_d)-representation of $\mathcal{B}$ for some $\mathcal{B} \in [\mathcal{A}]$. The set of all formal arrays that are (C, u)-representations of d-dimensional arrays $\mathcal{A}$ from V^{*d} is denoted by $\langle V^{*d}(C, u) \rangle$.*

Remark 4.1 *In contrast to $\langle V^{*d} \rangle$, which, as already mentioned in Example 3.1, has no finite representation of the form $M^{\odot*}c$ for some finite $M \subset \langle V^{*d} \rangle$ and some finite $C \subset Z_0^d$, for $\langle V^{*d}(C, u) \rangle$ we obtain such a finite representation by taking $M = \{\langle ua \rangle \mid a \in V\}$, i.e., $\langle V^{*d}(C, u) \rangle = \{\langle ua \rangle \mid a \in V\}^{\odot*}_C$.*

The following result is an immediate consequence of the definitions:

Lemma 4.1 *If $card(C) = 1$, then – if it exists – the (C, u)-representation (C-representation) of the array $\mathcal{A}$ ($[\mathcal{A}]$) is uniquely determined.*

Definition 4.2 *For any well-defined formal array $\langle v_i a_i \rangle_{i=1}^k$ the string image of $\langle v_i a_i \rangle_{i=1}^k$ is $a_1 \ldots a_k$, and we write $string(\langle v_i a_i \rangle_{i=1}^k) = a_1 \ldots a_k$; the string image of the empty array Λ_d is the empty word λ. Moreover, we define $string(\left[\langle v_i a_i \rangle_{i=1}^k \right]) = a_1 \ldots a_k$ as well as $string([\langle \emptyset_d \rangle]) = \lambda$.*

Example 4.1 *Consider the one-dimensional array $\{((0), a), ((1), b), ((2), c)\} \in \{a, b, c\}^{*1}$. This array has the unique $(\{(1), (-1)\}, (0))$-representation $\langle (0)a(1)b(1)c \rangle$, whereas we find two $(\{(1), (-1), (2)\}, (0))$-representations $\langle (0)a(1)b(1)c \rangle$ and $\langle (0)a(2)c(-1)b \rangle$. The corresponding string images are abc and acb.*

According to Lemma 4.1, each connected one-dimensional array $[\mathcal{A}]$ (for the notion of connectedness, e.g., see [5]) has a uniquely determined $(\{(1)\})$-representation and the corresponding string image is a natural representation of the array $[\mathcal{A}]$. In general, for $card(C) = 1$ with $C = \{v\}$, $v \in Z_0^d$, and each array $\mathcal{A} \in V^{*d}$ ($[\mathcal{A}] \in \left[V^{*d}\right]$) that has a – uniquely determined – (C, u)-representation $\langle ua_1 \ldots va_k\rangle$ (C-representation $[\langle \Omega_d a_1 \ldots va_k\rangle]$), $a_1 \ldots a_k$ is the string representation of $\mathcal{A}$ ($[\mathcal{A}]$), which allows us to recover the (C, u)-representation $\langle ua_1 \ldots va_k\rangle$ (C-representation $[\langle \Omega_d a_1 \ldots va_k\rangle]$) as well as the array $\mathcal{A}$ ($[\mathcal{A}]$) itself. Obviously, this possibility of recovering the original arrays is not given any more if $card(C) > 1$. On the other hand, Lemma 4.1 can be extended in the following way:

Lemma 4.2 *Let C be a positive subset of Z_0^d, i.e., every component of each vector $v \in C$ is non-negative, and let $u \in Z^d$. Then – if it exists – the (C, u)-representation (C-representation) of the array $\mathcal{A} \in V^{*d}$ ($[\mathcal{A}] \in \left[V^{*d}\right]$) is uniquely determined.*

Proof. On the contrary, let us assume that $\langle v_1 a_1 \ldots v_k a_k\rangle$ and $\langle v_1' a_1' \ldots v_k' a_k'\rangle$ with $v_1 = v_1' = u$ and $a_1 = a_1'$ are two different formal arrays which both are (C, u)-representations of the same array $\mathcal{A} \in V^{+d}$. Hence,

$$\mathcal{A} = \{(\sum_{j=1}^{m} v_j, a_m) \mid 1 \le m \le k\} = \{(\sum_{j=1}^{m} v_j', a_m') \mid 1 \le m \le k\}.$$

Both sequences of vectors $\langle \sum_{j=1}^{m} v_j\rangle_{m=1}^{k}$ and $\langle \sum_{j=1}^{m} v_j'\rangle_{m=1}^{k}$ are strictly increasing, and obviously,

$$\{\sum_{j=1}^{m} v_j \mid 1 \le m \le k\} = \{\sum_{j=1}^{m} v_j' \mid 1 \le m \le k\}.$$

As $\langle v_1 a_1 \ldots v_k a_k\rangle$ and $\langle v_1' a_1' \ldots v_k' a_k'\rangle$ are two different formal arrays both representing the same array $\mathcal{A}$, there must be some n with $1 \le n \le k$ such that $(v_j, a_j) = (v_j', a_j')$ for $1 \le j \le n-1$ and $(v_N, a_N) \ne (v_N', a_N')$; $a_N \ne a_N'$ but $v_N = v_N'$ is impossible, because in that case the array $\mathcal{A}$ would have different symbols at the position $\sum_{j=0}^{n} v_j$, hence, we must have $v_N \ne v_N'$ for some $n \ge 2$.

As $\{\sum_{j=1}^{m} v_j \mid 1 \leq m \leq k\} = \{\sum_{j=1}^{m} v'_j \mid 1 \leq m \leq k\}$,

$$\sum_{j=1}^{n} v_j \in \{\sum_{j=1}^{m} v_j \mid 1 \leq m \leq k\} \text{ and } \sum_{j=1}^{n} v'_j \in \{\sum_{j=1}^{m} v_j \mid 1 \leq m \leq k\};$$

moreover, $\sum_{j=1}^{n} v'_j = (\sum_{j=1}^{n-1} v_j) + v'_N$. Now, without loss of generality, let us assume

$$(\sum_{j=1}^{n-1} v_j) + v_N < (\sum_{j=1}^{n-1} v_j) + v'_N.$$

As $(\sum_{j=1}^{n-1} v_j) < (\sum_{j=1}^{n-1} v_j) + v_N$, in the strictly increasing sequence of vectors $\langle \sum_{j=1}^{m} v'_j \rangle_{m=1}^{k}$ the vector $\sum_{j=1}^{n} v_j$ could not appear any more, which is a contradiction to the fact that $\{\sum_{j=1}^{m} v_j \mid 1 \leq m \leq k\} = \{\sum_{j=1}^{m} v'_j \mid 1 \leq m \leq k\}$. Hence, we conclude that the (C, u)-representation of any array $\mathcal{A} \in V^{+d}$ is uniquely determined. The observation that Λ_d has a unique (C, u)-representation completes the proof.   □

Remark 4.2 *As an immediate consequence of the preceding lemma we observe the following. Let C be a positive subset of Z_0^d and let $u \in Z^d$; if the (C, u)-representation $\mathcal{B}$ of the array $\mathcal{A} \in V^{*d}$ exists, then there exists exactly one string image of $\mathcal{A}$ with resprect to (C, u), i.e., string$(\mathcal{B})$.*

We now define the specific classes of regular array grammars to be considered in this paper:

Definition 4.3 *((C, u)-regular array grammar) Let C be a finite subset of Z_0^d and $u \in Z^d$. A (C, u)-regular array grammar is a d-dimensional array grammar G, $G = (d, V_N, V_T, \#, P, v_0, S)$, such that $v_0 = u$ and every array production in P is of the form $A \to a$ with $A \in V_N$, $a \in V_T$, or of the form $\{(\Omega_d, A), (v, \#)\} \to \{(\Omega_d, a), (v, B)\}$ with $A, B \in V_N$, $a \in V_T$, and $v \in C$ (we also write $A \to avB$); finally, we also allow the array production $S \to \#$, if for any array production in P of the form $A \to avB$, we have $B \neq S$. If, in addition, for every array production $A \to avB$ in P with $A, B \in V_N$, $a \in V_T$, and $v \in C$ we have $A \to av'B \in P$ for every $v' \in C$, then G is called maximal.*

The d-dimensional regular array grammars G, $G = (d, V_N, V_T, \#, P, v_0, S)$, usually considered in the literature (e.g., see [5]), can be seen as d-dimensional (U_d, v_0)-regular array grammars, where $U_d = \{(i_1, \ldots, i_d) \mid \sum_{k=1}^{d} |i_k| = 1\}$. In contrast to the standard definition of a derivation in a d-dimensional array grammar with the array productions working on d-dimensional arrays, for a (C, u)-regular array grammar G, $G = (d, V_N, V_T, \#, P, u, S)$, we now define a *terminal derivation* as a sequence $\langle \mathcal{A}_i \rangle_{i=0}^{n}$ of formal arrays $\mathcal{A}_i$ such that

1. $\mathcal{A}_0 = \langle uS \rangle = \langle v_0 X_0 \rangle$,

2. for every i with $0 < i < n$,

   $\mathcal{A}_i = \langle v_0 a_0 \ldots v_{i-1} a_{i-1} v_i X_i \rangle$ is a well-defined formal array and $X_{i-1} \to a_{i-1} v_i X_i \in P$ as well as $v_i \in C$, $a_{i-1} \in V_T$ and $X_i \in V_N$; finally,

3. $\mathcal{A}_N = \langle v_0 a_0 \ldots v_{n-1} a_{n-1} \rangle$ is a well-defined formal array, $X_{n-1} \to u_{n-1} \in P$ and $a_{n-1} \in V_T$.

If $S \to \# \in P$, then $\langle \langle uS \rangle, \langle \emptyset_d \rangle \rangle$ is a terminal derivation, too, which derives the empty array (observe that in this case the non-terminal symbol S cannot appear in another terminal derivation $\langle \mathcal{A}_i \rangle_{i=0}^{n}$ in an $\mathcal{A}_i$ with $i \geq 1$). Any subsequence $\langle \mathcal{A}_i \rangle_{i=0}^{m}$ of a terminal derivation $\langle \mathcal{A}_i \rangle_{i=0}^{n}$, $0 \leq m \leq n$, is called a *derivation* in G. Moreover, from the definitions we immediately infer that $\mathcal{A}_i \in \langle (V_N \cup V_T)^{*d}(C, u) \rangle$, $0 \leq i \leq n-1$, and $\mathcal{A}_N \in \langle V_T^{*d}(C, u) \rangle$.

Obviously, the terminal derivation of formal arrays $\langle \mathcal{A}_i \rangle_{i=0}^{n}$ corresponds with the sequence of arrays $\langle array(\mathcal{A}_i) \rangle_{i=0}^{n}$, which represents the usual derivation of the terminal array $array(\mathcal{A}_N) \in V_T^{*d}$, i.e.,

$$array(\mathcal{A}_0) \Longrightarrow_G array(\mathcal{A}_1) \Longrightarrow_G \ldots \Longrightarrow_G array(\mathcal{A}_N).$$

Hence, we restrict ourselves to consider the set of formal arrays that can be generated by G and thus we define

$$L(G) = \{\mathcal{A} \mid \mathcal{A} \in \langle V_T^{*d}(C, u) \rangle, \langle \mathcal{A}_i \rangle_{i=0}^{n} \text{ is a terminal derivation in } G,$$
$$\text{and } \mathcal{A}_N = \mathcal{A}\}.$$

The family of (array) languages that can be generated by (maximal) d-dimensional (C, u)-regular array grammars with the terminal alphabet V_T now is denoted by $L(d\text{-}(C, u)\text{-}reg(V_T))$ (and $L(d\text{-}(C, u)\text{-}reg_{max}(V_T))$); moreover, we define $[L(d\text{-}C\text{-}reg(V_T))] = [L(d\text{-}(C, \Omega_d)\text{-}reg(V_T))]$ as well as $[L(d\text{-}C\text{-}reg_{max}(V_T))] = [L(d\text{-}(C, \Omega_d)\text{-}reg_{max}(V_T))]$.

We now define the specific classes of regular array sets that correspond with some special classes of specific (C, u)-regular (C-regular) array languages.

Definition 4.4 (*C-regular array sets*) *Let C be a finite subset of Z_0^d. A set of formal arrays $R \subseteq \left[\langle V^{*d} \rangle \right]$ is called a C-regular array set over V, if it can be obtained in a finite number of steps using the following rules:*

1. $\emptyset$ *is C-regular.*

2. $\{[\langle \emptyset_d \rangle]\}$ *is C-regular.*

3. $\{[\langle \Omega_d a \rangle]\}$ *is C-regular for every $a \in V$.*

4. *If X and Y are C-regular, then $X \cup Y$ is C-regular.*

5. *If X and Y are C-regular, then $X \odot_d Y$ is C-regular for every non-empty $D \subseteq C$.*

6. *If X is C-regular, then $X^{\odot_d^*}$ is C-regular for every non-empty $D \subseteq C$.*

If we restrict the catenation operations to $D = C$, then the set of formal arrays R is called maximal. *The set of all (maximal) C-regular array sets over V is denoted by $[REGA_d(V, C)]$ ($[REGA_{d,max}(V, C)]$).*

For subsets of $\langle V^{*d} \rangle$ the situation is a little bit more difficult than for subsets of $\left[\langle V^{*d} \rangle \right]$, because in this case all the arrays have to start from the same origin:

Definition 4.5 (*(C, u)-regular array sets*) *Let C be a finite subset of Z_0^d and let $u \in Z^d$. A set of formal arrays $R \subseteq \langle V^{*d} \rangle$ is called a (C, u)-regular array set over V, if it can be obtained in a finite number of steps using the following rules:*

1. $\emptyset$ *is (C, u)-regular.*

2. $\{\langle \emptyset_d \rangle\}$ *is (C, u)-regular.*

3. $\{\langle ua \rangle\}$ *is (C, u)-regular for every $a \in V$.*

4. *If X and Y are (C, u)-regular, then $X \cup Y$ is (C, u)-regular.*

5. *If X and Y are (C, u)-regular, then $X \odot_d Y$ is (C, u)-regular for every non-empty $D \subseteq C$.*

6. *If X is (C, u)-regular, then $X^{\odot_d^*}$ is (C, u)-regular for every non-empty $D \subseteq C$.*

If we restrict the catenation operations to $D = C$, then the set of formal arrays R is called maximal. *The set of all (maximal) (C, u)-regular array sets over V is denoted by $REGA_d(V, C, u)$ ($REGA_{d,max}(V, C, u)$).*

According to the construction rules given in the definitions above, (C, u)-regular array sets can be represented as specific finite *regular expressions* using the operations union, catenation, and iterated catenation of sets of formal arrays. In the following we shall not distinguish between the notions (C, u)-regular array set and (C, u)-regular array expression.

Remark 4.3 *Obviously, $REGA_d(V, C, u) \subset \langle V_T^{*d}(C, u) \rangle$; moreover,*

$$[REGA_d(V, C)] = [REGA_d(V, C, \Omega_d)] \ \text{and}$$
$$[REGA_{d,max}(V, C)] = [REGA_{d,max}(V, C, \Omega_d)] \,.$$

If $C = \{w\}$ for some $w \in Z_0^d$, then, by definition, all C-regular and (C, u)-regular array sets are maximal:

$$REGA_d(V, \{w\}, u) = REGA_{d,max}(V, \{w\}, u) \ \text{and}$$
$$[REGA_d(V, \{w\})] = [REGA_{d,max}(V, \{w\})] \,.$$

Finally, $L(d\text{-}(\emptyset, u)\text{-}reg(V_T))$, $REGA_d(V, \emptyset, u)$, $L(d\text{-}(\emptyset, u)\text{-}reg_{max}(V_T))$, and $REGA_{d,max}(V, \emptyset, u)$ characterize the finite family of subsets of $\{\langle ua \rangle \mid a \in V_T\} \cup \{\langle \emptyset_d \rangle\}$, and $[L(d\text{-}\emptyset\text{-}reg(V_T))]$, $REGA_d(V, \emptyset)$, $L(d\text{-}\emptyset\text{-}reg_{max}(V_T))$, and $REGA_{d,max}(V, \emptyset)$ characterize the finite family of subsets of $\{[\langle \Omega_d a \rangle] \mid a \in V_T\} \cup \{[\langle \emptyset_d \rangle]\}$.

More interesting features of specific families of C-regular and (C, u)-regular array sets will be established in the succeeding section.

5 Regular Array Sets with Algebraic Representation

As already mentioned earlier in this paper, for U_d-regular array grammars with $d \geq 2$ it is undecidable whether the array language generated by a given d-dimensional U_d-regular array grammar is empty, finite or infinite. This follows from the constructions used in [5], where it was shown how every recursively enumerable string language can be represented by an at least two-dimensional #-context-free array grammar. Using this construction we might show that for any recursively enumerable string language L there exists a U_2-regular array grammar G such that $card(L) = card(L(G))$. For $d > 2$ we may consider $i_{2,d}(G)$ yielding $i_{2,d}(L(G))$. In sum, we obtain the following result (we have to omit the proof which would go much beyond the scope of this paper).

Proposition 5.1 *Let $d \geq 2$, $C \supseteq U_d$ be a finite subset of Z_0^d, and $u \in Z^d$. Then it is undecidable whether for a given d-dimensional (C,u)-regular array grammar G the language of formal arrays $L(G)$ is empty, finite or infinite.*

In the one-dimensional case we obtain the following results, which mostly are consequences of well-known results (e.g., see [4]):

Proposition 5.2 *For any terminal alphabet V_T,*

$$[REGA_1(V_T, \{(1)\})] = [REGA_1(V_T, \{(-1)\})] =$$
$$[L(1\text{-}\{(1)\}\text{-}reg(V_T))] = [L(1\text{-}\{(-1)\}\text{-}reg(V_T))];$$

moreover, $string([L(1\text{-}\{(1)\}\text{-}reg(V_T))]) = L(reg(V_T))$.

We now prove these results in a more general context for arbitrary singleton sets $C \subset Z_0^d$:

Theorem 5.1 *For any terminal alphabet V_T as well as $w \in Z_0^d$ and $u \in Z^d$,*

1. $REGA_d(V_T, \{w\}, u) = L(d\text{-}(\{w\}, u)\text{-}reg(V_T)) =$
   $REGA_{d,max}(V_T, \{w\}, u) = L(d\text{-}(\{w\}, u)\text{-}reg_{max}(V_T)),$

2. $[REGA_d(V_T, \{w\})] = [L(d\text{-}\{w\}\text{-}reg(V_T))] =$
   $[REGA_{d,max}(V_T, \{w\})] = [L(d\text{-}\{w\}\text{-}reg_{max}(V_T))]$

3. $string(L(d\text{-}(\{w\}, u)\text{-}reg(V_T))) = L(reg(V_T)).$

4. $[REGA_d(V_T, \{-w\})] = [L(d\text{-}\{-w\}\text{-}reg(V_T))] =$
   $[L(d\text{-}\{w\}\text{-}reg(V_T))] = [REGA_{d,max}(V_T, \{-w\})] =$
   $[L(d\text{-}\{-w\}\text{-}reg_{max}(V_T))]$

Proof. According to Remark 4.3, $(\{w\}, u)$-regular array sets are already maximal. Moreover, obviously any $(\{w\}, u)$-regular array grammar is maximal, too. Therefore we need not take care of the feature of maximality throughout the rest of the proof.

$REGA_{d,max}(V_T, \{w\}, u) = L(d\text{-}(\{w\}, u)\text{-}reg_{max}(V_T))$ will be shown later in Theorem 5.2, hence we obtain 1 and 2.

Let G, $G = (d, V_N, V_T, \#, P, u, S)$, be a d-dimensional $(\{w\}, u)$-regular array grammar. Then consider the string grammar G', $G' = (V_N, V_T, P', S)$, with $P' = P'_1 \cup P'_2 \cup P'_3$ and

$$P'_1 = \{A \to a \mid A \to a \in P, A \in V_N, a \in V_T\},$$
$$P'_2 = \{A \to aB \mid A \to awB \in P, A, B \in V_N, a \in V_T\},$$
$$P'_3 = \{S \to \lambda \mid S \to \# \in P\}.$$

Clearly, $string(L(G)) = L(G')$. On the other hand, because of the simple structure of the formal arrays it contains, $L(G)$ is easily reconstructible from $L(G')$, i.e., given the regular string grammar G', $G' = (V_N, V_T, P', S)$, we immediately obtain the corresponding d-dimensional $(\{w\}, u)$-regular array grammar G, $G = (d, V_N, V_T, \#, P, u, S)$, with $string(L(G)) = L(G')$ by taking $P = P_1 \cup P_2 \cup P_3$ and

$$P_1 = \{A \to a \mid A \to a \in P', A \in V_N, a \in V_T\},$$
$$P_2 = \{A \to awB \mid A \to aB \in P', A, B \in V_N, a \in V_T\},$$
$$P_3 = \{S \to \# \mid S \to \lambda \in P'\}.$$

Therefore we conclude that (3) is valid.

Finally, for any string language L, let L^r denote the mirror image of L. Let G be a d-dimensional $(\{-w\}, u)$-regular array grammar. Then consider the string grammar G_s for $string(L(G))$. $L(reg(V_T))$ is closed under mirror image, i.e., from G_s we can effectively construct a string grammar G_r such that $L(G_r) = (L(G_s))^r$. According to the construction above, from G_r we obtain a $(\{w\}, u)$-regular array grammar G' with $string(L(G')) = (string(L(G)))^r$, and moreover, $[L(G')] = [L(G)]$. Hence, we conclude $[L(d\text{-}\{-w\}\text{-}reg(V_T))] = [L(d\text{-}\{w\}\text{-}reg(V_T))]$, which completes the proof. $\qquad \square$

Remark 5.1 *In addition to the results proved above we should like to mention that for any* $(\{w, -w\}, u)$-*regular array grammar* G, $G =$

$(d, V_N, V_T, \#, P, u, S)$, we can construct a $(\{w\}, u)$-regular array grammar G' such that $[L(G')] = [L(G)]$, i.e.,

$$[L(d\text{-}\{w, -w\}\text{-}reg(V_T))] = [L(d\text{-}\{w\}\text{-}reg(V_T))].$$

This follows from the fact that $[L(G)]$ simply is the union $[L(G)] = [L(G_1)] \cup [L(G_2)]$ of a language $[L(G_1)] \in [L(d\text{-}\{w\}\text{-}reg(V_T))]$ and a language $[L(G_2)] \in [L(d\text{-}\{-w\}\text{-}reg(V_T))]$, where $G_i = (d, V_N, V_T, \#, P_i, u, S)$ with

$$P_i = P - \{A \to a((-1)^{2-i}w)B \mid A \to a((-1)^{2-i}w)B \in P,$$
$$A, B \in V_N, a \in V_T\}, \ i \in \{1, 2\}.$$

The correspondence of array languages in $L(d\text{-}(\{w\}, u)\text{-}reg(V_T))$ with regular string languages immediately yields the following decidability results.

Corollary 5.1 Let $w \in Z_0^d$ and $u \in Z^d$. Then emptiness and finiteness of $L(G)$ are decidable for any $(\{w\}, u)$-regular array grammar G.

Proof. According to the proof of Theorem 5.1, the string grammar for $string(L(G))$ can effectively be constructed from G, and $L(G)$ is empty/finite if and only if $string(L(G))$ is empty/finite.   □

Corollary 5.2 Let C be a finite subset of Z_0^d and $u \in Z^d$. Then emptiness and finiteness of $L(G)$ are decidable for any maximal (C, u)-regular array grammar G.

Proof. Let G, $G = (d, V_N, V_T, \#, P, u, S)$, be a d-dimensional maximal (C, u)-regular array grammar and $w \in C$. Then consider the d-dimensional maximal $(\{w\}, u)$-regular array grammar G_w, $G_w = (d, V_N, V_T, \#, P_w, u, S)$, where

$$P_w = P - \{A \to avB \mid A \to avB \in P \text{ and } v \neq w\}.$$

Obviously, $L(G_w) \subseteq L(G)$ and $L(G)$ is empty/finite if and only if $L(G_w)$ is empty/finite, which is decidable by Corollary 5.1.   □

In the second part of this section we elaborate the equivalence of characterizing specific d-dimensional regular array languages by d-dimensional maximal (C, u)-regular array grammars and by d-dimensional maximal (C, u)-regular expressions. We also show how d-dimensional maximal

(C, u)-regular array sets can be obtained as the fixpoint solution of a linear system of equations directly constructed from the corresponding d-dimensional maximal (C, u)-regular array grammar.

Our main result can be formulated in the following way:

Theorem 5.2 *For any terminal alphabet V_T, $u \in Z^d$, and any non-empty finite subset C of Z_0^d,*

$$REGA_{d,max}(V_T, C, u) = L(d\text{-}(C, u)\text{-}reg_{max}(V_T)).$$

The inclusion $\subseteq$ can be proved by showing an even more general result:

Lemma 5.1 *For any terminal alphabet V_T, $u \in Z^d$, and any non-empty finite subset C of Z_0^d,*

1. *$REGA_d(V_T, C, u) \subseteq L(d\text{-}(C, u)\text{-}reg(V_T))$ and*

2. *$[REGA_d(V_T, C)] \subseteq [L(d\text{-}C\text{-}reg(V_T))]$ as well as*

3. *$REGA_{d,max}(V_T, C, u) \subseteq L(d\text{-}(C, u)\text{-}reg_{max}(V_T))$ and*

4. *$[REGA_{d,max}(V_T, C)] \subseteq [L(d\text{-}C\text{-}reg_{max}(V_T))]$.*

Proof. Let $L \in REGA_d(V_T, C, u)$.

If $L \in \{\emptyset, \{\langle \emptyset_d \rangle\}\} \cup \{\{\langle ua \rangle\} \mid a \in V_T\}$, then obviously L can be generated by a d-dimensional even maximal (C, u)-regular array grammar.

We now show that $L(d\text{-}(C, u)\text{-}reg(V_T))$ and $L(d\text{-}(C, u)\text{-}reg_{max}(V_T))$ are closed under union as well as catenation and iterated catenation with respect to C. Let $L_i \in L(d\text{-}(C, u)\text{-}reg(V_T))$, $i \in \{1, 2\}$, i.e., there are two d-dimensional (C, u)-regular array grammars $G_i = (d, V_N^{(i)}, V_T, \#, P_i, u, S_i)$, such that $L(G_i) = L_i$, $i \in \{1, 2\}$. Without loss of generality we may assume that $V_N^{(1)}$ and $V_N^{(2)}$ are disjoint sets.

1. If L is obtained as the union $L_1 \cup L_2$, we define the new d-dimensional (C, u)-regular array grammar G, $G = (d, V_N, V_T, \#, P, u, S)$, such that S is a new symbol, $V_N = V_N^{(1)} \cup V_N^{(2)} \cup \{S\}$, $P = ((P_1 \cup P_2) - \{S_1 \to \#, S_2 \to \#\}) \cup P' \cup P''$, where $P' = \{S \to \alpha \mid S_1 \to \alpha \in P_1\} \cup \{S \to \beta \mid S_2 \to \beta \in P_2\}$ and $P'' = \{S \to \#\}$ if either $S_1 \to \# \in P_1$ or $S_2 \to \# \in P_2$ and $P'' = \emptyset$ otherwise. Clearly, G is a d-dimensional (C, u)-regular array grammar and $L(G) = L_1 \cup L_2$. If G_1 and G_2 are maximal, then G is maximal, too.

2. Now let us consider $L = L_1 \odot_D L_2$ for some $D \subseteq C$; moreover, assume that the empty d-dimensional formal array $\langle \emptyset_d \rangle$ does not belong to $L_1 \cup L_2$. Then we define $G = (d, V_N, V_T, \#, P, u, S_1)$, $V_N = V_N^{(1)} \cup V_N^{(2)}$, $P = (P_1 - \{X \to a \mid X \to a \in P_1\}) \cup P_2 \cup P'$, where $P' = \{X \to awS_2 \mid X \to a \in P_1 \text{ and } w \in D\}$. One can easily verify that G is a d-dimensional (C, u)-regular array grammar and $L(G) = L_1 \odot_D L_2$. If G_1 and G_2 are maximal and $D = C$, then G is maximal, too.

   The cases where the empty d-dimensional formal array $\langle \emptyset_d \rangle$ is in L_1 or in L_2 or in both of them are left to the reader.

3. Let $L = L_1^{\odot D}$ for some $D \subseteq C$. Then we define the d-dimensional (C, u)-regular array grammar $G = (d, V_N^{(1)} \cup \{S\}, V_T, \#, P_1', u, S)$ such that S is a new symbol, $P_1' = (P_1 - \{S_1 \to \#\}) \cup P' \cup P'' \cup \{S \to \#\}$, where $P' = \{S \to \alpha \mid S_1 \to \alpha \in P_1\}$ and $P'' = \{X \to awS_1 \mid X \to a \in P_1 \text{ and } w \in D\}$. Obviously, G is a d-dimensional (C, u)-regular array grammar and $L(G) = L_1^{\odot D}$. If G_1 is maximal and $D = C$, then G is maximal, too.

All the closure properties proved above for $L(d\text{-}(C, u)\text{-}reg(V_T))$ and $L(d\text{-}(C, u)\text{-}reg_{max}(V_T))$ are valid for the corresponding families $[L(d\text{-}C\text{-}reg(V_T))]$ and $[L(d\text{-}C\text{-}reg_{max}(V_T))]$, too, i.e., in these cases we simply may consider d-dimensional (C, Ω_d)-regular array grammars. □

In order to prove the inclusion $\supseteq$ in Theorem 5.2, we need some additional notations and results (e.g., see [7]).

Remark 5.2 *For any terminal alphabet V_T, $u \in Z^d$, and any non-empty finite subset C of Z_0^d, let $A(d, V_T, C, u)$ and $[A(d, V_T, C)]$ denote the powerset of $\langle V_T^{*d}(C, u)\rangle$ and $[\langle V_T^{*d}(C)\rangle]$ $(= [\langle V_T^{*d}(C, u)\rangle])$, respectively. Then $(A(d, V_T, C, u), \cup, \odot_C, \emptyset, \{\langle \emptyset_d \rangle\})$ and $(A(d, V_T, C), \cup, \odot_C, \emptyset, \{[\langle \emptyset_d \rangle]\})$ is an ω-complete semiring with first element $\emptyset$.*

Remark 5.3 *Let $\mathcal{S} = (S, +, \odot, 0, 1)$ be an ω-complete semiring with first element 0. Let $Y = \{Y_i \mid 1 \le i \le m\}$ be a set of variables and let $\mathcal{E}$ be a rational system of equations over $\mathcal{S}$, i.e., $\mathcal{E}$ is a set of equations of the form*

$$Y_i = \alpha_{i1} \odot Y_1 + \ldots \odot \alpha_{im} \odot Y_m + \beta_i,$$

where $\alpha_{ij} \in S$, $\beta_i \in S$, $1 \le i \le m$, $1 \le j \le m$.
   It is well-known (see [7]) that the system $\mathcal{E}$ has a solution, and moreover, each component of the solution of $\mathcal{E}$ is a rational set in S.

Example 5.1 *For a terminal alphabet V_T, $u \in Z^d$, and a non-empty finite subset C of Z_0^d, consider the equation $Y = \alpha \odot_C Y \cup \beta$, where α and β are subsets of $\langle V^{*d}(C, u) \rangle$ or $[\langle V^{*d}(C) \rangle]$. Then the minimal fixpoint of this equation is $\alpha^{\odot_C^*} \odot_C \beta$ according to Kleene's theorem.*

For any d-dimensional maximal (C, u)-regular array grammar we now define a rational system of equations $\mathcal{E}_G$ such that $L(G)$ is a component of the (minimal) fixpoint solution of $\mathcal{E}_G$.

Definition 5.1 *Let $G = (d, V_N, V_T, \#, P, u, Y_1)$ be a d-dimensional maximal (C, u)-regular array grammar. The rational system of equations $\mathcal{E}_G$ associated with G is defined as follows:*

1. *The elements in V_N, $V_N = \{Y_i \mid 1 \leq i \leq m\}$, are the variables of the system $\mathcal{E}_G$.*

2. *The equations forming $\mathcal{E}_G$ are*

$$Y_i = \alpha_{i1} \odot_C Y_1 \cup \ldots \alpha_{im} \odot_C Y_m \cup \beta_i,$$

   where $\alpha_{ij} = \{\langle ua \rangle \mid Y_i \to avY_j \in P \text{ for all } v \in C\}$, $1 \leq j \leq m$, $1 \leq i \leq m$, and $\beta_i = \{\langle ua \rangle \mid Y_i \to a \in P\}$, $1 \leq i \leq m$; if $Y_1 \to \# \in P$, then we also add $\langle \emptyset_d \rangle$ to β_1.

Using these notations (similar notations can be introduced for C-regular array grammars in an obvious way) we state the following important result:

Theorem 5.3 *Let $G = (d, V_N, V_T, \#, P, u, Y_1)$ be a d-dimensional maximal (C, u)-regular array grammar and let $\mathcal{E}_G$ be the rational system of equations associated with G as defined in Definition 5.1. Let L_{Y_i} be the i-th component $(Y_i$-component) of the minimal fixpoint solution of $\mathcal{E}_G$. Then,*

$$L_{Y_i} = L(G_{Y_i}), \text{ where } G_{Y_i} = (d, V_N, V_T, \#, P, u, Y_i),$$

and therefore $L(G) = L_{Y_1}$.

Proof. First we show that if the formal array $\mathcal{A}$ is the result of a terminal derivation in G_{Y_i} starting from $\langle uY_i \rangle$, then it is an element of the Y_i-component L_{Y_i} of the solution of $\mathcal{E}_G$, too, which can be shown by induction with respect to the length of terminal derivations in the grammars G_{Y_i}.

If the terminal derivation is $\langle\langle uY_1\rangle, \langle\emptyset_d\rangle\rangle$, then by definition, $Y_1 \to \# \in P$ and therefore $\langle\emptyset_d\rangle \in \beta_1$. If the terminal derivation is $\langle\langle uY_i\rangle, \langle ua\rangle\rangle$ for some $a \in V_T$, then by definition, $Y_i \to a \in P$ and therefore $\langle ua\rangle \in \beta_i$. Hence, in any case, for a terminal derivation starting with $\langle uY_i\rangle$ of length 1, the result of a terminal derivation in G_{Y_i} is an element of the Y_i-component L_{Y_i} of the solution of $\mathcal{E}_G$, too. Now assume that the sequence $\langle\mathcal{A}_k\rangle_{k=0}^n$ of formal arrays $\mathcal{A}_k$ is a terminal derivation in G_{Y_i} with $n > 1$ such that:

1. $\mathcal{A}_0 = \langle uY_i\rangle = \langle v_0X_0\rangle$,

2. for every k with $0 < k < n$, $\mathcal{A}_k = \langle v_0a_0 \ldots v_{k-1}a_{k-1}v_kX_k\rangle$ is a well-defined formal array and $X_{k-1} \to a_{k-1}v_kX_k \in P$ as well as $v_k \in C$, $a_{k-1} \in V_T$, and $X_k \in V_N$; finally,

3. $\mathcal{A}_N = \langle v_0a_0 \ldots v_{n-1}a_{n-1}\rangle$ (is a well-defined formal array), $X_{n-1} \to a_{n-1} \in P$ and $a_{n-1} \in V_T$.

Then we consider the sequence $\langle\mathcal{B}_k\rangle_{k=1}^n$ with:

1. $\mathcal{B}_1 = \langle uX_1\rangle = \langle v_1'X_1\rangle$;

2. for every k with $1 < k < n$, $\mathcal{B}_k = \langle v_1'a_1 \ldots v_{k-1}'a_{k-1}v_k'X_k\rangle$, $v_1' = u$, $v_l' = v_l$, $1 < l \le k$;

3. $\mathcal{B}_N = \langle v_1'a_1 \ldots v_{n-1}'a_{n-1}\rangle = \langle v_1a_1 \ldots v_{n-1}a_{n-1}\rangle$.

The formal arrays $\mathcal{B}_k$, $1 \le k \le n$, are well-defined, and $\langle\mathcal{B}_k\rangle_{k=1}^n$ is a terminal derivation in G_{X_1} of length $n - 1$. By the induction hypothesis, $\mathcal{B}_N \in L_{X_1}$. According to the construction of the system $\mathcal{E}_G$, $X_0 \to a_0v_1X_1 \in P$ implies that $\langle ua_0\rangle \in \alpha_{ij}$, where j is the index such that $X_1 = Y_j$. Hence, by the definition of the fixpoint solution, $\mathcal{A}_N = \langle ua_0\rangle \odot_{v_1} \mathcal{B}_N$ and $\mathcal{A}_N \in L_{Y_i}$.

For the converse inclusion, the proof is by induction with respect to the number of Kleene-operations necessary to obtain a formal array $\mathcal{A}$ as an element of the component L_{Y_i}:

If $\mathcal{A}$ belongs to the first iteration, this means that $\mathcal{A}$ has to be an element of β_i. According to the definition of $\mathcal{E}_G$, $\mathcal{A}$ then is the result of a terminal derivation of length 1 in G_{Y_i}. Now assume that for all formal arrays coming up in the $(n-1)$-th Kleene-iteration, $n > 1$, there exists a terminal derivation in G_{Y_i} of length $(n-1)$ yielding $\mathcal{A}$. Now consider a formal array $\mathcal{A}$ coming up in the n-th Kleene-iteration of the component Y_i, i.e., $\mathcal{A} = \langle ua_0\rangle \odot_{v_1} \mathcal{B}$, where $\mathcal{B}$ is a formal array having come up in the $(n-1)$-th Kleene-iteration of the component Y_j for some index j and $\langle ua_0 \rangle \in \alpha_{ij}$. By the induction hypothesis, $\mathcal{B}$ is the result of a terminal derivation $\langle \mathcal{B}_k \rangle_{k=1}^n$ of length $n-1$ in G_{Y_j}. Moreover, by the construction of the system $\mathcal{E}_G$, $Y_i \to a_0 v_1 Y_j \in P$; therefore, the terminal derivation $\langle \mathcal{B}_k \rangle_{k=1}^n$ of $\mathcal{B}$ can be extended to a terminal derivation $\langle \mathcal{A}_k \rangle_{k=0}^n$ of $\mathcal{A}$ of length n in G_{Y_i} with $\mathcal{A}_0 = \langle uY_i \rangle$ and $\mathcal{A}_k = \langle ua_0 \rangle \odot_{v_1} \mathcal{B}_k$, $1 \le k \le n$; consequently, $\mathcal{A} \in L(G_{Y_i})$. □

The proof of Theorem 5.2 is now complete, it follows from Lemma 5.1 and Theorem 5.3. The results of Theorem 5.2 and its consequences can be summed up in the following way:

Corollary 5.3 *For any terminal alphabet V_T, $u \in Z^d$, and any (non-empty) finite subset C of Z_0^d,*

1. $REGA_{d,max}(V_T, C, u) = L(d\text{-}(C, u)\text{-}reg_{max}(V_T))$;

2. $[REGA_{d,max}(V_T, C)] = [L(d\text{-}C\text{-}reg_{max}(V_T))]$;

3. $L(d\text{-}(C, u)\text{-}reg_{max}(V_T))$ $([L(d\text{-}C\text{-}reg_{max}(V_T))])$ *coincides with the family of rational languages in $A(d, V_T, C, u)$ (in $A(d, V_T, C)$).*

6 Future Research Topics

This paper can be seen as a starting point for further investigations in the field of algebraic representations of d-dimensional array languages. One major research topic will be the investigation of possible algebraic representations of d-dimensional regular array languages generated by d-dimensional (C, u)-regular array grammars that are not maximal as well as of possible algebraic representations of d-dimensional array languages generated by specific (#-)context-free d-dimensional array grammars. Moreover, all the notions introduced in this paper can also be introduced in the framework of Cayley grids as proposed in [2], where the underlying grid is based on the more general structure of Cayley graphs of finitely generated groups; many of the results obtained in this paper can be carried over into this more general framework, although some of them are not valid any more in the form presented in this paper.

References

[1] C. R. Cook, P. S.-P. Wang, A Chomsky hierarchy of isotonic array grammars and languages, *Computer Graphics and Image Processing*, **8** (1978), 144–152.

[2] E. Csuhaj-Varjú, R. Freund, V. Mitrana, *Arrays on Cayley Grids*, TUCS Report, 1998.

[3] H. Fernau, R. Freund, M. Holzer, The generative power of d-dimensional #-context-free array grammars, *Proceedings MCU'98*, Vol. II (M. Margenstern, ed.), Metz, 1998, 43–56.

[4] R. Freund, Gh. Păun, One-dimensional matrix array grammars, *J. Inform. Process. Cybernet.*, EIK, **29**, 6 (1993), 1–18.

[5] R. Freund, Control mechanisms on #-context-free array grammars, in *Mathematical Aspects of Natural and Formal Languages* (Gh. Păun, ed.), World Scientific Publ., Singapore, 1994, 97–136.

[6] C. Kim, I. H. Sudborough, The membership and equivalence problems for picture languages, *Theoretical Computer Science*, **52** (1987), 177–191.

[7] W. Kuich, A. Salomaa, *Semirings, Automata, Languages*, Springer-Verlag, Berlin, 1986.

[8] K. Morita, Y. Yamamoto, K. Sugata, The complexity of some decision problems about two-dimensional array grammars, *Information Sciences*, **30** (1983), 241–262.

[9] A. Rosenfeld, *Picture Languages*, Academic Press, Reading, MA, 1979.

[10] A. Salomaa, *Formal Languages*, Academic Press, Reading, MA, 1973.

[11] I. H. Sudborough, E. Welzl, Complexity and decidability for chain code picture languages, *Theoretical Computer Science*, **36** (1985), 173–202.

[12] P. S.-P. Wang, Some new results on isotonic array grammars, *Information Processing Letters*, **10** (1980), 129–131.

[13] Y. Yamamoto, K. Morita, K. Sugata, Context-sensitivity of two-dimensional regular array grammars, in *Array Grammars, Patterns and Recognizers* (P. S.-P. Wang, ed.), WSP Series in Computer Science, Vol. 18, World Scientific Publ., Singapore, 1989, 17–41.

Rough Set Processing of Vague Information Using Fuzzy Similarity Relations

Salvatore Greco, Benedetto Matarazzo

Faculty of Economics, University of Catania
Corso Italia 55, 95129 Catania, Italy
E-mail: salgreco@vm.unict.it

Roman Slowinski

Institute of Computing Science, Poznan University of Technology
Piotrowo 3a, 60-965 Poznan, Poland
E-mail: slowinsk@sol.put.poznan.pl

Abstract. The rough sets theory has proved to be a very useful tool for analysis of information tables describing objects by means of disjoint subsets of condition and decision attributes. The key idea of rough sets is approximation of knowledge expressed by decision attributes using knowledge expressed by condition attributes. From a formal point of view, the rough sets theory was originally founded on the idea of approximating a given set represented by objects having the same description in terms of decision attributes, by means of an indiscernibility binary relation linking pairs of objects having the same description by condition attributes. The indiscernibility relation is an equivalence binary relation (reflexive, symmetric and transitive) and implies an impossibility to distinguish two objects having the same description in terms of the condition attributes. It produces crisp granules of knowledge that are used to build approximations. In reality, due to vagueness of the available information about objects, small differences are not considered significant. This situation may be formally modelled by similarity or tolerance relations instead of the indiscernibility relation. We are using a similarity relation which is only reflexive, relaxing therefore the properties of symmetry and transitivity.

Moreover, the credibility of our similarity relation is gradual,
i.e., it is a fuzzy similarity relation. As a consequence, the granules of knowledge produced by this relation are fuzzy and their
credibility can change gradually from any finite degree to an
infinitely small degree, thus giving different credibility of rough
approximations.

1   Introduction

The rough sets theory proposed by Pawlak [11], [12] was originally founded
on the idea of approximating a given set by means of an indiscernibility
binary relation which was assumed to be an equivalence binary relation
(reflexive, symmetric and transitive). It produces crisp granules of knowledge that are used to built approximations. With respect to the basic
rough set idea, three main theoretical developments have been proposed
for handling different types of the available information:

1. extensions to a fuzzy description of objects (e.g., [2], [3], [18], [19],
   [21]);

2. extensions of the indiscernibility relation to more general binary relations modelling similarity between objects (e.g., [7], [8], [14], [16],
   [23], [24], [25]);

3. extensions to preference-ordered scales of attributes (criteria) and to
   multicriteria decision problems, using dominance relations and pairwise comparison tables (e.g., [5]).

   In this paper, we put together extensions 1) and 2), considering the approach proposed in [23], [24] within a fuzzy context. More specifically, we
propose to approximate a given fuzzy set by means of reflexive fuzzy similarity relations. This extension is motivated by vagueness of the available
information about objects. As a consequence of considering fuzzy similarity relations instead of the classical indiscernibility relation, the granules
of knowledge produced by this relation are fuzzy and their credibility can
change gradually from any finite degree to an infinitely small degree, thus
giving different credibility of rough approximations.

   The paper is structured as follows. In Section 2, we recall some basic
concepts concerning rough sets, fuzzy sets, approximations by similarity
relations and rough fuzzy sets. In Section 3, we introduce the rough approximation by fuzzy similarity relations. In Section 4, we discuss the rule

induction from rough approximations by fuzzy similarity relations. In Section 5, we present an application of the proposed approach to an exemplary problem. Section 6 groups conclusions.

2 Basic Elements of Fuzzy Sets and Rough Sets Theories

2.1 Elements of the Rough Sets Theory

The rough set concept proposed by Pawlak [11], [12] is founded on the assumption that with every object of the universe of discourse there is associated some information (data, knowledge). For example, if objects are firms submitted to a bankruptcy risk evaluation, their financial, economic and technical characteristics form information (description) about the firms. Objects characterized by the same description are indiscernible in view of the available information about them. The *indiscernibility relation* generated in this way is the mathematical basis of the rough sets theory.

Any set of indiscernible objects is called an *elementary set* and forms a crisp granule (atom) of knowledge about the universe. Any subset Y of the universe can either be expressed precisely, in terms of the granules, or roughly only. In the latter case, subset Y can be characterized by two ordinary sets called *lower* and *upper approximations*. The two approximations define the rough set. The lower approximation of Y consists of all elementary sets included in Y, whereas the upper approximation of Y consists of all elementary sets having a non-empty intersection with Y. Obviously, the difference between the upper and the lower approximation constitutes the *boundary region* including objects which cannot be properly classified as belonging or not to Y, using the available information. Cardinality of the boundary region says, moreover, how exactly we can describe Y in terms of available information.

For algorithmic reasons, knowledge about objects will be represented in the form of an information table. The rows of the table are labelled by *objects*, whereas columns are labelled by *attributes* and entries of the table are *attribute-values*. Formally, by an *information table* we understand the 4-tuple $S = \langle U, Q, V, f \rangle$, where U is a finite set of objects, Q is a finite set of attributes, $V = \bigcup_{q \in Q} V_q$ and V_q is a domain of the attribute q, and $f : U \times Q \longrightarrow V$ is a total function such that $f(x, q) \in V_q$ for every $q \in Q, x \in U$, called an *information function* (see [12]).

An information table can be seen as a *decision table* assuming that the set of attributes Q is equal to $C \cup D$ such that $C \cap D = \emptyset$, where set C contains so-called *condition attributes*, and D, *decision attributes*.

From the decision table a set of *decision rules* can be induced and expressed as logical statements "*if... then...*" relating descriptions of condition and decision classes. The rules are exact or approximate depending whether a description of a condition class corresponds to a unique decision class or not. Different procedures for derivation of decision rules have been presented (see, e.g., [6], [27], [10], [22]).

2.2   Elements of the Fuzzy Sets Theory

We use the following basic concepts with respect to fuzzy sets theory. For a more extensive review see, e.g., [1], [4]. Each set X in a universe of discourse U is represented by its membership function $\mu_X : U \longrightarrow [0,1]$, where $\mu_X(x)$ is the grade of membership of x in X, from full nonmembership ($\mu_X(x) = 0$) to full membership ($\mu_X(x) = 1$), through all intermediate values. Within fuzzy logic the conjunction operator "and" is represented by a triangular norm or t-norm $T(x,y)$ while the disjunction operator "or" is represented by a triangular conorm or t-conorm $T^*(x,y)$ ([9], [15], [26]). More precisely,

- a t-norm is a function $T : [0,1] \times [0,1] \longrightarrow [0,1]$ satisfying the following properties:
  $T(1,x) = x$, for all $x \in [0,1]$,
  $T(x,y) = T(y,x)$, for all $x,y \in [0,1]$,
  $T(x,y) \leq T(u,v)$, for all $x \leq u, y \leq v$,
  $T(x,T(y,z)) = T(T(x,y),z)$, for all $x,y,z \in [0,1]$;

- a t-conorm is a function $T^* : [0,1] \times [0,1] \longrightarrow [0,1]$ satisfying the following properties:
  $T^*(0,x) = x$, for all $x \in [0,1]$,
  $T^*(x,y) = T^*(y,x)$, for all $x,y \in [0,1]$,
  $T^*(x,y) \leq T^*(u,v)$, for all $x \leq u, y \leq v$,
  $T^*(x,T^*(y,z)) = T^*(T^*(x,y),z)$, for all $x,y,z \leq [0,1]$.

The powers of the function T and T^* are defined as follows:

$$T^1(x_1, x_2) = T(x_1, x_2),$$
$$T^{*1}(x_1, x_2) = T^*(x_1, x_2),$$

for all $x_1, x_2 \in [0,1]$, and

$$T^n(x_1, \ldots, x_{n+1}) = T(T^{n-1}(x_1, \ldots, x_n), x_{n+1}),$$

$$T^{*n}(x_1, \ldots, x_{n+1}) = T^*(T^{*n-1}(x_1, \ldots, x_n), x_{n+1}),$$

for all $x_1, \ldots, x_2 \in [0,1]$ and for all $n \geq 2$.

We adopt the following notation:

$$T^{n-1}(x_1, \ldots, x_n) = T(x_1, \ldots, x_n) = T_{i=1}^n x_i = T_{x \in A} x,$$
$$T^{*n-1}(x_1, \ldots, x_n) = T^*(x_1, \ldots, x_n) = T_{i=1}^{*n} x_i = T_{x \in A}^* x,$$

where $A = \{x_1, \ldots, x_n\}$ and $x_1, \ldots, x_n \in [0,1]$.

Let us observe that for each t-norm T and for each t-conorm T^* the following conditions hold for all $x, y \in [0,1]$

$$T(x, y) \leq \min x, y, \tag{1}$$
$$T^*(x, y) \geq \max x, y. \tag{2}$$

In the context of fuzzy logic the negation operator $N(\cdot)$ is a nonincreasing function $N : [0,1] \longrightarrow [0,1]$ such that $N(0) = 1$ and $N(1) = 0$. A negation is strict if $N(\cdot)$ is a strictly decreasing continuous function. Often the negation N is required to satisfy also the involutory condition

$$N(N(x)) = x, \quad \text{for all } x \in [0,1]. \tag{3}$$

Given a fuzzy set X in universe U, whose membership function is $\mu_X(x)$ the complement of X, denoted by $U - X$, is the fuzzy set on U having the membership function $\mu_{U-X}(x) = N(\mu_X(x))$.

Given a t-norm T, a t-conorm T^* and a strict negation N, (T, T^*, N) is a De Morgan triplet iff $N(T^*(x, y)) = T(N(x), N(y))$.

In fuzzy logic several implication operators have been proposed. Let us remember that an implication is a function $I^{\rightarrow}(x, y) : [0,1] \times [0,1] \longrightarrow [0,1]$ satisfying the following properties ([4]):

$$I^{\rightarrow}(x, y) \geq I^{\rightarrow}(z, y), \quad \text{for all } x \leq z \text{ and for all } y,$$
$$I^{\rightarrow}(x, y) \geq I^{\rightarrow}(t, y), \quad \text{for all } x \leq t \text{ and for all } y,$$
$$I^{\rightarrow}(0, x) = 1, \quad \text{for all } x,$$
$$I^{\rightarrow}(x, 1) = 1, \quad \text{for all } x,$$
$$I^{\rightarrow}(1, 0) = 0.$$

In this paper we consider a T^*-implication $I^{\rightarrow}_{T^*, N}$, i.e., a function $I^{\rightarrow}_{T^*, N} : [0,1] \times [0,1] \longrightarrow [0,1]$ associated with a t-conorm T^* and a negation N defined by $I^{\rightarrow}_{T^*, N}(x, y) = T^*(N(x), y))$.

A fuzzy binary relation is a function $R : U \times U \longrightarrow [0,1]$. It is

- reflexive if $R(x, x) = 1$, for each $x \in U$,

- symmetric if $R(x, y) = R(y, x)$, for all $x, y \in U$,

- T-transitive if $T(R(x, y), R(y, z)) \leq R(x, z)$, for all $x, y, z \in U$.

A fuzzy binary relation which is reflexive, symmetric and T-transitive is a fuzzy equivalence binary relation. In fuzzy sets theory, a fuzzy equivalence relation is called fuzzy similarity; however, we prefer to keep the term of fuzzy similarity for fuzzy binary relations that are only reflexive.

2.3   Fuzzy Rough Sets

Let $X \subseteq U$. Let also R be a binary equivalence relation defined in U, and $[x]_R$ the equivalence class including $x \in U$. The lower approximation of X, denoted by $\underline{R}(X)$, and the upper approximation of X, denoted by $\overline{R}(X)$, are defined as

$$\underline{R}(X) = \{x \in U \mid [x]_R \subseteq X\}, \tag{4}$$
$$\overline{R}(X) = \{x \in U \mid [x]_R \cap X \neq \emptyset\}. \tag{5}$$

Let X be a fuzzy set defined in a finite universe U, $\mu_X(x)$ the membership function of X, R an equivalence relation defined on U, $[x]_R$ the equivalence class including $x \in U$. The lower approximation of X can be defined as a fuzzy set whose membership function associated to each $x \in U$ is equal to the credibility that "each $y \in [x]_R$ belongs to X", i.e.,

$$\mu(x, \underline{R}(X)) = T_{y \in [x]_R} \mu_X(y). \tag{6}$$

Analogously, the upper approximations of X can be defined as a fuzzy set whose membership function associates to each $x \in U$ the credibility that "there is at least one $y \in [x]_R$ belonging to X", i.e.,

$$\mu(x, \overline{R}(X)) = T^*_{y \in [x]_R} \mu_X(y). \tag{7}$$

Let $X \subseteq U$, and let R be an equivalence fuzzy binary relation defined on U. In this case the lower approximation of X can be defined as a fuzzy set whose membership function associates to each $x \in U$ the credibility that "for each $y \in X$, y is not in the relation R with x", i.e.,

$$\mu(x, \underline{R}(X)) = T_{y \notin X} N(R(y, x)). \tag{8}$$

Analogously the upper approximations of X can be defined as a fuzzy set whose membership function associates to each $x \in U$ the credibility that "there is at least one $y \in X$ which is in the relation R with x", i.e.,

$$\mu(x, \overline{R}(X)) = T^*_{y \in X} R(y, x). \tag{9}$$

Finally, let X be a fuzzy set defined in a finite universe U. Moreover, let $\mu_X(x)$ be the membership function of X and R be an equivalence fuzzy binary relation defined on U. In this case, the lower approximation of X can be defined as a fuzzy set whose membership function associates with each $x \in U$ the credibility that "for each $y \in U$, y is not in the relation R with x and/or y belongs to X", i.e.,

$$\mu(x, \underline{R}(X)) = T_{y \in U} T^*(N(R(y, x), \mu_X(y))). \tag{10}$$

The upper approximation of X can be defined in turn as a fuzzy set whose membership function associates to each $x \in U$ the credibility that "there is at least one $y \in U$ which is in the relation R with x and which belongs to X", i.e.,

$$\mu(x, \overline{R}(X)) = T^*_{y \in U} T(R(y, x), \mu_X(y)). \tag{11}$$

Let us observe that

a) definitions (5) and (6) are special cases of definitions (10) and (11), when the fuzzy equivalence binary relation R becomes a crisp equivalence binary relation,

b) definitions (8) and (9) are special cases of definitions (10) and (11), when the fuzzy set X becomes a crisp set,

c) the classical definitions (5) and (5) represent special cases of definitions (10) and (11), when the fuzzy equivalence binary relation R becomes a crisp equivalence binary relation and the fuzzy set X becomes a crisp set.

2.4   Similarity

As observed above, indiscernibility implies an impossibility to distinguish two objects of U having the same description in terms of the attributes from Q. This relation induces equivalence classes on U, which constitute the crisp granules of knowledge. In reality, due to uncertainty and imprecision of data describing the objects (vague information), small differences are often not considered significant for the purpose of discrimination. This situation may be formally modelled by considering similarity or tolerance relations (see e.g., [7], [8], [14], [16], [23], [24], [25]).

In general, the similarity relations R do not generate partitions of U; the information regarding similarity may be represented using *similarity classes* for each object $x \in U$. Precisely, the similarity class of x, denoted by $R(x)$, consists of the set of objects which are similar to x:

$$R(x) = \{y \in U \mid yRx\}.$$

It is obvious that an object $y \in R(x)$ may be similar to another object $z \in U$, and $z \notin R(x)$. The similarity relation is of course reflexive (each object is similar to itself). In [23], [24] one proposes a similarity relation which is only *reflexive,* relaxing therefore the properties of symmetry and transitivity. The abandoning of the transitivity requirement is easily justi-fiable, remembering – for example – Luce's paradox of the cups of tea. As for the symmetry, one should notice that yRx, which means "y is similar to x", is directional; there is a subject y and a referent x, and in general this is not equivalent to the proposition "x is similar to y". This is quite immediate when the similarity relation is defined in terms of a percent-age difference between evaluations of the objects compared on a numerical attribute, calculated with respect to the evaluation of the referent object. Therefore, the symmetry of the similarity relation should not be imposed and it makes sense to consider the inverse relation of R, denoted by R^{-1}, where $xR^{-1}y$ means again "y is similar to x"; $R^{-1}(x), x \in U$, is then the class of referent objects to which x is similar:

$$R^{-1}(x) = \{y \in U \mid xRy\}.$$

Given a subset $X \subseteq U$ and a similarity relation R on U, an object $x \in U$ is said to be *non-ambiguous* in each of the two following cases:

- x clearly belongs to X, that is $x \in X$ and $R^{-1}(x) \subseteq X$; such objects are called *positive*;

- x clearly does not belong to X, that is $x \in U - X$ and $R^{-1}(x) \subseteq U - X$ (or $R^{-1}(x) \cap X = \emptyset$); such objects are called *negative*.

The objects which are neither positive nor negative are said to be *am-biguous*.

A more general definition of lower and upper approximation may thus be offered (see [24]). Let $X \subseteq U$ and R be a reflexive binary relation defined on U; the lower approximation of X, denoted by $\underline{R}(X)$, and the upper approximation of X, denoted by $\overline{R}(X)$, are defined, respectively, as:

$$\underline{R}(X) = \{x \in U \mid R^{-1}(x) \subseteq X\},$$
$$\overline{R}(X) = \bigcup_{x \in X} R(x).$$

It may be demonstrated that the key property $\underline{R}(X) \subseteq X \subseteq \overline{R}(X)$ still holds and that

$$\underline{R}(X) = U - \overline{R}(U - X) \text{ (complementarity property) and}$$
$$\overline{R}(X) = \{x \in U \mid R^{-1}(x) \cap X \neq \emptyset\}.$$

Moreover, the definitions proposed are the only ones which correctly characterize the set of positive objects (lower approximation) and the set of positive or ambiguous objects (upper approximation) when a similarity relation is reflexive, but not necessarily symmetric nor transitive.

Using a similarity relation one is able to induce decision rules from a decision table. The syntax of a rule is the following:

"if $f(x, q_1)$ is similar to r_{q_1} and $f(x, q_2)$ is similar to r_{q_2} and $\ldots f(x, q_p)$ is similar to r_{q_p}, then x belongs to Y_{j_1} or Y_{j_2} or $\ldots Y_{j_k}$",

where $\{q_1, q_2, \ldots, q_p\} \subseteq C, (r_{q_1}, r_{q_2}, \ldots, r_{q_p}) \in V_{q_1} \times V_{q_2} \times \ldots \times V_{q_p}$ and $Y_{j-1}, Y_{j_2}, \ldots, Y_{j_k}$ are some classes of the considered classification (D-elementary sets). If $k = 1$, then the rule is *exact*, otherwise it is *approximate* or *uncertain*. Procedures for generation of decision rules adapt the general induction scheme.

3 Rough Approximations by Reflexive Fuzzy Similarity Relations

3.1 Basic Concepts

Let U be a finite non-empty set of objects, called universe, and R a fuzzy reflexive binary relation defined on U, which represents a certain form of similarity. Let X be a fuzzy set in U and let also $\mu_X : U \longrightarrow [0, 1]$ be the membership function of X. Given $x \in U$ we say that:

1. the membership degree of x to the set of positive objects with respect to X, denoted by $Pos(x, X)$, is the credibility that "for each $y \in U$,

y is not similar to x and/or y belongs to X", i.e.,

$$Pos(x, X) = T_{y \in U} T^*(N(R(x, y)), \mu_X(y));$$

2. the membership degree of x to the set of negative objects with respect to X, denoted by $Neg(x, X)$, is the credibility that "for each $y \in U$, y is not similar to x and/or y does not belong to X", i.e.,

$$Neg(x, X) = T_{y \in U} T^*(N(R(x, y)), N(\mu_X(y))).$$

Let us remark that, remembering the definition of T^*-implication, we can write

$$Pos(x, X) = T_{y \in U} I_{\overrightarrow{T^*}, N}(R(x, y), \mu_X(y)), \tag{12}$$
$$Neg(x, X) = T_{y \in U} I_{\overrightarrow{T^*}, N}(N(R(x, y)), N(\mu_X(y))). \tag{13}$$

On the basis of (12), $Pos(x, X)$ can be seen as the credibility that "for each $y \in U$ the similarity of x to y implies that y belongs to X". Analogously, from (13), $Neg(x, X)$ can be seen as the credibility that "for each $y \in U$ the 'non-similarity' of x to y implies that y does not belong to X".

Considering a subset $X \subseteq U$ and a reflexive binary relation R defined on U, the lower approximation of X, denoted by $\underline{R}(X)$, and the upper approximation of X, denoted by $\overline{R}(X)$, are fuzzy subsets of U whose membership functions are respectively defined as

$$\mu(x, \underline{R}(X)) = T_{y \in U} T^*(N(R(x, y)), \mu_X(y))),$$
$$\mu(x, \overline{R}(X)) = T^*_{y \in U} T(R(x, y), \mu_X(y)).$$

Theorem 3.1 *For each* $x \in U$ *we have:* $\mu(x, \underline{R}(X)) \leq \mu_X(x) \leq \mu(x, \overline{R}(X))$.

Proof. From property (1) we have for each $x \in U$

$$\mu(x, \underline{R}(X)) = T_{y \in U} T^*(N(R(x, y)), \mu_X(y))$$
$$\leq \min_{y \in U} T^*(N(R(x, y)), \mu_X(y)). \tag{14}$$

Since $\min_{y \in U} T^*(N(R(x, y)), \mu_X(y)) \leq T^*(N(R(x, x)), \mu_X(x))$ and from the reflexivity of R we have

$$T^*(N(R(x, x)), \mu_X(x)) = T^*(N(1), \mu_X(x)) = T^*(0, \mu_X(x)) = \mu_X(x),$$

from (14) we obtain

$$\mu(x, \underline{R}(X)) \le \mu_X(x).$$

Analogously, from property (2), we have: $\mu(x, \overline{R}(X)) = T^*_{y \in U} T(R(x, y),$
$\mu_X(y)) \ge \max_{y \in U} T(R(x, y), \mu_X(y)) \ge T(R(x, x), \mu_X(x)) = T(1, \mu_X(x)) = \mu_X(x).$
$\square$

Theorem 3.1 can be read as the fuzzy counterpart of the fact that set X includes its P-lower approximation and is included in its P-upper approximation.

Let us observe that $\mu(x, \underline{R}(X)) = Pos(x, X)$ while $\mu(x, \overline{R}(X))$ can be seen as the credibility that "there is at least one $y \in U$ such that y belongs to X and x is similar to y".

Theorem 3.2 *Considering a fuzzy set X in U and a fuzzy binary reflexive relation R defined on U, if N is involutory and (T^*, T, N) is a De Morgan triple, then $\overline{R}(X)$ is the complement of the set of negative objects with respect to X, i.e., $\mu(x, \overline{R}(X)) = N(Neg(x, X))$.*

Proof. From De Morgan law we have $\mu(x, \overline{R}(X)) = T^*_{y \in U} T(R(x, y), \mu_X(y)) = T^*_{y \in U} N(T^*(N(R(x, y)), N(\mu_X(y)))) = N(T_{y \in U} T^*(N(R(x, y)), N(\mu_X(y)))) = N(Neg(x, X)).$
$\square$

Theorem 3.3 *Considering a fuzzy set X in U and a fuzzy binary reflexive relation R defined on U, if (T^*, T, N) is a De Morgan triple, then $\underline{R}(X)$ is the complement of the set of $\overline{R}(x, U - X)$, i.e., $\mu(x, \underline{}(X)) = N(\mu(x, \overline{R}(U - X)).$*

Proof. From De Morgan law we have $\mu(x, \underline{R}(X)) = T_{y \in U} T^*(N(R(x, y)), \mu_X(y)) = N(T^*_{y \in U} N(T^*(N(R(x, y)), \mu_X(y)))) = N(T^*_{y \in U} T(R(x, y), N(\mu_X(y)))) = N(\mu(x, \overline{R}(U - X)).$
$\square$

Theorem 3.3 expresses in fuzzy terms the following well-known complementarity property of the rough sets theory: the P-lower (P-upper) approximation of set X is the complement of the P-upper (P-lower) approximation of its complementary set $U - X$.

3.2   Modelling Fuzzy Similarity Relations

In this section, for each attribute $q \in Q$ we consider a valued binary relation R_q, i.e., a function $R_q : U \times U \longrightarrow [0, 1]$, where for all $x, y \in U, R_q(x, y)$

represents the intensity or degree of similarity of x to y with respect to attribute q. More precisely, for each $q \in Q$ and for all $x, y \in U$ we have:

$R_q(x, y) = 0$ means that there is no similarity of x to y with respect to attribute q, for all $x, y \in U$,

$R_q(x, y) = 1$ means that x is definitely similar to y with respect to attribute q, for all $x, y \in U$,

$R_q(x, y) \geq R_q(w, z)$ means that with respect to attribute q the similarity of x to y is at least as credible as the similarity of w to z, for all $x, y, w, z \in U$.

The similarity classes defined by the valued binary relation R_q correspond to fuzzy granules of knowledge; their credibility varies gradually between a finite degree equal to one and an infinitely small degree, thus giving different credibility of rough approximations.

The similarity R_q should satisfy the following mimimal condition:

$$[f(x, q) = f(y, q)] \implies R_q(x, y) = 1,$$

for all $x, y \in U$ and for each $q \in C$.

If $V_q = (v_{q^*}, v_q^*) \subseteq R$, then the similarity R_q should satisfy also the condition

$$[f(x, q) \leq f(y, q) \leq f(w, q) \leq f(z, q)] \Leftrightarrow R_q(y, w) \leq R_q(x, z),$$

for all $x, y, w, z \in U$ and for each $q \in C$.

We propose to model the similarity binary relation R_q in the following way for each $q \in Q$. Let ε_{q1} and ε_{q2} be two functions such that $\varepsilon_{q1} : V_q \longrightarrow R^+$, $\varepsilon_{q2} : V_q \longrightarrow R^+$ and $\varepsilon_{q1}(f(x, q)) \leq \varepsilon_{q2}(f(x, q))$ for each $x \in U$. Given $x, y \in U$

$$R_q(x, y) = \begin{cases} 1, & \text{if } |f(x, q) - f(y, q)| \leq \varepsilon_{q1}(f(y, q)), \\ 0, & \text{if } |f(x, q) - f(y, q)| > \varepsilon_{q2}(f(y, q)), \\ \text{linear}, & \text{between the two bounds.} \end{cases}$$

The following form of the functions $\varepsilon_{q1}, \varepsilon_{q2}$ can be adopted:

$$\varepsilon_{q1}(f(x, q)) = \alpha_{q1} f(x, q) + \beta_{q1}, \tag{15}$$

$$\varepsilon_{q2}(f(x, q)) = \alpha_{q2} f(x, q) + \beta_{q2}. \tag{16}$$

To model the comprehensive similarity of a to $b \in U$ with respect to $P = \{q_1, q_2, \ldots, q_p\} \subseteq Q$, denoted by $R_P(a, b)$, we consider the credibility of the proposition "$f(a, q_1)$ is similar to $f(b, q_1)$ with respect to q_1 and $f(a, q_2)$ is similar to $f(b, q_2)$ with respect to q_2 and $\ldots f(a, q_p)$ is similar to $f(b, q_p)$ with respect to q_p". Thus, we obtain

$$R_P(x, y) = T_{q \in P} R_q(x, y).$$

Theorem 3.4 *For all $x, y \in U$ and for each $P \subseteq T \subseteq C, R_P(x, y) \geq R_T(x, y)$.*

Proof. Let us suppose that $T = P \cup \{q\}$ with $q \in C - P$. Thus, from property (1) of T-norms, we have, for all $x, y \in U$, $R_T(x, y) = T(R_P(x, y), R_q(x, y)) \leq \min\{R_P(x, y), R_q(x, y)\} \leq R_P(x, y)$. Iterating the reasoning we can prove the result for each $P \subseteq T \subseteq C$. $\qquad\square$

Theorem 3.5 *For each $x \in U$, for each $X \subseteq U$ and for each $P \subseteq T \subseteq C$, we have*

$$\mu(x, \underline{R}_P(X)) \leq \mu(x, \underline{R}_T(X)) \text{ and } \mu(x, \overline{R}_P(X)) \geq \mu(x, \overline{R}_T(X)).$$

Proof. From the monotonicity properties of T-norms, T-conorms and negation and from Theorem 3.4, we have for each $x \in U$, for each $X \subseteq U$ and for each $P \subseteq T \subseteq C$,

$$\mu(x, \underline{R}_P(X)) = T^*_{y \in U}(N(R_P(x, y)), \mu_X(y)))$$
$$\leq T^*_{y \in U}(N(R_T(x, y)), \mu_X(y)) = \mu(x, \underline{R}_T(X)),$$
$$\mu(x, \overline{R}_P(X)) = T^*_{y \in U} T(R(x, y), \mu_X(y))$$
$$\geq T^*_{y \in U} T(R(x, y), \mu_X(y)) = \mu(x, \overline{R}_T(X)),$$

which proves the theorem. $\qquad\square$

Theorem 3.5 expresses in a fuzzy context the following property of the classical rough sets: using greater sets of attributes, it is possible to obtain more accurate approximations of X: thus, while in the classical rough sets theory the lower approximation becomes greater (more precisely, not smaller) and the upper approximation becomes smaller (more precisely, not greater), in the fuzzy case the membership of the lower approximation increases (more precisely, it does not decrease) and the membership of the upper approximation decreases (more precisely, it does not increase).

4   Rule Induction

Using the fuzzy similarity relation one is able to induce decision rules from a decision table. In the following we consider *certain* and *possible decision rules*. The syntax of a *certain decision rule* is the following:

> "if $f(x, q_1)$ is similar to r_{q_1} and $f(x, q_2)$ is similar to r_{q_2} and
> $\ldots f(x, q_p)$ is similar to r_{q_p}, then x <u>belongs</u> to X",

where $\{q_1, q_2, \ldots, q_p\} \subseteq C, (r_{q_1}, r_{q_2}, \ldots, r_{q_p}) \in V_{q_1} \times V_{q_2} \times \ldots \times V_{q_p}$ and $X \subseteq U$.

The syntax of a *possible decision rule* is the following:

> "if $f(x, q_1)$ is similar to r_{q_1} and $f(x, q_2)$ is similar to r_{q_2} and
> $\ldots f(x, q_p)$ is similar to r_{q_p}, then x <u>could belong</u> to X",

where $\{q_1, q_2, \ldots, q_p\} \subseteq C, (r_{q_1}, r_{q_2}, \ldots, r_{q_p}) \in V_{q_1} \times V_{q_2} \times \ldots \times V_{q_p}$ and $X \subseteq U$. To each decision rule r there is associated a credibility $\aleph(r)$.

A statement r: "if $f(x, q_1)$ is similar to r_1 and $f(x, q_2)$ is similar to $r_2 \ldots$ and $f(x, q_p)$ is similar to r_p, then x belongs to X", is accepted as a certain decision rule with a credibility equal to $\aleph(r)$ if there is at least one $w \in U$ such that $f(w, q_1) = r_1$ and $f(w, q_2) = r_2, \ldots, f(w, q_p) = r_p$, and

$$T(\mu(w, \underline{R}_P(X)), T_{y \in U}^* T^*(N(R_P(y, w)), \mu_X(y))) = \aleph(r) > 0,$$

where $P = \{q_1, q_2, \ldots, q_P\}$.

Let us observe that $\aleph(r)$ can be interpreted as the credibility of the proposition "w belongs to the P-lower approximation of X and all the objects similar to it also belong to X".

A statement r: "if $f(x, q_1)$ is similar to r_{q_1} and $f(x, q_2)$ is similar to r_{q_2} and $\ldots f(x, q_p)$ is similar to r_{q_p}, then x could belong to X", is accepted as a possible decision rule with a credibility equal to $\aleph(r)$ if there is at least one $w \in U$ such that $f(w, q_1) = r_1$ and $f(w, q_2) = r_2, \ldots, f(w, q_p) = r_p$, and

$$
\begin{aligned}
&T(\mu(w, \overline{R}_P(X)), T_{y \in U}^* T(R_P(y, w), \mu_X(y))) \\
&= T(\mu(w, \overline{R}_C(X)), T_{y \in U}^* T(R_C(y, w), \mu_X(y))) = \aleph(r) > 0,
\end{aligned}
$$

where $P = \{q_1, q_2, \ldots, q_p\}$.

Let us observe that, in this case, $\aleph(r)$ can be interpreted as the credibility of the proposition "w belongs to the upper approximation of X and there is at least one object similar to it that also belongs to X".

The different formulation of accepted certain and possible decision rules depends on Theorem 3.5. In simple words, it says that the credibility that an object $x \in U$ belongs to the upper approximation of a set $X \subseteq U$ does not increase, and very often decreases, when considering a larger set of attributes. Thus, considering a smaller set of attributes there is the possibility of overestimating the credibility that an object $x \in U$ belongs to the upper approximation of a set $X \subseteq U$. Therefore, an accepted possible decision rule may be built on object $w \in U$ from an upper approximation with respect to set $P \subseteq C$; however, its credibility cannot be greater than the credibility of a possible decision rule built on the same object $w \in U$ from an upper approximation with respect to set C.

A certain decision rule r: "if $f(x, q_1)$ is similar to r_{q_1} and $f(x, q_2)$ is similar to r_{q_2} ... and $f(x, q_p)$ is similar to r_{q_p} then $x \in X$", whose credibility is equal to $\aleph(r)$, is called *minimal* if there is no other certain decision rule s: "if $f(x, h_1)$ is similar to s_1 and $f(x, h_2)$ is similar to s_2 ... and $f(x, h_k)$ is similar to s_k then $x \in X$", whose credibility is $\aleph(s)$, such that $\{h_1, \ldots, h_k\} \subseteq \{q_1, \ldots, q_p\}$, $r_{h_1} = s_{h_1}, r_{h_2} = s_{h_2}, \ldots, r_{h_k} = s_{h_k}$ and $\aleph(s) \geq \aleph(r)$.

A possible decision rule r: "if $f(x, q_1)$ is similar to r_1 and $f(x, q_2)$ is similar to r_2 ... and $f(x, q_p)$ is similar to r_p then $x \in X$", whose credibility is $\aleph(r)$ is called *minimal* if there is no other possible or certain decision rule s: "if $f(x, h_1)$ is similar to s_1 and $f(x, h_2)$ is similar to s_2 ... and $f(x, h_k)$ is similar to s_k then $x \in X$", whose credibility is $\aleph(s)$, such that $\{h_1, \ldots, h_k\} \subseteq \{q_1, \ldots, q_p\}$, $r_{h_1} = s_{h_1}, r_{h_2} = s_{h_2}, \ldots, r_{h_k} = s_{h_k}$ and $\aleph(s) \geq \aleph(r)$.

Let us observe that, since each decision rule is an implication, the minimal decision rules represent the implications such that there is no other implication with an antecedent at least of the same weakness and a consequent of at least the same strength.

5   An Illustrative Example

In order to illustrate the methodology proposed in this paper, let us consider a simple example. First, we will solve this example using the classical rough set approach based on indiscernibility and crisp granules of knowledge. Then, we will solve it using the extended rough set approach based on fuzzy similarity relations and fuzzy granules of knowledge. The example is based on a decision table describing six firms which have got an approximately equal credit in a bank. The firms are characterized by three condition attributes: A1 = value of fixed capital, A2 = value of sales in the year preceding the application, A3 = kind of activity. Attributes A1 and A2 are numerical ones, while attribute A3 is a qualitative one with three possible values.

Decision attribute d makes a partition of the firms as follows: $d =$ "yes" if the firm paid back its debt, and $d =$ "no" otherwise.

The decision table is shown in Table 1.

Table 1: Original decision table

Firm	A1	A2	A3	d
F1	120	60	A	yes
F2	140	72	A	no
F3	170	80	A	yes
F4	99	49.5	B	no
F5	101	50.5	B	yes
F6	90	45	B	no

In order to apply the classical rough set approach to the above decision table the values of numerical attributes A1 and A2 must be translated into some nominal terms, e.g., low, medium or high. This translation, called coding, involves a division of the original domain of the numerical attributes into subintervals to which nominal terms are assigned. In our example the following boundary values were adopted for the subintervals:

fixed capital < 100 $\implies$ fixed capital is low,
$100 <$ fixed capital < 150 $\implies$ fixed capital is medium,
fixed capital ≥ 150 $\implies$ fixed capital is high,
values of sales < 50 $\implies$ value of sales is low,
$50 <$ values of sales < 75 $\implies$ value of sales is medium,
value of sales ≥ 100 $\implies$ value of sales is high.

Therefore, the coded decision table shown in Table 2 was obtained.

Table 2: Coded decision table

Firm	A1	A2	A3	d
F1	medium	medium	A	yes
F2	medium	medium	A	no
F3	high	high	A	yes
F4	low	low	B	no
F5	medium	medium	B	yes
F6	low	low	B	no

Denoting by X_Y the set of firms which paid back the debt and by X_N the set of firms which did not pay back the debt, the lower and the upper approximations of the set of firms paying back the debt are $\underline{C}(X_Y) = \{F3, F5\}$ and $\overline{C}(X_Y) = \{F1, F2, F3, F5\}$, while the lower and the upper approximations of the set of firms not paying back the debt are $\underline{C}(X_N) = \{F4, F6\}$ and $\overline{C}(X_N) = \{F1, F2, F4, F6\}$.

Let us observe that F1 and F2 belong to the upper approximations of X_Y and X_N, i.e., to the boundary between X_Y and X_N because, in the coded decision table, they have the same description with respect to the condition attributes from C but they belong to different classes according to the decision attribute d, i.e., F4 paid back and F6 did not pay back the debt. Let us observe, moreover, that even if in the coded decision table, F4 and F5 have quite different description, the original decision table shows that F4 and F5 have quite similar values on the two first condition attributes. This is the result of the "boundary effect" of discretization due to the fact that the thresholds between the intervals of low fixed capitals and medium fixed capitals is between the corresponding evaluations of F4 and F5 and the same happens with respect to the intervals of low values of sales and medium values of sales.

This boundary effect can be avoided when considering the similarity relation instead of the indiscerniblity relation. For instance, let us assume that: x is similar to y with respect to A1 (we write xR_1y) iff

$$\frac{|\text{fixed capital of } x - \text{fixed capital of } y|}{\text{fixed capital of } y} \leq 10\%,$$

x is similar to y with respect to A2 (we write xR_2y) iff

$$\frac{|\text{value of sales of } x - \text{value of sales of } y|}{\text{value of sales of } y} \leq 10\%,$$

and x is similar to y with respect to A3 (we write xR_3y) iff x and y have the same kind of activity.

Furthermore, we state that x is similar to y with respect to C, and denote this relation by xR_Cy iff xR_1y, xR_2y, and xR_3y.

Therefore, we have $R_C = \{(F1, F1), (F2, F2), (F3, F3), (F4, F4),$ $(F4, F5),$ $(F4, F6),$ $(F5, F4),$ $(F5, F5),$ $(F6, F4),$ $(F6, F6)\}$ and $R_C^{-1}(F1) = \{F1\}$, $R_C^{-1}(F2) = \{F2\}$, $R_C^{-1}(F3) = \{F3\}$, $R_C^{-1}(F4) = \{F4, F5, F6\}$, $R_C^{-1}(F5) = \{F4, F5\}$, $R_C^{-1}(F6) = \{F4, F6\}$. The lower and upper approximations of the set of firms paying back the debt are $\underline{R}_C(X_Y) = \{F1, F3\}$ and $\overline{R}_C(X_Y) = \{F1, F3, F4, F5\}$, while the lower

and the upper approximations of the set of firms not paying back the debt
are $\underline{R}_C(X_N) = \{F2, F6\}$ and $\overline{R}_C(X_N) = \{F2, F4, F5, F6\}$. Let us observe
that in this case F4 is similar to F5 and F5 is similar to F4, with the consequence that F4 and F5 belong to the boundary between X_Y and X_N rather
than to the lower approximation of X_N and X_Y, respectively. This result
was obtained because the use of a similarity relation permitted avoiding
the boundary effect typical for discretization of numerical attributes and
the use of the indiscerniblity relation.

However, let us observe that a more subtle boundary effect is hidden in
the proposed example. In fact F4 is similar to F6, but F5, which in turn
is very similar to F4, is not similar to F6. It is enough of a small change
in the definition of similarity with respect to A1 and A2 to obtain rather
different results. For instance, let us suppose that $x R_1 y$ iff

$$\frac{|\text{fixed capital of } x - \text{fixed capital of } y|}{\text{fixed capital of } y} \leq 12\%,$$

and $x R_2 y$ iff

$$\frac{|\text{value of sales of } x - \text{value of sales of } y|}{\text{value of sales of } y} \leq 12\%.$$

We obtain $R_C = \{(F1, F1), (F2, F2), (F3, F3), (F4, F4), (F4, F5),$
$(F4, F6), (F5, F4), (F5, F5), (F6, F4), (F6, F5), (F6, F6)\}$, $R_C^{-1}(F1) =$
$\{F1\}$, $R_C^{-1}(F2) = \{F2\}$, $R_C^{-1}(F3) = \{F3\}$, $R_C^{-1}(F4) = \{F4, F5, F6\}$,
$R_C^{-1}(F5) = \{F4, F5\}$, $R_C^{-1}(F6) = \{F4, F5, F6\}$. The lower and upper
approximations of the set of firms paying back the debt are $\underline{R}_C(X_Y) =$
$\{F1, F3\}$ and $\overline{R}_C(X_Y) = \{F1, F3, F4, F5, F6\}$, while the lower and the
upper approximations of the set of firms not paying back the debt are
$\underline{R}_C(X_N) = \{F2\}$ and $\overline{R}_C(X_N) = \{F2, F4, F5, F6\}$. Let us observe that in
this case F5, which belongs to X_Y, is similar to F6, which belongs to X_N,
and, therefore, F6 does not belong to the lower approximation of X_Y as in
the previous case but, instead, it belongs to the boundary of X_Y and X_N.
The rather large sensitivity of the rough approximations to small changes
in the definition of the similarity relation suggests introducing a graduality
in the concept of similarity and thus justifies a rough approximation based
on a fuzzy similarity.

Therefore, let us consider the following definition of fuzzy similarity: x

is similar to y with respect to A1 with credibility $R_1(x, y)$, where

$$R_1(x, y) = \begin{cases} 1, & \text{if } \dfrac{|\text{fixed capital of } x - \text{fixed capital of } y|}{\text{fixed capital of } y} \leq 10\%, \\ 0, & \text{if } \dfrac{|\text{fixed capital of } x - \text{fixed capital of } y|}{\text{fixed capital of } y} > 15\%, \\ \text{linear}, & \text{between the two bounds} \end{cases}$$

x is similar to y with respect to A2 with credibility $R_2(x, y)$, where

$$R_2(x, y) = \begin{cases} 1, & \text{if } \dfrac{|\text{value of sales of } x - \text{value of sales of } y|}{\text{value of sales of } y} \leq 10\%, \\ 0, & \text{if } \dfrac{|\text{value of sales of } x - \text{value of sales of } x|}{\text{value of sales of } y} > 15\%, \\ \text{linear}, & \text{between the two bounds} \end{cases}$$

Furthermore, we consider the following fuzzy logical operators:

$$T(a, b) = \min(a, b), \text{ for all } a, b \in [0, 1],$$
$$T^*(a, b) = \max(a, b), \text{ for all } a, b \in [0, 1],$$
$$N(a) = 1 - a, \text{ for each } a \in [0, 1].$$

The following results were obtained. The fuzzy similarity $R_C(x, y)$ is presented in Table 3. We calculated the lower and upper approximation of X_Y and X_N obtaining the results presented in Table 4.

Table 3: Fuzzy binary relation $R_C(x, y)$

	F1	F2	F3	F4	F5	F6
F1	1	0	0	0	0	0
F2	0	1	0	0	0	0
F3	0	0	1	0	0	0
F4	0	0	0	1	1	1
F5	0	0	0	1	1	0.56
F6	0	0	0	1	0.82	1

Table 4: Lower and upper approximations with respect to {A1, A2, A3}

Firm	$\mu(x, \underline{R}(X_Y))$	$\mu(x, \underline{R}(X_N))$	$\mu(x, \overline{R}(X_Y))$	$\mu(x, \overline{R}(X_N))$
F1	1	0	1	0
F2	0	1	0	1
F3	1	0	1	0
F4	0	0	1	1
F5	0	0	1	1
F6	0	0.18	0.82	1

From the decision table in Table 1 the following exact decision rules were induced (between parentheses there is the credibility of the decision rule):

1. if fixed capital is similar to 140, then the firm does not pay back its debt (credibility 0.86);

2. if fixed capital is similar to 90, then the firm does not pay back its debt (credibility 0.18);

3. if fixed capital is similar to 120, then the firm pays back its debt (credibility 0.86);

4. if fixed capital is similar to 170, then the firm pays back its debt (credibility 1);

5. if value of sales is similar to 45, then the firm does not pay back its debt (credibility 0.18);

6. if value of sales is similar to 60, then the firm does not pay back its debt (credibility 1).

Furthermore, from the decision table in Table 1 the following possible decision rules were induced:

7. if fixed capital is similar to 140, then the firm could not pay back its debt (credibility 1);

8. if fixed capital is similar to 99, then the firm could not pay back its debt (credibility 1);

9. if fixed capital is similar to 99, then the firm could pay back its debt (credibility 1);

10. if fixed capital is similar to 90, then the firm could not pay back its debt (credibility 1);

11. if fixed capital is similar to 90, then the firm could pay back its debt (credibility 0.18);

12. if fixed capital is similar to 120, then the firm could pay back its debt (credibility 1);

13. if fixed capital is similar to 101, then the firm could pay back its debt (credibility 1);

14. if fixed capital is similar to 101, then the firm could not pay back its debt (credibility 1);

15. if value of sales is similar to 72, then the firm could not pay back its debt (credibility 1);

16. if value of sales is similar to 49.5, then the firm could not pay back its debt (credibility 1);

17. if value of sales is similar to 49.5, then the firm could pay back its debt (credibility 1);

18. if value of sales is similar to 45, then the firm could not pay back its debt (credibility 1);

19. if value of sales is similar to 45, then the firm could pay back its debt (credibility 0.18);

20. if value of sales is similar to 80, then the firm could not pay back its debt (credibility 1);

21. if value of sales is similar to 50.5, then the firm could pay back its debt (credibility 1);

22. if value of sales is similar to 50.5, then the firm could not pay back its debt (credibility 1);

23. if kind of activity is A, then the firm could not pay back its debt (credibility 1);

24. if kind of activity is A, then the firm could pay back its debt (credibility 1);

25. if kind of activity is B, then the firm could not pay back its debt (credibility 1);

26. if kind of activity is B, then the firm could pay back its debt (credibility 1).

Let us remark that the credibility of each decision rule must be interpreted in relation with the different syntax of the exact and approximate decision rule. For example, decision rule 1 and 7 have the same condition part ("if fixed capital is similar to 140") and a very similar decision part (rule 1: "then the firm does not pay back its debt", and rule 7: "then the firm could not pay back its debt"). However, the credibility of the two rules is different: 0.86 for rule 1 and 1.0 for rule 7. This means that the implication of rule 1, being an exact decision rule, is not fully credible. This is because there is some firm which at least partly satisfies the condition part of the rule without belonging to the decision class suggested by the rule: more precisely, with respect to rule 1, firm F1 has a fixed capital partly similar to 140, but it paid back its debt. In contrast, the implication of the decision rule 7, being a possible decision rule, is fully credible. This is because there is at least one firm which satisfies the condition part of the rule and belongs to the decision class suggested by the rule: more precisely, with respect to rule 7, firm F2 has a fixed capital similar (equal) to 140 and it does not pay back its debt.

6   Conclusions

We introduced rough approximations of fuzzy sets by means of similarity relations defined as reflexive fuzzy binary relations. The general framework proposed represents a theoretical extension of the rough set approach into a fuzzy context and also a generalization of the indiscernibility relations to more general binary relations. The similarity classes defined by the fuzzy similarity relation correspond to fuzzy granules of knowledge; their credibility varies gradually between a finite degree equal to one and an infinitely small degree, thus giving different credibility of the rough approximations. Decision rules induced from these rough approximations have a more general syntax and deal more naturally with numerical attributes. Due to flexibility introduced by fuzzy sets and similarity relations, this new approach to rough set analysis gives more credible results, as shown by an exemplary problem.

Acknowledgement: The research of the first two authors has been supported by the Italian Ministry of University and Scientific Research (MURST). The third author wishes to acknowledge financial support from State Committee for Scientific Research, KBN research grant no. 8 T11C 013 13, and from CRIT 2 Esprit Project no. 20288.

References

[1] D. Dubois, H. Prade, *Fuzzy Sets and Systems: Theory and Applications*, Academic Press, New York, 1980.

[2] D. Dubois, H. Prade, Rough fuzzy sets and fuzzy rough sets, *Int. J. of General Systems*, **17** (1990), 191–200.

[3] D. Dubois, H. Prade, Putting rough sets and fuzzy sets together, in *Intelligent Decision Support, Handbook of Applications and Advances of the Rough Sets Theory* (R. Slowinski, ed.), Kluwer, Dordrecht, 1992, 203–233.

[4] J. Fodor, M. Roubens, *Fuzzy Preference Modelling and Multicriteria Decision Support*, Kluwer, Dordrecht, 1994.

[5] S. Greco, B. Matarazzo, R. Slowinski, The use of rough sets and fuzzy sets in MCDM, Chapter 14 in *Advances in Multiple Criteria Decision Making* (T. Gal, T. Hanne, T. Stewart, eds.), Kluwer, Dordrecht, 1999, 14.1–14.59.

[6] J. W. Grzymala-Busse, LERS – a system for learning from examples based on rough sets, in *Intelligent Decision Support. Handbook of Applications and Advances of the Rough Sets Theory* (R. Slowinski, ed.), Kluwer, Dordrecht, 1992, -3-18.

[7] T. Lin, Neighborhood systems and approximation in database and knowledge base systems, in *Proc. 4th International Symposium on Methodologies for Intelligent Systems*, 1989.

[8] S. Marcus, Tolerance rough sets, Cech topologies, learning processes, *Bull. of the Polish Academy of Sciences, Technical Sciences*, **42** (1994), 471–487.

[9] K. Menger, Probabilistic theories of relations, *Proc. Nat. Acad. Sci. (Math.)*, **37** (1951), 178–180.

[10] R. Mienko, J. Stefanowski, K. Toumi, D. Vanderpooten, Discovery-oriented induction of decision rules, *Cahier du LAMSADE* 141, Université de Paris Dauphine, Paris, 1996.

[11] Z. Pawlak, Rough sets, *Intern. J. of Information and Computer Sciences*, **11** (1982), 341–356.

[12] Z. Pawlak, *Rough Sets. Theoretical Aspects of Reasoning about Data*, Kluwer, Dordrecht, 1991.

[13] Z. Pawlak, Rough sets and fuzzy sets, *Fuzzy Sets and Systems*, **17** (1985), 99–102.

[14] L. Polkowski, A. Skowron, J. Zytkow, Tolerance based rough sets, in *Soft Computing: Rough Sets, Fuzzy Logic, Neural Networks, Uncertainty Management, Knowledge Discovery* (T. Y. Lin, A. Wildberger, eds.), Simulation Councils, Inc., San Diego, CA, 1995, 55–58.

[15] B. Schweizer, A. Sklar, *Probabilistic Metric Spaces*, North-Holland, Amsterdam, 1983.

[16] A. Skowron, J. Stepaniuk, Generalized approximation spaces, in *Soft Computing: Rough Sets, Fuzzy Logic, Neural Networks, Uncertainty Management, Knowledge Discovery* (T. Y. Lin, A. Wildberger, eds.), Simulation Councils, Inc., San Diego, CA, 1995, 18–21.

[17] R. Slowinski, A generalization of the indiscernibility relation for rough set analysis of quantitative information, *Rivista di matematica per le scienze economiche e sociali*, **15** (1993), 65–78.

[18] R. Slowinski, Rough set processing of fuzzy information, in *Soft Computing: Rough Sets, Fuzzy Logic, Neural Networks, Uncertainty Management, Knowledge Discovery* (T. Y. Lin, A. Wildberger, eds.), Simulation Councils, Inc., San Diego, CA, 1995, 142–145.

[19] R. Slowinski, J. Stefanowski, Rough classification in incomplete information systems, *Math. Comput. Modelling*, **2**, 10/11 (1989), 1347–1357.

[20] R. Slowinski, J. Stefanowski, Handling various types of uncertainty in the rough set approach, in *Rough Sets, Fuzzy Sets and Knowledge Discovery* (W. P. Ziarko, ed.), Springer-Verlag, London, 1993.

[21] R. Slowinski, J. Stefanowski, Rough set reasoning about uncertain data, *Fundamenta Informaticae*, **27** (1996), 229–243.

[22] R. Slowinski, J. Stefanowski, Rough family – software implementation of the rough set theory, in *Rough Sets in Knowledge Discovery* (L. Polkowski, A. Skowron, eds.), vol. 2, Physica-Verlag, Heidelberg, New York, 1998, 581–586.

[23] R. Slowinski, D. Vanderpooten, Similarity relation as a basis for rough approximations, in *Advances in Machine Intelligence and Soft-Computing* (P. P. Wang, ed.), vol.IV, Duke University Press, Durham, NC, 1997, 17–33.

[24] R. Slowinski, D. Vanderpooten, A generalized definition of rough approximation based on similarity, *IEEE Transactions on Data and Knowledge Engineering*, 1999 (to appear).

[25] Y. Yao, S. Wong, Generalization of rough sets using relationships between attribute values, *Proc. 2nd Annual Joint Conference on Information Sciences*, Wrightsville Beach, NC, 1995, 30–33.

[26] S. Weber, A general concept of fuzzy connectives, negations and implications based on t-norms and t-conorms, *Fuzzy Sets and Systems*, **11** (1983), 115–134.

[27] W. Ziarko, D. Golan, D. Edwards, An application of DATALOGIC/R knowledge discovery tool to identify strong predictive rules in stock market data, *Proc. AAAI Workshop on Knowledge Discovery in Databases*, Washington D.C., 1993, 89–101.

Relativized Code Concepts
and Multi-Tube DNA Dictionaries

Tom Head

Mathematics Department, Binghamton University
Binghamton, New York 13902-6000, USA
E-mail: tom@math.binghamton.edu

Abstract. Concepts from the theory of variable length codes are relativized to a language L so that demands are made only for those strings that belong to L. Decision procedures and regularity results appropriate for the relativized forms are given. This suggests the definition of a sequence of code types: the join codes of level k $(1 \leq k)$. At level 1 we have precisely the familiar comma free codes. For a join code of level k, a procedure for segmenting messages into code words is observed that consists of k steps, with each step being carried out in the elementary comma free manner. An illustration of this segmentation process is given using an idealization of the Watson-Crick base pairing of complementary strands of DNA. In this process the code words appear as ssDNA molecules in k distinct test tubes with each test tube content being washed over each incoming ssDNA complemented message strand. After k washings the sequence of attached code words provides the required segmentation.

Preamble

The interplay between finiteness and infinity is particularly clear in studies of the forms and the essence of computational processes. The specification and manipulation of infinite sets of formal objects is desired. In order to maintain confidence in the precision of thought finite displays of symbols are created that can be interpreted to generate or specify the desired infinite sets. Infinite sets can often be compared by comparing the finite schemes

175

that specify them. It is tempting to say that finite creatures are able in
this way to interact with the infinite. Perhaps this is the essential spirit of
mathematics. In the present article we use finite automata and finite sets of
code words to manipulate members from an infinite sets of strings – to each
of which we may ascribe a specific meaning. In this way an infinite number
of choices of distinct meanings is provided by the use of finite schemes. At
this point the goals of mathematics and linguistics fuse.

1   Introduction

We allow any subset of the set A^* of all strings over a finite set A to be
called a *code*. In this context we refer to A as an *alphabet*. Any subset
of A^* may also be called a *language*. Whether we choose to call a subset
of A^* a code or a language is determined by the role we intend it to play
in our investigations. Elements of codes will be called *words* and elements
of languages will be called *strings*. For C a code, elements of C^* will be
called *messages*. In some expositions of the theory of codes, a subset C of
A^* is not called a code unless it is uniquely decipherable as defined here in
Section 2. Among the references for this article, only [10] and [1] require
unique decipherability before a set of strings is allowed to be called a code.

By linking the relativization of code properties [6] with the behaviours of
DNA molecules [7], a sequence of generalizations of the concept of a comma
free code [1] is developed and illustrated. The codes of this sequence are
the join codes of various levels. The comma free codes are precisely the
join codes of the first level. The join codes of level k allow the segmentation of messages to be made in a sequence of k steps for which each step
has the simplicity of a comma free segmentation. This segmentation process is illustrated in Section 6 by using an idealization of the annealing of
complementary single stranded DNA molecules.

As background for the present article, only familiarity with the elementary aspects of the theory of regular languages and finite automata is
required. For this material see [15] or selected chapters in either [10] or
[1]. If references for the theory of codes are desired, see [10], [1], and [9].
Regarding DNA behaviour, see [12]. In the two closing paragraphs of this
section our choice of automata notation is established and used to confirm
a regularity result concerning the set of factors of a language.

Let $M = (A, Q, q_0, F, E)$ be a deterministic automaton where: Q is the
set of states of M; q_0 is the initial state of M; F is the set of final states;
and E is the set of labeled edges of M, i.e., E is a subset of $Q \times A \times$

Q. Such an automaton recognizes the language consisting of all strings in A^* that are the labels of paths that initiate at q_0 and terminate at a state in F. The language recognized by the automaton (A, Q, q_0, F, E) will be denoted $L(A, Q, q_0, F, E)$. Each such language is regular and each regular language is recognized by such an automaton. Decision procedures concerning regular languages commonly use an automaton recognizing the language.

A string w in A^* is a *factor* of a language L if there are strings x, y in A^* for which xwy is in L. $\mathrm{Fac}(L)$ will denote the set of all factors of L. When L is regular, $\mathrm{Fac}(L)$ is also regular since, for $L = L(A, Q, q_0, F, E)$, $\mathrm{Fac}(L)$ is the union of the set of all languages $L(A, Q', p, \{q\}, E)$, where Q' is the set of all states in Q from which F is accessible and p, q are in Q'.

2 Relative Unique Decipherability

Definition 2.1 *A code C is* uniquely decipherable *if, for each message w in C^*, there is only one non-negative integer n and one finite sequence $c_1, c_2, \ldots c_{n-1}, c_n$ of words in C for which $w = c_1 c_2 \ldots c_{n-1} c_n$.*

Definition 2.2 *A code C is* uniquely decipherable relative to a language L *if, for each string w in L, there is at most one non-negative integer n and one finite sequence $c_1, c_2, \ldots c_{n-1}, c_n$ of words in C for which $w = c_1 c_2 \ldots c_{n-1} c_n$.*

Note that a code C over an alphabet A is uniquely decipherable if it is uniquely decipherable relative to either A^* or C^* and that a uniquely decipherable code is uniquely decipherable relative to every language contained in A^*. More generally, if C is uniquely decipherable relative to a language L and S is a subset of L then C is also uniquely decipherable relative to S.

Observe that the code $C = \{ab, a, ba\}$ is uniquely decipherable relative to each of the first three of the following four languages, but not to the fourth: $(ab)^*$, $a^* + (ab + ba)^*$, $(a + ab)^* + (ab + ba)^*$, $(a + ab)^* + (a + ba)^*$.

The following proposition was demonstrated in [6], where an algorithm based on deciding single-valuedness of a-transducers was given.

Proposition 2.1 *For a given regular code C and regular language L, a procedure exists for deciding whether C is uniquely decipherable relative to L.*

Algorithm. Let C be a regular code and L a regular language. Let $M = (A, Q, q_0, F, E)$ be the minimal deterministic automaton recognizing

L. For each p in Q let $I(p) = L(A, Q, q_0, \{p\}, E)$. For each p, q in Q let $L(p, q) = L(A, Q, p, \{q\}, E)$. For each q in Q let $F(q) = L(A, Q, q, F, E)$. Observe that C fails to be uniquely decipherable relative to L if and only if there are p, q, r in Q for which each of the following seven regular languages is not empty:

1. $C^* \cap I(p)$,

2. $C^\cap F(r)$, which we call $T(r)$ below,

3. $C \cap L(p, q)$, which we call $C(p, q)$ below,

4. $C \cap L(p, r)$, which we call $C(p, r)$ below,

5. $L(q, r) \cap A^+$, which we call $N(q, r)$ below,

6. $C^* \cap N(q, r)T(r)$, which we call $T(q, r)$ below,

7. $C(p, q)T(q, r) \cap C(p, r)T(r)$.

Relative unique decipherability can therefore be decided by deciding the emptiness of these intersections. $\qquad\square$

The following known result is a special case of Proposition 2.1.

Corollary 2.1 *For a regular code C, a procedure exists for deciding whether C is uniquely decipherable.*

3 Relative Solidity

Definition 3.1 *A code C is a solid code if:*

1. *u and puq in C can hold only if pq is null; and*

2. *pu and uq in C and u non-null, can hold only if pq is null.*

Definition 3.2 *A code C is solid relative to a language L if:*

1. *$w = ypuqz$ in L with both u and puq in C can hold only if pq is null; and*

2. *$w = ypuqz$ in L with pu and uq in C and u non-null, can hold only if pq is null.*

Note that a code C is solid if and only if C is solid relative to A^* and that a solid code is solid relative to every language contained in A^*. More generally, if C is solid relative to a language L and S is a subset of L then C is also solid relative to S.

Observe that the code $C = \{ab, c, ba\}$ is not solid, but it is solid with respect to the language $(abc + cba)^*$. It is not solid with respect to $C^* = (ab + c + ba)^*$.

Proposition 3.1 *For a given regular code C and a regular language L, a procedure exists for deciding whether C is solid relative to L.*

Algorithm. Let X be the union of A^+CA^* and A^*CA^+. (1) of the definition of relative solidity holds if and only if the intersection of the three regular languages $\mathrm{Fac}(L)$, C, and X is empty. Since this intersection is regular, (1) can be decided by deciding the emptiness of this intersection. We decide (2) of the definition only in the case that (1) holds. For the purpose of deciding (2) we use the minimal automaton $M = (A, Q, q_0, F, E)$ recognizing C. Consider each ordered pair (s, t) of states in Q in turn. Associate with each (s, t) in Q the four languages $I(s) = L(A, Q, q_0, \{s\}, E)$, $T(s) = L(A, Q, s, F, E)$, $I(t) = L(A, Q, q_0, \{t\}, E)$, and $T(t) = L(A, Q, t, F, E)$. For each of the ordered pairs (s, t), proceed as follows.

Decide whether the intersection of $I(s)$ and $T(t)$ contains a non-null string. If so, call this intersection $U(s, t)$; if not, do not consider (s, t) further and proceed to an ordered pair not yet treated, if one remains. If none remains, (2) holds and we terminate. If $U(s, t)$ has been defined, decide whether the intersection of the regular languages $\mathrm{Fac}(L)$ and $I(t)U(s, t)T(s)$ contains a string not in C. If so, (2) fails and we terminate; if not, proceed to an ordered pair not yet treated. If none remains, (2) holds and we terminate. □

Observe that if C is a code that is solid relative to C^* then C is necessarily uniquely decipherable. If C is not uniquely decipherable then there is a violation of (1), of the definition of relative solidity, with p null and q non-null. The concept of a comma free code is much older than the concept of a solid code. The following definition, which is convenient in the present context, is equivalent to the traditional definition:

Definition 3.3 *A code C is* comma free *if it is solid relative to C^*.*

Thus the new concept of relative solidity covers two classes of codes, each of which has a considerable literature: solid codes [9], Section 11, and

comma free codes [1]. The following known results are special cases of
Proposition 3.1.

Corollary 3.1 *For a given regular code C, a procedure exists for deciding
whether C is a solid code.*

Corollary 3.2 *For a given regular code C, a procedure exists for deciding
whether C is a comma free code.*

4   Joins Relative to a Language

Definition 4.1 *A string w in A^* is a join relative to a language L if w is
a factor of L and for every u, v in A^* for which uwv is in L, both u and v
are also in L.*

Proposition 4.1 *The set J of joins relative to a regular language L is a
regular language.*

Proof. Let L be a regular language over the alphabet A. Let $M = (A, Q, q_0, F, E)$ be the minimal deterministic automaton recognizing L. J
is regular since it is the complement in $\mathrm{Fac}(L)$ of the union of the set of lan-
guages recognized by the automata $(A, Q, p, \{q\}, E)$ where p, q are in Q and
the following two conditions hold: (1) F is accessible from q; and (2) either
p is not in F or the language recognized by the automaton (A, Q, q, F, E)
is not contained in L.   □

Definition 4.2 *A word w in a code C is a* join *in C if w is a join relative
to C^*. $J(C)$ will denote the set of all joins in the code C.*

Corollary 4.1 *The subset $J(C)$ of all join words in a regular code C is a
regular language.*

Observe that if C is uniquely decipherable then no word in $J(C)$ can be
a factor of any other word in C. If $v = puq$ were in C with u in $J(C)$ and
pq non-null then, since both p and q would necessarily lie in C^*, $v = puq$
would contradict the unique decipherability of C.

Proposition 4.2 *If a code C is uniquely decipherable, then $J(C)$ is solid
relative to C^*.*

Proof. Let C be a uniquely decipherable code. That condition (1), of the definition of relative solidity, holds is a consequence of the stronger observation immediately above. We consider condition (2). Suppose that $w = ypuqz$ is in C^* with both pu and uq in $J(C)$ and u non-null. Then y, yp, qz, z must all be in C^*. Consequently, in C^* we have the factorizations $w = (y)(pu)(qz) = (yp)(uq)(z)$. From pu & uq in C and u non-null, it follows from the unique decipherability of C that pq is null. $\square$

Corollary 4.2 *For each uniquely decipherable code C, $J(C)$ is comma free.*

5 Reading Messages over a Join Code

Let C be a uniquely decipherable code over the alphabet A. Let $C_0 = C$. We define recursively a descending chain of subsets C_i $(0 \leq i)$ of C_0. With C_i defined and $J(C_i)$ not empty, we define $C_{i+1} = C_i \backslash J(C_i)$. If a non-negative integer k arises for which $J(C_k)$ is empty, then $C = C_0$ is the union of the pairwise disjoint subsets: $J(C_0), J(C_1), \ldots, J(C_{k-1}), C_k$. If no such integer k arises, then $C = C_0$ is the union of an infinite sequence of pairwise disjoint subsets: $J(C_0), J(C_1), \ldots, J(C_i), \ldots$ $(0 \leq i)$, and a possibly non-empty subset D which would be the complement in C of the union of all the subsets $J(C_i)$, $(0 \leq i)$. If C is regular then, by Proposition 4.1, all the sets arising in the decomposition process are regular, with the possible exception of D.

Definition 5.1 *A uniquely decipherable code C for which the decomposition procedure carried out above yields either a non-negative integer k with C_k defined and empty, or an infinite descending chain C_i $(0 \leq i)$ with D empty, will be called a* join code *of level k, in the first case, and of* infinite level, *in the second case.*

Note that the join codes of level one are precisely the comma free codes. For these codes we have C_1 empty, which is equivalent to $C = C_0 = J(C)$. The join code concept enlarges the comma free concept into a hierarchy of code classes, each of which allows a parallel distributed segmentation process that extends the natural segmentation process applicable to the comma free codes.

We continue with the notations set in the first paragraph of this section and explain this segmentation process. For maximum clarity we consider first the case in which $C = C_0$ is a *finite* uniquely decipherable code. In this case a non-negative integer k is certain to arise that provides a partition of C into *finite* subsets $J(C_0), J(C_1), \ldots, J(C_{k-1}), C_k$. Note that it follows from Proposition 4.2 that, for each i $(0 \leq i \leq k-1)$, $J(C_i)$ is solid relative to C_i^*. We may imagine the words of the code C as being listed in a dictionary consisting of $k + 1$ successive volumes: $J(C_0), \ldots, J(C_{k-1}), C_k$. A message w can be segmented into words as follows. Distributed agents, perhaps one at each letter occurrence of the message, examine w independently and in parallel using only volume 0 of the dictionary until all occurrences in w of words in volume 0 have been enclosed in parentheses. There will be no overlapping of these occurrences by Proposition 4.2. Agents now examine independently the remaining *unparenthesized subsegments* of w using only volume 1 of the dictionary – inserting parentheses bounding occurrences of words that appear in volume 1. Similar processes takes place in succession using volumes 2 through $k - 1$. *If C is a join code of level k then C_k is empty and w has been completely segmented into its code words in a parallel distributed manner that involves only local features of the message.* On the other hand, if C_k is not empty then, at this point, we have performed all of the segmentation that can be done purely locally by non-communicating agents. The remainder of the segmentation must be done by standard methods – the most primitive of which is to work from, say, the left end of each remaining subsegment attempting matches with words in C_k – backtracking when necessary.

Consider now the case of an *infinite* uniquely decipherable code $C = C_0$. We will consider only the case in which C *is a regular join code of infinite level*. The case in which C is a regular join code of finite level will be clear after a discussion of the case in which the level is infinite. The purpose of the regularity assumption is to allow the construction of the subsets C_i and $J(C_i)$ $(0 \leq i)$. The restriction to join codes demands that D be empty and assures that the method proposed here for segmentation of messages will terminate. In these circumstances a segmentation process is possible that, conceptually at least, differs very little from the case in which C is finite.

Let $C = C_0$ be a uniquely decipherable regular join code of infinite level. Each of subsets C_i, $J(C_i)$ $(0 \leq i)$ is regular and any finite number of these can be constructed (in the form of recognizing automata) whenever needed. Suppose now that we have a message w that we wish to segment into words in C. We first note the length n of the message w. No word in C of length greater than n can occur in w. For each i $(0 \leq i)$, let $J'(C_i)$ be a listing of the words in $J(C_i)$ having length at most n. Each of these finite lists $J'(C_i)$ is constructible from an automaton recognizing $J(C_i)$. We regard each of the lists $J'(C_i)$ as volume i of a multiple volume dictionary. To segment w we construct, as the need arises, a finite number of the finite volumes $J'(C_0)$, $J'(C_1), \ldots, J'(C_i), \ldots$ $(0 \le i)$ required for segmenting w exactly as in the previously treated case when C is finite; except that the use of $J(C_i)$ as volume i is now replaced by the use of $J'(C_i)$ as volume i. That this process terminates in the complete segmentation of w is guaranteed by the fact that there are only finitely many code words of length at most n, and each of these must lie in one of the subsets $J'(C_i)$, since D is empty.

Examples of join codes of each level are easily given for each alphabet having at least two symbols. We begin with an example of a finite join code of level three. Let the alphabet be $A = \{a, b, c\}$ and the code be $C = C_0 = \{a, bab, cbabc\}$. For this code C the following are easily confirmed: $J(C_0) = \{cbabc\}$; $C_1 = \{a, bab\}$; $J(C_1) = \{bab\}$; $C_2 = \{a\}$; $J(C_2) = C2 = \{a\}$; and C_3 is empty. If a two symbol alphabet $A = \{0, 1\}$ is desired, then the a, b, and c may be replaced by 101, 1001, and 10001, respectively. An analogous example of an infinite join code D of infinite level over alphabet $A = \{0, 1\}$ is $D = \{101, 10011011001, 100011001101100110001, \ldots\}$. A finite join code E of any desired level k is obtained by choosing the first k code words listed in D.

6   A Conceptual Illustration Suggested by DNA Complementary Base Pairing

As alphabet we will use the four symbols $\{A, C, G, T\}$ that are commonly used in biochemistry for denoting the four deoxyribonucleotides based on Adenine, Cytosine, Guanine, and Thymine. We consider here how we might communicate by transferring single stranded (ss) DNA molecules. We would like to derive the meaning represented by an incoming ssDNA molecule, m, by allowing the attachment to the message of (usually much shorter) complementary ssDNA code words. These attachments will be expected to form spontaneously through the classical Watson-Crick complementary base pairings which are held together by hydrogen bonds. (Recall that the pairings are A/T, C/G, G/C, and T/A.) The code molecules may have easily recognizable (possibly phosphorescent) labels attached. The meaning conveyed by the message molecule m is expressed by the sequence of (labels of) the code word molecules that attach through hydrogen bonds to the message molecule. How can we choose our code words to maximize the coherence of this proposed decoding scheme? One method is to choose the code word molecules in such a way that they form a join code.

Suppose that C_0 is a join code of level three over the alphabet $\{A, C, G, T\}$. Then C_0 is the union of $J(C_0)$, $J(C_1)$, $J(C2)$, with C_3 empty. We regard the words in each of the $J(C_i)$ as labeled ssDNA molecules. Let the molecules in each $J(C_i)$ be stored in aqueous solution in a test tube T_i. When a message (a ssDNA molecule) arrives that is to be read (decoded), we assume that it is the Watson-Crick complement of the concatenation of a sequence of code word molecules lying in C_0. Thanks to the join code feature of C_0 the decoding process can be carried out in the following sequence of three ultra-parallel steps: hold the message molecule(s) permanently. Wash message with solution from T_0. Wash with solution from T_1. Wash with solution from T_2. What is the result expected after these three washes?

We make the assumption that the code words attach where and only where the Watson-Crick matching is perfect. During the wash with T_0, code words from T_0 attach (globally in parallel) in each of a fully unique set of locations, since $J(C_0)$ is solid relative to C_0. (These code words are assumed to remain attached during the two remaining washes.) During the wash with T_1, code words from T_1 attach in each of a fully unique set of locations, since $J(C_1)$ is solid relative to C_1. (These code words are assumed to remain attached during the remaining wash.) During the wash with T_2, code words from T_2 attach in each of a fully unique set of locations, since $J(C_2)$ is comma free. Thus, after the three washes, all bases of the message molecule are paired with bases of the attached code word molecules. The meaning of the original message is expressed by the sequence of (labels of) these attached code word molecules.

From the considerations above it is clear that (under hypotheses that idealize the biochemical conditions) the decoding of ssDNA messages based on a finite join code of level k can be carried out with k washes done in the proper sequence. *The multi-volume dictionaries of Section 5 have become multi-tube DNA dictionaries of this section.* The parallel dictionary look up using volume i in Section 5 has become the i-th wash in this section.

7 Questions

In Sections 2 and 3 the concepts of unique decipherability and solidity were relativized to regular languages. Which of the many additional code concepts relativize in an interesting way? Is multiset decipherability [11], [13], [8], [4] relative to a regular language decidable? Can the theory of decipherability with respect to a variety of monoids, which has been developed by F. Guzman [5] and by E. Gumustop [3], be relativized in an interesting manner? Does the homophonic extension [14] of the code concept relativize meaningfully?

Can an algorithm be given that will decide whether any given uniquely decipherable infinite regular code is: (1) A join code? (2) A join code of finite level?

In the first paragraph of Section 5 a subset D of a uniquely decipherable infinite regular code C arose. When is D regular? Which (regular) subsets can arise in the role of D?

One may wish to extend results of Sections 4 and 5 in either of two ways. One way would be to allow context to be considered and the other would be to relativize the join code concept with respect to a language L. Formal results in Sections 4 and 5 can probably be generalized in either of these ways, but their biomolecular interpretation, in the manner of Section 6, may be less satisfying.

For a fascinating new example of the use of DNA molecules as message carriers, see [2].

Acknowledgments

The results in this article arose in part from an effort to analyze theoretically the potential of biomolecular operations used in various sticker approaches to DNA computing [12]. Partial support for this research through NSF CCR- 9509831 and DARPA/NSF CCR-9725021 is gratefully acknowledged.

References

[1] J. Berstel, D. Perrin, *Theory of Codes*, Academic Press, Orlando, FL, 1985.

[2] C. T. Clelland, V. Risca, C. Bancroft, Hiding messages in DNA microdots, *Nature*, **399** (1999), 533–534.

[3] E. R. Gumustop, *Varieties of Codes and Codifiable Varieties of Monoids*, Doctoral Dissertation in Mathematics, Binghamton University – State University of New York, Binghamton, NY, 1997.

[4] F. Guzman, A complete list of small proper MSD and SD codes (manuscript).

[5] F. Guzman, Decipherability of codes, *J. Pure and Applied Algebra* (to appear).

[6] T. Head, Unique decipherability relative to a language, *Tamkang J. Math.*, **11** (1980) 59–66.

[7] T. Head, Gh. Păun, D. Pixton, Language theory and molecular genetics: generative mechanisms suggested by DNA recombination, in Vol. II of *Handbook of Formal Languages* (G. Rozenberg, A. Salomaa, eds.), Springer-Verlag, Berlin, 1997.

[8] T. Head, A. Weber, Deciding multiset decipherability, *IEEE Trans. Information Theory*, **41** (1995), 291–297.

[9] H. Jurgensen, S. Konstantinidis, Codes, in Vol. I of *Handbook of Formal Languages* (G. Rozenberg, A. Salomaa, eds.), Springer-Verlag, Berlin, 1997.

[10] G. Lallement, *Semigroups and Combinatorial Theory*, Wiley, New York, NY, 1979.

[11] A. Lempel, On multiset decipherable codes, *IEEE Trans. Information Theory*, **32** (1986), 714–716.

[12] Gh. Păun, G. Rozenberg, & A. Salomaa, *DNA Computing. New Computing Paradigms*, Springer Verlag, Berlin, 1998.

[13] A. Restivo, A note on multiset decipherable codes, *IEEE Trans. Information Theory*, **35** (1989), 662–663.

[14] A. Weber, T. Head, The finest homophonic partition and related code concepts, *IEEE Trans. Information Theory*, **42** (1996), 1569–1575.

[15] S. Yu, Regular languages, in Vol. I of *Handbook of Formal Languages* (G. Rozenberg, A. Salomaa, eds.), Springer-Verlag, Berlin, 1997.

Uniformly Scattered Factors[1]

Lucian Ilie[2]

Turku Centre for Computer Science (TUCS)
Lemminkaisenkatu 14A, Data City 4th floor
FIN-20520, Turku, Finland
E-mail: `lucili@utu.fi`

Ion Petre

Turku Centre for Computer Science (TUCS) and
Department of Mathematics, University of Turku
FIN-20520, Turku, Finland
E-mail: `ipetre@cs.utu.fi`

Grzegorz Rozenberg

Department of Computer Science, Leiden University
P.O.Box 9512, 2300 RA Leiden, The Netherlands, and
Department of Computer Science
University of Colorado at Boulder
Boulder, CO 80309, USA
E-mail: `rozenber@wi.leidenuniv.nl`

Abstract. A word u appears as a factor of another word v as it is: in one piece. When u is a subword of v, u may be scattered as several factors. We consider the case in between and put some restrictions on the number of factors as to which u is allowed to be scattered. A large class of partial orders which are generalizations of factors and subwords is obtained. Investigating the borderline between their finite and infinite antichains, we are able to fully characterize the property of being well partial order. The result generalizes Higman's theorem.

---

[1]This research was partially done during the first author's visit to Leiden University.
[2]Research supported by the Academy of Finland, Project 137358. On leave of absence from Faculty of Mathematics, University of Bucharest, Str. Academiei 14, R-70109, Bucharest, Romania.

1   Introduction

Two of the most important partial orders on words are the factor and the subword partial orders. If a word u is a factor of a word v, then u appears inside v exactly as it is. If u is a subword of v, then u may appear inside v scattered as several factors. It naturally arises the idea of considering restrictions on the number of factors as to which u is allowed to be scattered. This idea is not new. It appeared first in a paper of Haines, cf. [2], who just mentioned the possibility of putting a fixed upper bound on the number of factors and notices that, in this way, one obtains a relation admitting infinite antichains (sets containing pairwise incomparable elements only). A deeper investigation of the problem is performed for the first time in [4], where some general restrictions are imposed and some necessary and sufficient conditions for the existence of the infinite antichains are given. In this paper, we consider a different approach to the problem. Our restrictions on the number of factors are similar to the ones of [4] but the way we consider the factors is different: all factors but the last one have the same length. In this way, all relations obtained are partial orders and, moreover, we can iterate the generalization process, obtaining a very large class of partial orders which are generalizations of factors and subwords. We then fully characterize those which are well partial orders. As we deal with well founded partial orders only, the problem consists of investigating the borderline between finite and infinite antichains. The result generalizes Higman's theorem.

2   Basic Partial Orders

In this section, we perform the first step of our generalization. We fix first some notations and concepts.

For an alphabet Σ, the set of all finite words over Σ is denoted Σ^*; the empty word is ε. For a word $w \in \Sigma^*$, $|w|$ denotes the length of w. For two words $u, v \in \Sigma^*$, we say that u is a *factor* of v, denoted $u \leq_f v$, if there are $x, y \in \Sigma^*$ such that $v = xuy$; u is a *subword* of v, denoted $u \leq_s v$, if $u = a_1 a_2 \cdots a_n$ for some $a_i \in \Sigma, 1 \leq i \leq n$, and $v = v_1 a_1 v_2 a_2 \cdots v_n a_n v_{n+1}$, for some $v_i \in \Sigma^*, 1 \leq i \leq n+1$.

Consider a set A and a partial order $\leq$ on A. A *chain* of $\leq$ is a set $B \subseteq A$ such that, for any $a, b \in B$, either $a \leq b$ or $b \leq a$. An *antichain* of $\leq$ is a set $B \subseteq A$ such that, for any $a, b \in B$, $a \not\leq b$. The partial order $\leq$ is called *well founded* if it has no infinite descending sequence $a_1 \leq^{-1} a_2 \leq^{-1} a_3 \leq^{-1} \cdots$

such that, for no $i \geq 1$, $a_i \leq a_{i+1}$, where $\leq^{-1}$ is the inverse of $\leq$; $\leq$ is called *well partial order* if it is well founded and has no infinite antichain.

It is easy to see that both the factor partial order $\leq_f$ and the subword partial order $\leq_s$ are well founded. Also, the factor partial order is not well partial order as soon as $\mathrm{card}(\Sigma) \geq 2$, since it has infinite antichains, e.g., $\{ab^n a \mid n \geq 1\}$. The famous theorem of Higman, cf. [3], states that this is not the case for $\leq_s$ (in fact Higman [3] proved a much more general result; Theorem 2.1 is a consequence of Higman's result in the case of words).

Theorem 2.1 (Higman [3]) *The subword partial order $\leq_s$ is well partial order.*

All the bounds we are to put on the number of factors as to which a certain word is allowed to be scattered are given by functions having as argument the length of the word. We now construct the family $\mathcal{G}$ of all these functions. It contains all $g : \mathbf{N} \longrightarrow \mathbf{R}$ with the following properties ($[r]$ denotes the integer part or r):

(i) $g(0) = 0, 1 \leq g(n) \leq n$, for any $n \geq 1$;

(ii) the function $[g(n)]$ is increasing;

(iii) the function $[\overline{g}(n)]$ is increasing, where $\overline{g}(0) = 0$ and $\overline{g}(n) = \frac{n}{g(n)}$, for any $n \geq 1$.

For a non-empty word $w \in \Sigma^+$ and a function $g \in \mathcal{G}$, we define the *g-factorization* of w as the factorization

$$w = w_1.w_2.\cdots.w_{[\overline{g}(|w|)]}.w_{[\overline{g}(|w|)]+1}$$

where

$$|w_1| = |w_2| = \cdots = |w_{[\overline{g}(|w|)]}| = [g(|w|)].$$

From the definition of the family $\mathcal{G}$, the g-factorization of a non-empty word is well defined. We may assume, for the sake of completeness, that the g-factorization of the empty word has just one factor.

Next, for a function $g \in \mathcal{G}$, we define the binary relation $\leq_{f,g}$ on Σ^* as follows.

(a) $\varepsilon \leq_{f,g} \varepsilon$; $\varepsilon \leq_{f,g} w$, for any $w \in \Sigma^*$; $w \leq_{f,g} \varepsilon$ implies $w = \varepsilon$;

(b) for two non-empty words $u, v \in \Sigma^+$, if their g-factorizations are, respectively,

$$u = u_1.u_2.\cdots.u_{[\overline{g}(|u|)]}.u_{[\overline{g}(|u|)]+1},$$
$$v = v_1.v_2.\cdots.v_{[\overline{g}(|v|)]}.v_{[\overline{g}(|v|)]+1},$$

then $u \leq_{f,g} v$ if and only if (i)–(iii) below are fulfilled:

(i) $|u| \leq |v|$;

(ii) there is a subsequence $1 \leq k_1 < k_2 < \cdots < k_{[\bar{g}(|u|)]} \leq [\bar{g}(|v|)]$ such that $u_i \leq_f v_{k_i}$, for any $1 \leq i \leq [\bar{g}(|u|)]$;

(iii) if $|u| = |v|$, then $u_{[\bar{g}(|u|)]+1} = v_{[\bar{g}(|v|)]+1}$.

Remark 2.1 For any $u, v \in \Sigma^*$, if $u \leq_{f,g} v$ and $|u| = |v|$, then $u = v$. Also, for any $g \in \mathcal{G}$, $\leq_{f,g}$ is well founded.

Consider next a few examples.

Example 2.1 Our first example shows that indeed our relations introduced above (and which are partial orders as proved below) are generalizations of the factor and subword partial orders. Consider the two functions

$$\mathbf{1}, \mathbf{1_N} : \mathbf{N} \longrightarrow \mathbf{N},$$

given by

$$\mathbf{1}(0) = 0, \quad \mathbf{1}(n) = 1, \quad \mathbf{1_N}(n) = n,$$

for any $n \geq 0$. It is clear that we have

$$\leq_{f,\mathbf{1}} = \leq_f, \quad \leq_{f,\mathbf{1_N}} = \leq_s .$$

Example 2.2 Consider the function $g_1(n) = \min(n, 2)$, for any $n \geq 0$. Then, for the words $u = aabab$ and $v = bbaaabba$ we have the g_1-factorizations respectively $aa.ba.b$ and $bb.aa.ab.ba$ and hence $u \leq_{f,g_1} v$. Also $aa.ba \nleq_{f,g_1} ba.ab.ab$.

Example 2.3 For the function

$$g_2 = \begin{cases} 1, & \text{if } 1 \leq n \leq 3, \\ \frac{n}{3}, & \text{if } n \geq 4, \end{cases}$$

we have that $aab.abb.aba.ab \leq_{f,g_2} baab.abbb.abaa$ but $ab.aa.bb.a \nleq_{f,g_2} ab.aa.bb.b$.

Example 2.4 Consider next a little bit stranger function, $g_3(n) = \sqrt{n}$. Clearly, $g_3 \in \mathcal{G}$. We have then, for instance, $aba.aba.bba.bbbaba \leq_{f,g_3} abba.abaa.baba.abba$.

Lemma 2.1 For any $g \in \mathcal{G}$, the relation $\leq_{f,g}$ is a partial order.

Proof. The reflexivity of the relation is obvious and the antisymmetry follows from Remark 2.1. For the transitivity, we just have to use the transitivity of the factor partial order $\leq_f$.   □

3 Characterization of Well Partial Orders

We give in this section a characterization of those partial orders introduced above which are well partial orders. As they are all well founded, we shall be concerned only with the finiteness of antichains.

We first notice that, for any $g \in \mathcal{G}$, since g is increasing, the limit $\lim_{n \to \infty} g(n)$ always exists in $\mathbf{R} \cup \{\infty\}$.

Theorem 3.1 *If $g \in \mathcal{G}$ such that $\lim_{n \to \infty} g(n) = \infty$, then $\leq_{f,g}$ is not well partial order as soon as* $\mathrm{card}(\Sigma) \geq 2$.

Proof. All we have to do is to find an infinite antichain for $\leq_{f,g}$. As $g(n)$ goes to infinity with n, it follows that there is a subsequence $(k_n)_{n \geq 1}$ of $(n)_{n \geq 1}$ such that

$$2 \leq g(k_1) < g(k_2) - 1 < g(k_3) - 2 < g(k_4) - 3 < \cdots$$

Consider then the following set of words over the two-letter alphabet $\{a, b\}$:

$$A = \{w_n = (ab^{[g(k_n)]-2}a)^{[\bar{g}(k_n)]}a^{k_n - [g(k_n)][\bar{g}(k_n)]} \mid n \geq 1\}.$$

We claim that A is an antichain of $\leq_{f,g}$. Assume that there are $1 \leq n < m$ such that $w_n \leq_{f,g} w_m$. For any $n \geq 1$, the g-factorizations of w_n is

$$w_n = ab^{[g(k_n)]-2}a.ab^{[g(k_n)]-2}a. \cdots .ab^{[g(k_n)]-2}a.a^{k_n - [g(k_n)][\bar{g}(k_n)]},$$

and so, from the fact that $w_n \leq_{f,g} w_m$, we get that $ab^{[g(k_n)]-2}a \leq_f ab^{[g(k_m)]-2}a$, which is impossible since $g(k_n) < g(k_m) - 1$ implies that $[g(k_n)] < [g(k_m)]$. The result is proved. □

We now prove that the converse of Theorem 3.1 hold true as well, thus obtaining a characterization of the relations in the family $(\leq_{f,g})_{g \in \mathcal{G}}$ which are well partial orders.

Theorem 3.2 *If $g \in \mathcal{G}$ such that $\lim_{n \to \infty} g(n) < \infty$, then $\leq_{f,g}$ is well partial order.*

Proof. Assume that $\lim_{n \to \infty} g(n) = d$, for some $d \in \mathbf{R}$. Therefore, as g is increasing, there are $c, n_0 \in \mathbf{N}, c \geq 1$, such that, for any $n \geq n_0$, $[g(n)] = c$ (in fact, $c \in \{[d], [d] - 1\}$).

We argue by contradiction. Assume that $\leq_{f,g}$ is not a well quasi order and consider an antichain of it, say $\{w_1, w_2, w_3, \ldots\}$. We may assume that $n_0 \leq |w_1| < |w_2| < |w_3| < \cdots$

Consider a new alphabet Δ which has as letters the words of length c over Σ, that is, $\Delta = \{w \in \Sigma^* \mid |w| = c\}$. For any $i \geq 1$, consider the g-factorizations of w_i $w_i = w_{i,1}.w_{i,2}.\cdots.w_{i,[\overline{g}(|w_i|)]}.w_{i,[\overline{g}(|w_i|)]+1}$ and denote $v_i = w_{i,1}w_{i,2}\cdots w_{i,[\overline{g}(|w_i|)]}$. Then the set $\{v_1, v_2, v_3, \ldots\} \subseteq \Delta^*$, where each word v_i is viewed as a word over Δ (its letters being the factors in the corresponding g-factorization), is an infinite antichain of the subword partial order $\leq_s$ on Δ^*. But this contradicts Higman's theorem. □

From Theorems 3.1 and 3.2 we obtain the characterization of the relations in the family $(\leq_{f,g})_{g \in \mathcal{G}}$ which are well partial orders.

Theorem 3.3 *For any $g \in \mathcal{G}$, $\leq_{f,g}$ is well partial order if and only if* $\lim_{n \to \infty} g(n) < \infty$.

4 Further Generalization

It is not difficult to see that in defining our general relations we may start with any partial order instead of the factor partial order $\leq_f$ and still the obtained relations are partial orders. In this way, a natural question arises: what are the well partial orders if we start with the subword partial order $\leq_s$ instead of the factor one $\leq_f$? More generally, we can iterate the process of defining our partial orders above. So, as $\leq_{f,g_1}$ is a partial order, for any $g_1 \in \mathcal{G}$, we can use it as a start instead of $\leq_f$. If the function of $\mathcal{G}$ we use is g_2, then denote the obtained partial order by $\leq_{f,g_1,g_2}$. We then iterate this process and get the most general partial orders which are of the form

$$\leq_{f,g_1,g_2,\ldots,g_n}, \tag{1}$$

where $n \geq 1$, $g_i \in \mathcal{G}$, for any $1 \leq i \leq n$. We notice that starting with $\leq_s$ is just a particular case here, namely, when $g_1(n) = \mathbf{1_N}$.

In this section we make all preliminaries for the main result of the paper, which characterizes those partial orders in (1) which are well partial orders. As all partial orders in (1) are well founded, we shall investigate only the existence of infinite antichains.

The following result on arbitrary partial orders is well known. We give here a short proof for the sake of completeness.

Lemma 4.1 *Let A be an infinite set and $\leq$ a partial order on A. Then A contains either an infinite antichain or an infinite chain of $\leq$.*

Proof. Consider the set M of all maximal elements of A with respect to $\leq$. M is antichain of $\leq$. If it is infinite, then we are done. If it is empty, then, clearly, A contains an infinite chain of $\leq$. Assume M is finite non-empty. Then there is $x \in M$ such that the set $N = \{y \in A \mid y \leq x\}$ is infinite. We now repeat the above reasoning with $N - \{x\}$ instead of A, namely, we consider the cardinality of the set of maximal elements of $N - \{x\}$. If the above reasoning can be repeated indefinitely, that is, at each step, the respective set of maximal elements is finite non-empty, then, again, we can construct an infinite chain of $\leq$ in A. $\qquad\square$

The next corollary follows immediately from Lemma 4.1.

Corollary 4.1 *Let A be an infinite set and $\leq$ a partial order on A. If $\leq$ is well partial order, then A contains an infinite chain of $\leq$.*

For any partial order $\leq$ on Σ^* and any $g \in \mathcal{G}$, we denote by $\leq_g$ the relation obtained as in Section 2, using $\leq$ instead of $\leq_f$.

For any positive integer $c \geq 1$, consider the function $\mathbf{c} : \mathbf{N} \longrightarrow \mathbf{N}$, defined by $\mathbf{c}(n) = \min(n, c)$, for any $n \geq 0$. Obviously, $\mathbf{c} \in \mathcal{G}$.

Lemma 4.2 *For any $c \geq 1$ and any partial order $\leq$ on Σ^*, the relation $\leq_\mathbf{c}$ is well partial order.*

Proof. Our idea is similar to the one used by Conway [1] for proving Higman's theorem and which is originally due to Nash-Williams, cf. [5]. Clearly, it is enough to prove that the relation

$$\leq'_\mathbf{c} = \leq_\mathbf{c} - \{(u, v) \mid u \leq_\mathbf{c} v, |u| \leq c\}$$

has no infinite antichain. We argue by contradiction. Assume that there is an infinite antichain of $\leq'_\mathbf{c}$, say

$$w_1, w_2, w_3, \ldots \qquad (2)$$

It follows that

$$\text{for any } i, j \geq 1, \text{if } i < j, \text{then } w_i \not\leq'_\mathbf{c} w_j. \qquad (3)$$

We may assume that (2) is an "earliest" sequence of words such that (3) is fulfilled, i.e., w_1 is a shortest word beginning a sequence satisfying (3), w_2

is a shortest word such that w_1, w_2 begin a sequence satisfying (3), and so on and so forth.

Now, there exists a word $w \in \Sigma^*, |w| = c$, such that infinitely many of the words of (2) start with w as a prefix. Assume that the indices of all such words are $(i_j)_{j \geq 1}$ such that $1 \leq i_1 < i_2 < i_3 < \cdots$ and put, for any $j \geq 1$, $w_{i_j} = wu_j$. Construct then the sequence

$$w_1, w_2, \ldots, w_{i_1-1}, u_1, u_2, \ldots \tag{4}$$

It is not difficult to check that the sequence (4) satisfies (3). Also, as $c \geq 1$, it is "earlier" than (2), a contradiction. The lemma is proved. $\square$

The next result generalizes Theorem 3.2.

Lemma 4.3 *For any partial order $\leq$ and any function $g \in \mathcal{G}$ such that $\lim_{n \to \infty} g(n) < \infty$, the relation $\leq_g$ is well partial order.*

Proof. Assume $\lim_{n \to \infty} g(n) = d$, for some $d \in \mathbf{R}$. Since g is increasing, there are $c, n_0 \in \mathbf{N}, c \geq 1$, such that, for any $n \geq n_0$, $[g(n)] = c$ (in fact, $c \in \{[d], [d] - 1\}$).

Assume that $\leq_g$ is not a well partial order. Then, there exists an infinite antichain for it, say $A = \{w_1, w_2, w_3, \ldots\}$. By possibly eliminating some of the words of A, we may assume that $n_0 \leq |w_1| < |w_2| < \cdots$.

Consider, for each $w_i, i \geq 1$, its g-factorization, say

$$w_i = w_{i,1}.w_{i,2}.\cdots.w_{i,[\bar{g}(|w_i|)]}.w_{i,[\bar{g}(|w_i|)]+1}$$

and denote, for any $i \geq 1$,

$$v_i = w_{i,1}w_{i,2}\cdots w_{i,[\bar{g}(|w_i|)]}.$$

Notice that $|w_{i,j}| = c$, for any $1 \leq j \leq [\bar{g}(|w_i|)]$.

Fix a letter $a \in \Sigma$. We then claim that

$$B = \{v_1a, v_2a, v_3a, \ldots\}$$

is an antichain of $\leq_{\mathbf{c}}$. To prove this, consider $v_ia, v_ja \in B$ and assume that $i < j$. Then, the **c**-factorizations of these two words are clearly

$$v_ia = w_{i,1}.w_{i,2}.\cdots.w_{i,[\bar{g}(|w_i|)]}.a,$$
$$v_ja = w_{j,1}.w_{j,2}.\cdots.w_{j,[\bar{g}(|w_j|)]}.a$$

and hence, using the fact that $|w_i| < |w_j|$, $v_i a \leq_c v_j a$ would imply $w_i \leq_g w_j$, a contradiction. Therefore, B is an infinite antichain of $\leq_c$, which contradicts Lemma 4.2. $\square$

For the rest of the paper, we shall make the following *assumption*: any partial order $\leq$ on words we shall consider has the property that $u \leq v$ implies $|u| \leq |v|$. Remark that all partial orders $\leq_{f,g_1,\ldots,g_n}$ do have this property.

Consider an alphabet Σ and a partial order $\leq$ on Σ^*. Denote by $\overline{\Sigma^*}$ the set of all tuples of words with equal length over Σ, that is,

$$\overline{\Sigma^*} = \{(w_1, w_2, \ldots, w_n) \mid n \geq 1, w_i \in \Sigma^*, |w_1| = |w_2| = \cdots = |w_n|\} \cup \{\bot\},$$

where $\bot$ stands for the empty tuple. We define the partial order $\overline{\leq}$ on $\overline{\Sigma^*}$ as follows: for any $x = (x_1, x_2, \ldots, x_n), y = (y_1, y_2, \ldots, y_m) \in \overline{\Sigma^*}$, we have $x \overline{\leq} y$ if and only if (i)–(iii) below are fulfilled (we assume that the length of the components of $\bot$ is zero):

(i) $n \leq m$;

(ii) $|x_1| \leq |y_1|$;

(iii) there is a subsequence $1 \leq k_1 < k_2 < \cdots < k_n \leq m$ such that $x_i \leq y_{k_i}$, for any $1 \leq i \leq n$.

It is easy to see that $\overline{\leq}$ is a partial order on $\overline{\Sigma^*}$. The next lemma shows that the property of being well is inherited from $\leq$ to $\overline{\leq}$.

Lemma 4.4 *If $\leq$ is a well partial order on Σ^*, then $\overline{\leq}$ is a well partial order on $\overline{\Sigma^*}$.*

Proof. By contradiction. Assume $\overline{\leq}$ is not a well partial order. We shall construct an infinite antichain of $\leq$, thus contradicting the hypothesis. This infinite antichain is constructed inductively, the main tool for the construction being the next claim.

Claim. Assume we have $B \subseteq \Sigma^*$ and $C \subseteq \overline{\Sigma^*}$ which verify $(*)$–$(***)$ below:

$(*)$ B is an antichain of $\leq$;

$(**)$ C is an infinite antichain of $\overline{\leq}$;

$(***)$ for any $u \in B$ and $v = (v_1, v_2, \ldots, v_n) \in C$, we have $u \not\leq v_i$, for any $1 \leq i \leq n$.

Then, there is $u \in \Sigma^*$ and $C' \subseteq \overline{\Sigma^*}$ such that $(*)$–$(***)$ above are fulfilled with $B' = B \cup \{u\}$ and C' instead of B and C, respectively.

Assuming that the claim holds, we construct an infinite antichain B of $\leq$ starting with $B = \emptyset$ and C an infinite antichain of $\widetilde{\leq}$ (there exists such a C by our assumption) and applying indefinitely the claim.

Proof of claim. Assume $C \subseteq \overline{\Sigma^*}$ is an infinite antichain of $\widetilde{\leq}$ and denote

$$C = \{w_n = (w_{n,1}, w_{n,2}, \ldots, w_{n,l(n)}) \mid n \geq 1\}.$$

By virtue of Lemma 4.2, we may assume that, for any $n < m$, $|w_{n,i}| < |w_{m,j}|$.

For any $n \geq 1$, denote by $i(n)$ the smallest $i, 1 \leq i \leq l(n)$, such that $w_{1,1} \leq w_{n,l(n)}$; if there is no such i, then set $i(n) = 0$ (of course, $i(1) = 1$).

If there are infinitely many n such that $i(n) = 0$, then $B' = B \cup \{w_{1,1}\}$ and $C' = \{w_n \in C \mid i(n) = 0\}$ are good choices. It is important to notice that when $l(1) = 1$, i.e., $w_1 = w_{1,1}$, then $i(n) = 0$, for all $n \geq 2$, and we can construct B' and C' as required.

If there are finitely many n with $i(n) = 0$, then consider the set

$$C_1 = \{(w_{n,1}, w_{n,2}, \ldots, w_{n,i(n)-1}) \mid i(n) \neq 0\}.$$

By Lemma 4.1, C_1 contains either an infinite antichain or an infinite chain of $\widetilde{\leq}$; denote it C_2 in either case. In the former case, take $B' = B \cup \{w_{1,1}\}$ and $C' = C_2$; clearly, B' and C' satisfy $(*)$–$(***)$.

Consider the latter case and put

$$C_3 = \{w_{n,i(n)} \mid (w_{n,1}, w_{n,2}, \ldots, w_{n,i(n)-1}) \in C_2\}.$$

Applying Corollary 4.1 to C_3 and $\leq$ we get an infinite chain of $\leq$ in C_3, say C_4. Therefore, the set

$$\begin{aligned} C_5 = {}& \{(w_{1,2}, w_{1,3}, \ldots, w_{1,l(1)})\} \\ & \cup \{(w_{n,i(n)+1}, w_{n,i(n)+2}, \ldots, w_{n,l(n)}) \mid w_{n,i(n)} \in C_4\} \end{aligned}$$

is an infinite antichain of $\widetilde{\leq}$ and also $(*)$–$(***)$ are fulfilled with C_5 instead of C. Notice the essential difference between C and C_5: $w_{1,1}$ has been eliminated. We next repeat the same reasoning with C_5 instead of C and continue in this way until either some good B' and C' are found or $w_{1,2}, w_{1,3}, \ldots, w_{1,l(1)-1}$ are all eliminated, one by one; in the latter case, B' and C' as required can again be constructed, as we already noticed. The proof is concluded. $\qquad\square$

Corollary 4.2 *For any well partial order $\leq$ on Σ^* and any $g \in \mathcal{G}$, the relation $\leq_g$ is well partial order.*

Proof. Assume that $\leq_g$ is not well partial order and consider an infinite antichain $\{w_n \mid n \geq 1\}$ of it. Without loss of generality we can assume that $|w_n| < |w_{n+1}|$ for any $n \geq 1$. If the g-factorization of w_n is

$$w_n = w_{n,1} \cdot w_{n,2} \cdots w_{n,[\overline{g}(|w_n|)]} \cdot w_{n,[\overline{g}(|w_n|)]+1},$$

then the set

$$\{(w_{n,1}, w_{n,2}, \ldots, w_{n,[\overline{g}(|w_n|)]}) \mid n \geq 1\}$$

is an infinite antichain of $\preceq$. This contradicts Lemma 4.4. □

5 The General Characterization

We are now ready to prove the main result of the paper: the characterization of the partial orders in (1) which are well partial orders. The part (i) in Theorem 5.1 is a generalization of Higman's theorem.

Theorem 5.1 *Let $m \geq 1$ and $g_1, \ldots, g_m \in \mathcal{G}$.*

(i) *If $\lim_{n \to \infty} g_k(n) < \infty$ for some $1 \leq k \leq m$, then $\leq_{f,g_1,\ldots,g_m}$ is well partial order.*

(ii) *If $\lim_{n \to \infty} g_k(n) = \infty$ for all $1 \leq k \leq m$, then $\leq_{f,g_1,\ldots,g_m}$ is not well partial order as soon as $\mathrm{card}(\Sigma) \geq 2$.*

Proof. (i) By Lemma 4.3 we obtain that $\leq_{f,g_1,\ldots,g_k}$ is well partial order. Then, by Corrolary 4.2, all relations $\leq_{f,g_1,\ldots,g_l}$ with $k \leq l \leq m$, are well partial orders.

(ii) For any $g \in \mathcal{G}$, let $[g]$ denote the function given, for any $n \geq 0$, by $[g](n) = [g(n)]$. It is clear from the hypothesis that

$$\lim_{n \to \infty} ([g_1] \circ \cdots \circ [g_m])(n) = \infty.$$

Then, there is a subsequence $(k_n)_{n \geq 1}$ of $(n)_{n > 1}$ such that $2 \leq ([g_1] \circ \cdots \circ [g_m])(k_1)$ and $([g_1] \circ \ldots \circ [g_m])(k_n) < ([g_1] \circ \ldots \circ [g_m])(k_{n+1}) - 1$ for all $n \geq 1$.
Denote $k_n^{(i)} = ([g_i] \circ \ldots \circ [g_m])(k_n)$ for any $n \geq 1$ and i, $1 \leq i \leq m$; put also $k_n^{(m+1)} = k_n$. We have then $2 \leq [g_1](k_1^{(2)}) < [g_1](k_2^{(2)}) - 1 < [g_1](k_3^{(2)}) - 2 < \cdots$ This, by the proof of Theorem 3.1, implies that the set

$$A_1 = \{(ab^{[g_1(k_n^{(2)})]-2}a)^{[\overline{g_1}(k_n^{(2)})]} a^{k_n^{(2)} - [g_1(k_n^{(2)})][\overline{g_1}(k_n^{(2)})]} \mid n \geq 1\}$$

is an antichain of $\leq_{f,g_1}$.

Assume that we have constructed an infinite antichain of $\leq_{f,g_1,\ldots,g_{i-1}}$, say $A_{i-1} = \{u_n \mid n \geq 1\}$, such that $|u_n| = k_n^{(i)}$, for all $n \geq 1$. We claim that we can construct an infinite antichain $A_i = \{v_n \mid n \geq 1\}$ of $\leq_{f,g_1,\ldots,g_i}$ such that $|v_n| = k_n^{(i+1)}$, for all $n \geq 1$. Notice that $1 \leq k_n^{(i)} = ([g_i] \circ [g_{i+1}] \circ \ldots \circ [g_m])(k_n) = [g_i](k_n^{(i+1)}) \leq k_n^{(i+1)}$.

We choose, for every $n \geq 1$, the word $v_n = (u_n)^{[k_n^{(i+1)}/k_n^{(i)}]} w_n$, where w_n is a prefix of u_n such that $|v_n| = k_n^{(i+1)}$ (actually, w_n can be any other word of the same length).

Suppose that A_i is not an antichain of $\leq_{f,g_1,\ldots,g_i}$, i.e., for some $r < s$ we have $v_r \leq_{f,g_1,\ldots,g_i} v_s$. Since $[g_i](|v_n|) = [g_i](k_n^{i+1}) = k_n^{(i)} = |u_n|$, for all $n \geq 1$, the g_i-factorizations of v_r and v_s are

$$v_r = u_r.\cdots.u_r.w_r, \qquad v_s = u_s.\cdots.u_s.w_s.$$

Thus, $v_r \leq_{f,g_1,\ldots,g_i} v_s$ implies that $u_r \leq_{f,g_1,\ldots,g_{i-1}} u_s$, which contradicts the fact that A_{i-1} is an antichain of $\leq_{f,g_1,\ldots,g_{i-1}}$.

For $i = m$ we obtain an infinite antichain A_m of $\leq_{f,g_1,\ldots,g_m}$, which concludes the proof of the theorem. $\qquad\qquad\square$

References

[1] J. H. Conway, *Regular Algebra and Finite Machines*, Chapman and Hall, 1971.

[2] L. Haines, On free monoids partially ordered by embedding, *J. Combin. Theory*, **6** (1969), 94–98.

[3] G. Higman, Ordering by divisibility in abstract algebras, *Proc. London Math. Soc.*, **2** (3) (1952), 326–336.

[4] L. Ilie, Generalized factors of words, *Fundamenta Inform.*, **33** (1998), 239–247.

[5] C. Nash-Williams, On well quasi-ordering finite trees, *Proc. Cambridge Philos. Soc.*, **59** (1963), 833–835.

Splicing Normalization and Regularity

Vincenzo Manca

Università di Pisa, Dipartimento di Informatica
Corso Italia, 40 - 56125 Pisa, Italy
E-mail: mancav@di.unipi.it

Abstract. Some canonic forms of splicing derivations are introduced and the notion of ω-splicing is used for proving that H systems with a finite number of splicing rules and a regular set of axioms generate regular languages.

1 Introduction

Splicing is one of the basic operations of DNA computing. It was introduced in [5], [6] as a formal representation of DNA recombinant behaviour: nucleic strands are cut in specific *sites*, by enzymes, so that pieces of different strands whose sticky ends match can be concatenated, producing new DNA molecules.

This mechanism suggested new generative systems in formal language theory, *H systems*, and introduced new perspectives in the combinatorial analysis of strings, languages, grammars, and automata.

More recently, going in the opposite direction, towards a Biological Mathematics, rather than a Mathematical Biology, it was discovered that a DNA soup can encode a combinatorial problem and can be transformed, by means of test tube genetic engineering techniques, into a final soup that encodes the solution of the problem [1], [11]. This approach disclosed new horizons in the search for new ideas, and applications. Thus, the biological trend, initiated by Kleene's finite state automata [9] and Lindenmayer systems [10], was continued, in a new direction, with H systems. Thereby, new computational models, inspired by biological metaphors, were introduced, and biochemical interpretations were found for concepts and results in formal language theory [19], [15], [16]. The field covering these subjects is now referred as DNA computing, or in wider perspectives, molecular computing, or natural computing. The benefits of this synergy (we could say *metasplicing*) between mathematics, computer science and biology are apparent and constitute an exceptional stimulus for a deep understanding of the combinatorial mechanism underlying splicing, the *catalyst* of all this process.

One of the most important mathematical properties of splicing, in its original formulation, was that the class of languages generated by splicing rules is a proper subclass of regular languages. Nevertheless, the proof of this result has a long story. It originates in [2], [3] and was developed in [17], in terms of a complex inductive construction of a finite automaton (in [16] is presented this proof). More general proofs, in terms of closure properties of abstract families of languages, are given in [18], [7], [15].

In this paper we show a different direct proof, as a natural consequence of some properties of normalization of splicing processes. We present a sort of geometrical representation of splicing derivations and introduce the notion of ω-splicing that allows us to clarify a crucial phenomenon on which regularity depends.

Consider an alphabet V and two symbols $\#, \$$ not in V. A *splicing rule* over V is a string $r = u_1 \# u_2 \$ u_3 \# u_4$, where $u_1, u_2, u_3, u_4 \in V^*$. For such a rule r and for $x, y, w \in V^*$ we define the (ternary) splicing relation $\Longrightarrow_r$

$$(x, y) \Longrightarrow_r w \quad \text{iff} \quad x = x_1 u_1 u_2 x_2,$$
$$y = y_1 u_3 u_4 y_2,$$
$$w = x_1 u_1 u_4 y_2, \text{ for some } x_1, x_2, y_1, y_2 \in V^*.$$

The string x is the *up premise* and the string y is the *down premise* of the *r-splicing step*; the string w is the *conclusion*. Strings u_1, u_2, u_3, u_4 are the *left up, right up, left down,* and *right down* components of the rule r; the pairs of strings (u_1, u_2), and (u_3, u_4) are the *up site* and the *down site* of the rule r. The string $x_1 u_1$ is the *r-head* of x and w; the string $y_1 u_3$ is the *r-head* of y. The string $u_4 y_2$ is the *r-tail* of y and w; while $u_2 x_2$ is the *r-tail* of x.

Therefore, when an *r*-splicing step is applied to two strings, they are cut in between the left and right components of sites of r, and then the *r-head* of the up premise is concatenated with the *r-tail* of the down premise. The resulting string is the conclusion of the step.

An *H system* system [16] Γ is given by: an alphabet V, a set A of strings over this alphabet, called axioms of the system, and a set R of splicing rules over this alphabet. The language $L(\Gamma)$ generated by Γ consists of the axioms and the strings that we can obtain starting from the axioms, by applying to them iteratively the splicing rules of Γ. If a terminal alphabet is considered, then we obtain an *extended* H system. We say *finitary* is an H system with a finite number of splicing rules. It is known that:

i) finitary H systems with finite axioms characterize a proper subfamily of regular languages;

ii) finitary H systems with regular sets of axioms characterize the regular languages;

iii) extended finitary H systems with finite axioms characterize the regular languages;

iv) extended H systems with regular sets of rules and finite axioms characterize the recursively enumerable languages;

v) and that extended finitary H systems with finite axioms plus certain controls on the use of rules or with certain distributed architectures characterize the recursively enumerable languages.

Comprehensive details can be found in [16].

2   Splicing Derivations

In [13] we introduced the notion of a *Derivation System* that allows us to analyze in an uniform way a great variety of symbolic systems, and to determine their common structure based on two main aspects: the combinatorial mechanism of the rules (e.g., replacement, parallel replacement, insertion/deletion, splicing, ...), and the regulation strategy that specifies the ways rules can be applied.

In this section, we consider derivations in the specific case of splicing systems, and show that we can express, in terms of splicing derivations, several interesting concepts useful in the analysis of splicing.

Definition 2.1 *Given an H system* $\Gamma = (V, A, R)$, *a splicing derivation* δ *of* Γ, *of length* n, *is a sequence of* n *strings and* n *labels, where each string associated with a corresponding label (written before an arrow that points to the string). A label is, either a triple (rule, string, string), or a special label indicated by* λ:

$$(\lambda \to \delta(1), \ l(1) \to \delta(2), \ \ldots, \ l(n-1) \to \delta(n))$$

where for $1 \leq i < n$:

- *if $l(i) \neq \lambda$ then $l(i) \in R \times \{\delta(1), \ldots, \delta(i-1)\} \times \{\delta(1), \ldots, \delta(i-1)\}$;*

- *if $l(i) = \lambda$, then $\delta(i+1) \in A$;*

- *if $l(i) = (r_i, \beta(i), \gamma(i))$, then $(\beta(i), \gamma(i)) \Longrightarrow_{r_i} \delta(i+1)$;*

- *$\forall i \ \exists j, \ i < j \leq n$ such that $\delta(i) = \beta(j)$;*

- *if $n \neq 1$, then $\delta(n) \notin A$.*

At each step i $\delta(i)$ is called the *current string* of that step. According to the definition, any current string, apart from the final one, has to be the conclusion of some splicing step with premises which are current strings of steps preceding i (no useless strings can occur in a derivation δ).

We indicate by $\Delta(\Gamma)$ the set of *(splicing) derivations* of a H system Γ.

If the last element $\delta(n)$ of a derivation δ is the string α, then we say that δ derives α and we write

$$\delta \vdash_\Gamma \alpha.$$

Two derivations of $\Delta(\Gamma)$ are said to be *equivalent* if they derive the same string.

A derivation $\delta \in \Delta(\Gamma)$ of length n is said to be *linear* when, for any $1 \leq i < n$, if the element $\delta(i+1)$ is not an axiom, then it is obtained by applying a rule of R with $\delta(i)$ as a premise.

Lemma 2.1 (Linearity Lemma) *Given a derivation $\delta \in \Delta(\Gamma)$ there is always a linear derivation equivalent to it.*

Proof. By induction on the length of derivations. For derivations where no rules are applied the linearity is trivial. Assume that for derivations of length smaller or equal to n we have linear derivations equivalent to them. Let δ be a derivation of length $n+1$. The string $\delta(n+1)$ has two premises $\delta(i), \delta(j)$ with $i, j \leq n$.

By the induction hypothesis there are two linear derivations δ', δ'' of length at most n that derive $\delta(i), \delta(j)$, respectively. Consider the concatenation of δ', δ'' where the steps of δ'' that are already in δ' are removed, then add to it the last splicing step of δ. This derivation is a linear derivation equivalent to δ. $\square$

A linear derivation can be represented in the following way:

$$(\lambda \rightarrow \delta(1), \ l(1) \rightarrow \delta(2), \ \ldots, \ l(n-1) \rightarrow \delta(n)),$$

where if at step i the label is different from λ, then it is $(r_i, \beta(i), p)$ with $\beta(i) \in \{\delta(1), \ldots, \delta(i-1)\}$, and $p \in \{0, 1\}$, in such a way that:

- if $l(i) = (r_i, \beta(i), 0)$, then $(\beta(i), \delta(i)) \Longrightarrow_{r_i} \delta(i+1)$;

- if $l(i) = (r_i, \beta(i), 1)$, then $(\delta(i), \beta(i)) \Longrightarrow_{r_i} \delta(i+1)$.

All the derivations we consider in the following are (tacitly) assumed to be linear derivations.

We introduce the notion of ω-splicing in order to consider infinite processes of splicing (ω stands for the set of natural numbers).

Given two (linear) derivations δ, δ' we say that δ is an *expansion* of δ', and we write $\delta < \delta'$, if δ' is obtained by interrupting δ at some step i, performing some further steps that derive a string, say β, and then, by continuing from β by applying the same labels of the steps that in δ are after the step i.

An ω-splicing is an infinite sequence of derivations:

$$\delta = (\delta^i \mid i \in \omega),$$

where, for every $i \in \omega$, δ^i is called a *component* of δ, and for every $i > 1$ there is a $j \le i$ such that $\delta^j < \delta^i$.

The set $\Delta^\omega(\Gamma)$ is the set of ω-splicings δ such that all their components belong to $\Delta(\Gamma)$. Any ω-splicing δ determines a language $L(\delta)$ constituted by the strings derived by its components.

The notion of ω-splicing allows us to give a necessary condition for the infinity of finitary H systems with finite axioms.

Lemma 2.2 (Infinity Lemma) *Let Γ be a finitary H system with finite axioms A. If the language $L(\Gamma)$ is infinite, then the set $\Delta^\omega(\Gamma)$ is not empty.*

Proof. Consider all the possible ways to apply the splicing rules starting from the axioms. We can arrange all these possibilities in a rooted tree where at first level we have the axioms and for each node its sons are obtained by applying a splicing step between it and some other node of a preceding level. Of course if $L(\Gamma)$ is infinite, then this tree is infinite. But, this tree is a finitary tree, therefore, by König's lemma it is infinite if it has an infinite path. This path is essentially an ω-splicing. □

An ω-splicing is linear iff all its components are linear. Now we introduce important classes of linear ω-splicings.

Given a step i of a linear derivation δ, it is k bounded if $\delta(i)$ is an axiom, or it is the conclusion of a splicing step where one of the two premises is the conclusion of the previous step, while the other one is at most k-old, that is, it is a string obtained at most k steps before that step. Splicing processes with an arithmetical control, labeling the *age* of strings in the derivation, were studied in a different context, and with different aims, in [12].

Let $\Gamma = (V, A, R)$ be an H system. The set $B_k(\Gamma)$ of k bounded derivations is the subset of derivations δ such that every splicing step of the derivations is k bounded. Two subclasses of $B_k(\Gamma)$ are the class $RD_k(\Gamma)$ of *right down* k bounded derivations and the class $LU_k(\Gamma)$ of k bounded *left up* derivations. In a derivation of $RD_k(\Gamma)$ the requirement of being at most k-old has to be satisfied for the down premise of every step, while in derivations of $LU_k(\Gamma)$ the same requirement has to hold for the up premise of every step.

The terms *right-down* and *left-up* are due to a graphical representation of splicing where the up premise is a (rectangular) frame, and the down premise another (rectangular) frame put under it.

In this manner the r-head of the superior frame (up premise) and the r-tail of the inferior frame (down premise), connected by an arrow, provide a representation of the result of the r-splicing. If the result of this step is the up premise of another step, we go in the *right down* direction (this kind of *staired* representation was adopted in [15]); otherwise if it is the down premise of a further step we go in the *left up* direction.

The following is the graphical representation of a (0 bounded) RD derivation with three steps.

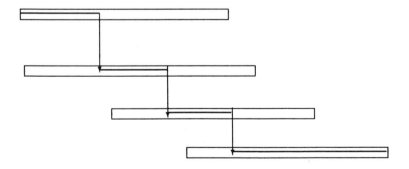

The following is the graphical representation of a (0 bounded) LU derivation with three steps.

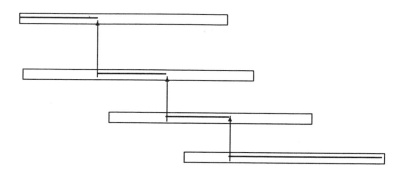

The classes $B_k^\omega(\Gamma)$, $RD_k^\omega(\Gamma)$, $LU_k^\omega(\Gamma)$ are constituted by ω-splicings whose components belong to $B_k(\Gamma)$, $RD_k(\Gamma)$, and $LU_k(\Gamma)$ respectively.

We can define the notion of RD and LU ω-splicing in terms of rewriting relations.

Given a splicing rule $r : u_1\#u_2\$u_3\#u_4$ over the alphabet V and $z \in V^*$, we write, for $x, y \in V^*$

$$x \Longrightarrow_{(r,z)}^{RD} y \quad \text{iff} \quad (x, z) \Longrightarrow_r y;$$

we say that $\Longrightarrow_{(r,z)}^{RD}$ is the RD (rewriting) rule associated to the splicing r with the index string z.

Analogously, $\Longrightarrow_{(r,z)}^{LU}$ is the LU (rewriting) rule associated to the splicing r with the index string z:

$$x \Longrightarrow_{(r,z)}^{LU} y \quad \text{iff} \quad (z, x) \Longrightarrow_r y.$$

Therefore, a 0 bounded RD derivation is obtained by applying RD rewriting rules indexed by the axioms, while a 0 bounded LU derivation is obtained by applying LU rewriting rules indexed by the axioms. Derivations of RD_0 or LU_0 are also called axiomatic (RD or LU) derivations.

Note that, in general, an axiomatic LU derivation cannot be transformed into an axiomatic RD derivation by changing the arrows in the staired representation. For example, given the splicing rule $\gamma\delta\#\beta\$\gamma\#\delta$ with $\gamma \neq \delta$, you can go in the left up direction any number of times, but only one step in the right down direction (analogously, a rule $\alpha\#\beta\$\gamma\#\alpha\beta$ with $\gamma \neq \alpha$ shows that, in general, axiomatic RD derivations cannot be transformed into LU axiomatic derivations by reversing the arrows).

Any derivation can be factorized in terms of different *nested levels* of axiomatic derivations. We can illustrate this factorization by using the following diagram, where horizontal lines indicate, sequence of steps and bullets represent start or end strings of these sequences. Strings α_2, α_3 are premises of a splicing step of conclusion α_4; and α_1, α_5 are premises of a splicing step of conclusion α_6.

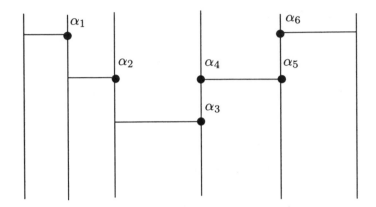

Given a derivation δ of length n, a *subderivation* of δ is a subsequence $\delta' = (l(j_1) \to \delta(j_1), ..., l(j_m) \to \delta(j_m))$, with $1 \le j_1 \le j_m \le n$, that is also a derivation. A component of δ is a *proper subderivation* if it is different from δ.

A derivation is *nonredundant* if it has no proper subderivation equivalent to it. For example, an H system with a splicing rule $u\#v\$u'\#v$ can generate derivations of any length that derive the same string. A less trivial situation of redundancy is illustrated by the following diagram, where frames are replaced by horizontal lines and only the parts remaining after splicing are indicated.

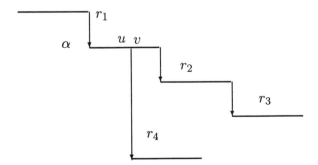

Consider the sequence of splicing steps r_1, r_2, r_3, r_4. Let α be the r_1-tail of the conclusion of the first splicing step. If the up site (u, v) of the fourth step is included in α, then the sequence of rules r_1, r_2, r_3, r_4 is equivalent to the sequence r_1, r_4, where the two steps r_2, r_3 are avoided.

An ω-splicing is *nonredundant* iff all its components are nonredundant.

In a linear splicing derivation, each splicing step is constituted by two substeps: the current string is cut in at a splicing point: a part of this string is kept (the left part in the case of an RD step, or the right part in the case of an LU step) and a string is appended to this point (on the right in an RD step, and on the left in an LU step). If at a step i, the splicing point is internal to a part of the current string that was appended in a previous step j, then we say that the step i is *counterdirectional* with respect to the step j (a splicing step is counterdirectional if it is counterdirectional with respect to some splicing step). For example, the derivation (r_1, r_2, r_3, r_4) represented by the diagram above the fourth splicing step is counterdirectional with respect to the second step and to the third step. In other words, a counterdirectional step removes, in the current string, the splicing point of a previous step.

In an RD axiomatic derivation, a counterdirectional step occurs when at some step the left up component of a rule is in the head of the conclusion of some previous splicing step. In an LU axiomatic derivation, in a counterdirectional step, the right down component of the rule is in the tail of the conclusion of some previous splicing step.

A step is *monotone* if it is not counterdirectional; a derivation is monotone if all its splicing steps are monotone. An ω-splicing is monotone if its components are monotone.

In any string we have a finite number of splicing points: moreover, by the definition above, all the splicing points of the counterdirectional steps of a given step are located in the same string. Therefore the following lemma holds.

Lemma 2.3 (Finite Counterdirectionality Lemma) *In a nonredundant ω-splicing the steps that are counterdirectional with respect to a given step i, if any, are a finite number.*

Monotone axiomatic RD or LU derivations can be represented in a schematic way with an oriented line going from the left up corner to the right down corner, or vice-versa. In this representation, any splicing step increases the line with a further piece.

There are nonredundant derivations that are not monotone. Consider the derivation (r_1, r_2, r_3, r_4) represented before in the diagram. If the right up component of the last splicing step r_4 is so long that it reaches the rightmost horizontal line, then we cannot reduce the derivation (r_1, r_2, r_3, r_4) to any proper subderivation of it.

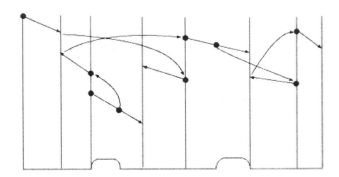

In conclusion, modulo elimination of redundant steps, we can factorize a derivation by composing RD and LU axiomatic derivations, by a diagram of the following type.

In this picture arrows represent monotone axiomatic RD (descending arrows) or LU (ascending arrows) subderivations. These subderivations are structured at different levels. At each level there is a current string of that level. The initial level is the most external level (the highest arrows). Vertical lines indicate the initial and the final splicing points of the subderivations.

When no counterdirectional step arises, a monotone axiomatic subderivation appends a string to (after in the RD case, or before in the LU case) the final splicing point of the current string of the superior level.

When a counterdirectional step occurs (in the picture, the arrows with two bullets), then the result of a monotone axiomatic subderivation is appended to the current string of the superior level, but at a splicing point that is not the last of that level. This produces a gap (in the picture, indicated by a bowed line) in the process of concatenating the substrings of the different levels.

The projections of arrows along the horizontal line indicate the appended strings. The horizontal line at the bottom represents the string globally derived.

Consider a derivation. There are two possibilities: at any step, we can go on by expanding it with further splicing steps, or we cannot continue anymore. This means that we can distinguish between derivations that cannot be extended after a certain number of steps and derivations that are components of ω-splicings. The next lemma explains the role of the derivations of the first type.

Lemma 2.4 (Finite Splicing Lemma) *Given a finitary H system Γ with a regular set of axioms, and a natural number k, the language of strings derived by derivations of Γ of length at most k is a regular language.*

Proof. It is a simple consequence of closure properties of noniterated splicing (cf. Lemmas 7.1, 7.2, and 7.3 in [16]). In fact, by using a more direct argument, we know that regular languages are closed with respect to prefix, subfix and concatenation. But any splicing step can be expressed by means of these operations applied to the premises. Therefore if we apply a splicing step to the axioms of Γ we get a regular language; if we continue this process k times, at the end we reach a regular language. □

This lemma can be easily extended by saying that, given a finitary H system Γ whose axioms belong to a class FL of formal languages, if $\Delta^\omega(\Gamma) = \emptyset$, then also $L(\Gamma)$ belongs to FL. Therefore, when Γ has finite axioms, if $L(\Gamma)$ is infinite, then $\Delta^\omega(\Gamma) \neq \emptyset$, as we already proved via the König lemma. If we consider only nonredundant ω-splicings, then we can reverse this implication, that is, $L(\Gamma)$ results to be infinite iff the set of nonredundant ω-splicings of $\Delta^\omega(\Gamma)$ is not empty.

3   Regularity

In this section we prove that finitary H systems with regular sets of axioms generate regular languages. This result, referred in [16] as *Regularity Preserving Lemma*, is a very important aspect of (nonextended) splicing. In the literature there are known indirect proofs [18], [16], [7], via algebraic closure properties of AFL (abstract families of languages), and direct proofs [2], [3], [17], [16] via a finite automaton that recognizes the language generated by a given H system. The proof we present here is another direct proof using the notion of ω-splicing, that makes evident a general aspect of finitary splicing that implies its regularity.

A crucial aspect of finitary H systems is that we can build arbitrarily long derivations, where an unbounded number of steps depend completely on a substring that occurred before in the derivation. We consider this aspect by using the notion of *locality* of a step.

Given a H system Γ we call the *left radius* of Γ the maximum length of left components of rules of Γ, and the *right radius* of Γ the maximum length of its right components (the maximum of these two lengths is the radius of Γ).

Let us call *locality* of a splicing step i, in a derivation of $\Delta(\Gamma)$, the substring $\theta = u'uvv'$ where uv is the site of the rule applied to $\delta(i)$ and $u'uvv'$ is the substring of $\delta(i)$, in a such a way that the length of $u'u$ equals the *left radius* of Γ, and the length of vv' equals the *right radius* of Γ.

At any step, if its splicing point will not be removed in a following step, what can happen after that step depends completely on its locality. In a finitary H system the localities are a finite number; therefore their lengths are bounded by some natural number, that is, in finitary H systems we have a sort of *local generation*. This phenomenon, illustrated by the following proof, resembles the notion of *locally testable language* [14], whose connection with splicing played a crucial role in the definition and analysis of this combinatorial mechanism [20], [4], [8].

Given two ω-splicings δ_1, δ_2, we can define a *generation* pre-ordering relation $\leq_L$ such that $\delta_1 \leq_L \delta_2$ if $L(\delta_1) \subseteq L(\delta_2)$. We say that an ω-splicing of $\Delta(\Gamma)$ is *maximal* if it is maximal with respect to this pre-ordering relation. Moreover, we say that two ω-splicings δ_1, δ_2 are *generation equivalent* if $L(\delta_1) = L(\delta_2)$. Let us indicate by $[\delta]$ the class of generation equivalence of the ω-splicing δ. The existence of maximal ω-splicings is a consequence of Zorn's Lemma (in an ordered set where each nonempty chain has the least upper bound, any chain has a maximal element too).

The essential tools of the following arguments are the *Pigeonhole principle* and the induction principle.

Lemma 3.1 (Infinity Gate Lemma) *Given a finitary H system Γ, there exists a number g_Γ such that in Γ any derivation of length greater than g_Γ is a component of an ω-splicing.*

Proof. Having in Γ a finite number of localities, we are sure that there exists a number g such that in g steps two occurrences of the same locality have to occur. If the second occurrence is not counterdirectional with respect to the first one, then $g_\Gamma = g$, because we can concatenate any number of times the string between the splicing points relative to the two occurrences of that locality. Otherwise, if we consider $2g, 3g, \ldots$ steps we get further occurrences of the same locality. We know from Lemma 2.3 that all these occurrences cannot be counterdirectional to the first one. Therefore, for some positive natural number k, $g_\Gamma = kg$.   □

Lemma 3.2 (Finite and Infinite Splicing Representation Lemma) *Given a finitary H system* Γ *with a regular set of axioms,*

$$L(\Gamma) = L_0 \cup \bigcup_{[\delta] \in M} L(\delta)$$

where M is the set of generation equivalence classes of maximal ω-splicings in $\Delta^\omega(\Gamma)$, and L_0 is a regular language.

Proof. Consider a derivation in $\Delta(\Gamma)$. If it can be carried on with an unbounded number of steps, then it is a component of an ω-splicing δ, thus what we can generate by means of it is generated by δ; otherwise, if it stops after some number of steps, then it can generate a string that we could not generate with an ω-splicing.

Let g_Γ be the number given by the previous lemma, and define L_0 to be the language of strings derived with at most g_Γ steps. As we have pointed out (Lemma 2.4), L_0 is regular; therefore, $L(\Gamma) = L_0 \cup L_1$, where L_1 is the set of strings derived with more than g_Γ steps. But L_1 consists of the strings generated by ω-splicings, and therefore the statement follows by the definitions of maximal ω-splicings and of generation equivalent ω-splicings. □

Theorem 3.1 (Regularity Preserving Lemma) *If Γ is a finitary H system with a regular set of axioms, then $L(\Gamma)$ is a regular language.*

Proof. By virtue of the previous lemmas, it is sufficient to prove that:

- The number of the generation equivalence classes of nonredundant and maximal ω-splicings in $\Delta^\omega(\Gamma)$ is finite.

- If δ is a nonredundant and maximal ω-splicing $\delta \in \Delta^\omega(\Gamma)$, then $L(\delta)$ is a regular language.

Consider a nonredundant and maximal (n.m.) ω-splicing δ. It has at least a rule that applies infinitely many times with the same locality, because rules and localities of Γ are finite, while δ is an ω-splicing and thus has infinitely many steps. Let (r, θ) be a pair *(rule, locality)* of this kind, that we call a *knot*. Consider the occurrence of a knot (r, θ) that we call *initial* because it is the first occurrence of (r, θ) in δ such that no splicing step of a further occurrence of (r, θ) can be counterdirectional with respect to (the step of) this occurrence. Such an initial occurrence must exist because otherwise, the occurrences of (r, θ) being infinitely many, we could have an infinite sequence of steps that are counterdirectional with respect to the same splicing step, contradicting Lemma 2.3. For the same reason, after the (first) initial occurrence of (r, θ) we have the second initial occurrence of (r, θ) such that no other further occurrences of (r, θ) could be counterdirectional with respect to it. Analogously we can define all the other initial occurrences (the third, the fourth, and so on).

The following picture represents the factorization of a n.m. ω-splicing δ with respect to the initial occurrences of (r, θ).

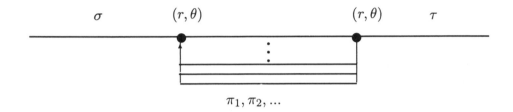

In this factorization $\sigma, \{\pi_1, \pi_2, \ldots\}, \tau$ are called the *seed*, the *power*, and the *termination* of δ. The power $\{\pi_1, \pi_2, \ldots\}$ contains all the sequences of steps (possibly infinitely many) that starting from an initial occurrence of the knot (r, θ) can reach again another initial occurrence of the same knot. The seed and the termination could be empty.

By the assumption on the initial occurrences of (r, θ), we know that if α is the string that we keep at the step of an initial occurrence of (r, θ), then when the next initial occurrence of (r, θ) occurs, some string β is added to α; beside, after appending α the same locality occurs, therefore, since δ is maximal, for every $n \in \omega$ in δ also β^n is appended to α.

Moreover, if some step in the termination would be counterdirectional, with respect to all the initial occurrences of the knot that can occur after its first initial occurrence, then the knot could be avoided, because only a finite number of occurrences of (r, θ) could be necessary in performing the termination. This would imply that δ should be redundant, but we assumed δ to be nonredundant.

Now, we proceed by induction on the number of knots in n.m. ω-splicings.

Consider a n.m. ω-splicing with only the knot (r, θ). In its seed, termination, and power no other knot can occur. In fact, any locality can occur a finite number of times: m_1 the first one, m_2 the second one, and so on. But the derivations that verify this condition are a finite number.

Therefore, we have only a finite number of possibilities for $\{\sigma, \pi_1, \pi_2, \ldots, \tau\}$, and even if we can arrange the elements of the power in different ways, all maximal nonredundant ω-splicings are equivalent inasmuch as they generate the same language:

$$L(\delta) = \alpha\{\beta_1, \beta_2, \ldots, \beta_n\}^*\gamma,$$

where α is the string that is kept at the first initial occurrence of (r, θ), $\{\beta_1, \beta_2, \ldots, \beta_n\}$ is the set of strings that can be appended to α by the (finite) elements of the power, and γ is the string that can be appended to α without producing an initial occurrence of the knot (r, θ). Of course, $L(\delta)$ is regular. Therefore, since the number of possible knots is finite, the ω-splicings with 1 knot generate a finite number of regular languages.

Now consider the inductive case. Assume that n.m. ω-splicings with at most k knots generate a finite number of regular languages. Let δ be a n.m. ω-splicing with $k+1$ knots. We could factorize δ in the same manner we did before: seed, power and termination; but in this case it can happen that among the seed, power or termination of δ there are infinitely many steps, because some knots can occur repetitively. Therefore,

$$L(\delta) = L_0(L_1 + L_2 + \ldots + L_m)^*L',$$

where, L_0 are the strings kept when the steps of σ finish, the languages $L_1, L_2, \ldots, L_m$ consist of strings appended to L_0 by $\{\pi_1, \pi_2, \ldots\}$, and L' is the language of strings appended by τ. By the induction hypothesis, $L_0, L_1, L_2, \ldots, L_m, L'$ are regular languages, because $\sigma, \pi_1, \pi_2, \ldots, \tau$ are n.m. ω-splicings with at most k knots. Therefore, $L(\delta)$ is regular.

Moreover, since the number of knots is finite, we have a finite number of possibilities for the first of $k+1$ knots; thus the n.m. ω-splicings with $k+1$ knots can generate a finite number of regular languages.

In conclusion, n.m. ω-splicings determine a finite number of generation equivalence classes, and every n.m. ω-splicing generates a regular language. This concludes the proof. □

If Γ is an H system with an infinite set of rules, then the argument of a finite set of localities does not apply.

The proof of this theorem is based on a sort of a *pumping* mechanism. It would be interesting to define a notion of *locally generated* languages where this pumping phenomenon is a direct consequence of general *locality conditions* that do not depend on the particular combinatorial mechanism underlying the derivations.

References

[1] L. Adleman, Molecular computation of solutions to combinatorial problems, *Science*, **266** (Nov. 1994), 1021–1024.

[2] K. Culik II, T. Harju, The regularity of splicing systems and DNA, in *Proc. ICALP 1989, Lecture Notes in Computer Science*, **372**, Springer-Verlag, 1989, 222–233.

[3] K. Culik II, T. Harju, Splicing semigroups of dominoes and DNA, *Discrete Applied Mathematics*, **31** (1991), 261–277.

[4] A. De Luca, A. Restivo, A characterization of strictly locally testable languages and its application to subsemigroups of a free semigroup, *Information and Control*, **44** (1980), 300–319.

[5] T. Head, Formal language theory and DNA: an analysis of the generative capacity of recombinant behaviors, *Bulletin of Mathematical Biology* **49** (1987), 737–759.

[6] T. Head, Splicing schemes and DNA, in: *Lindenmayer Systems: Impacts on Theoretical Computer Science, Computer Graphics, and Developmental Biology*, (G. Rozenberg, A. Salomaa, eds.), Springer-Verlag, Berlin, Heidelberg, 1992, 371–383.

[7] T. Head, Gh. Păun, D. Pixton, Language theory and molecular genetics, Chapter 7 in *Handbook of Language Theory* (G. Rozenberg, A. Salomaa, eds.), vol 2., Springer-Verlag, Berlin, Heidelberg, 1997, 295–360.

[8] T. Head, Splicing representations of strictly locally testable languages, *Discrete Applied Mathematics*, **87**, 1-3 (1998), 139–147.

[9] S. C. Kleene, Representation of events in nerve nets and finite automata, in *Automata Studies*, Princeton Univ. Press, Princeton, NJ, 1956.

[10] A. Lindenmayer, Mathematical models for cellular interaction in development, I and II, *J. Theoret. Biol.*, **18** (1968), 280–315.

[11] R. Lipton, DNA solution of hard computational problems, *Science*, **268** (April 1995), 542–545.

[12] V. Manca, Gh. Păun, Arithmetically controlled H systems, *Computer Science Journal of Moldova*, **6**, 2(17) (1998), 103–118.

[13] V. Manca, Logical string rewriting, *Theoretical Computer Science*, to appear.

[14] R. McNaughton, S. Papert, *Counter-Free Automata*, MIT Press, Cambridge, MA, 1971.

[15] Gh. Păun, On the splicing operation, *Discrete Applied Mathematics*, **70**, 1 (1996), 57–79.

[16] Gh. Păun, G. Rozenberg, A. Salomaa, *DNA Computing. New Computing Paradigms*, Springer-Verlag, Berlin, 1998.

[17] D. Pixton, Regularity of splicing languages, *Discrete Applied Mathematics*, **69**, 1-2 (1996), 101–124.

[18] D. Pixton, Splicing in abstract families of languages, *Technical Report* 26/97 of SUNY, Univ. of Binghamton, New York, 1997.

[19] G. Rozenberg, A. Salomaa, eds., *Handbook of Language Theory*, 3 Volumes., Springer-Verlag, Berlin, Heidelberg, 1997.

[20] M. P. Schützenberger, Sur certaines operations de fermeture dans les languages rationels, *Sympos. Math.*, **15** (1975), 245–253.

Infinitely Many Infinities

Carlos Martín-Vide

Research Group on Mathematical Linguistics and Language Engineering
Rovira i Virgili University
Pl. Imperial Tarraco, 1
43005 Tarragona, Spain
E-mails: cmv@astor.urv.es, cmv@tinet.fut.es

[The Country, the Olympic year 1992, sunny though a bit cloudy – fine weather, as ever; P and Q seated on the ground amongst the pines; tulips, cypresses and rosemary nearby.]

P: Always, we come here to spend the day. Always, the weather is the same, hot and dry. Always, we become thirsty by mid-morning. Meteorologists call this fine weather! It must be fine for them. To me, it's a torture. Above all, because of how monotonous it is. When it's raining, there exists atmospheric activity. When the sun is shining as today, it doesn't...

Q: And now it happens that you say that meteorologists are conservative!

P: Of course! Their television says, for example, that the weather is, will be or was fine. I don't have to remain impassive when listening to a statement of that kind. I like atmospheric uprising, simply that something is occurring in the sky. And I'm about to take accelerated classes for learning to pray for rain. Afterwards, I'll set up a company together with some old colleagues from the seminary...

Q: Your tendency to excess, to go too far, to say the first thing that crosses your mind appears again. If you had some more sense, your affairs in life would have gone better.

P: They have not gone so bad to me. After all, I can waste time as I've been doing the last five years, since I left that old building which welcomed us every morning reluctantly, and where we came in with the same spirit and the same little passion. Let me get a bit excited now, when my passions

217

can no longer bother anybody, as there is nobody. Well, yes, you. But my moods don't affect you, do they?

Q: You say everything. Of course, they affect me, the same as it must happen to the same chickens... I mean to the same birds... that come to meet us every morning. But I cannot allow myself a passion for this.

P: Why cannot you allow yourself a passion?

Q: Because it's not a part of my programme and gets me out of my rails.

P: Well, be derailed, my dear fellow. At your age, have you not already realized that passions move the world?

Q: You are a disbeliever.

P: Of course. And it has cost me a great deal. I prefer to be that than conceited, or than a believer.

Q: Well, bah! Don't start again with the bore of last Sunday. It is obvious that you like war, conflict, risk, struggle. I prefer placidity and contemplation. For nearly two hundred consecutive days, we take the train at exactly the same time and come to exactly the same place. On top of that, you cannot want us to talk about the same topics always.

P: There exists a rather small variety of topics to talk about. Notice that I always bring the same book.

Q: Yes, I noticed it. Have you not yet read it?

P: I never read it. I think that I haven't even opened it since the pretty girl pupil in the school gave it to me last year.

Q: Not even you'll open it.

P: Not even I'll open it.

Q: And what is it about?

P: About infinity. It's a book that Cantor wrote about infinite sets last century. It looks very interesting.

Q: And if it looks so to you, why not glance through it at least?

P: Well, look, because I think that, if I open it, I'll finally find out everything. And I prefer to remain in my ignorance, since it means to me a kind of secret.

Q: How very strange things you think!

P: I read only encyclopedias. Alphabetically.

Q: Alphabetically?

P: Yes, because nothing is of more interest to me than any other thing.

Q: And don't you believe that, at your age, you ought to have already chosen?

P: Yes, look at what happened to me with Marilin...

Q: Well... keep apart your conjugal problems, as I'd have much to say about them. What is surprising to me is that for a mathematician like you...

P: ...a wretched teacher of mathematics until his retirement...

Q: ...well, so for a wretched teacher of mathematics like you everything is interesting. What is common is that just the opposite happens: that one is only interested in mathematics.

P: No, look. Shortly after I began to teach, I realized that my colleague teaching French and I were doing not so very different tasks: both were teaching languages. Our pupils did not perceive so, but I was seeing it more and more in this way.

Q: As much as doing the same...!

P: Definitely yes. That is why I began to be interested in languages. In the plenty of time I had after lunch and before coming back to the school, I took the habit to sit down in the yard of the opposite faculty and to start to read some stuff that I was lent from the library of the university. I was reading without any order things by Martinet, Vossler, Harris and a lot of other people. I felt them very close to the way of arguing I was used to. After a time, some day I decided to stand up from my seat and to look at the offices for some professor who could guide me. I was successful: he was an old, bearded but kind professor, who was with me for a good while, until I had to run from my loved school.

Q: And did he take you seriously?

P: Yes, in a discoloured filing card that he gave to me, he noted down in his own way a brief guide to reading that I should do. Look, I got some very precious information by chance. Good things of life occur by chance: love, lottery and everything else!

Q: God doesn't play dice, my dear.

P: And how! I've just read a quotation from Hawking in the newspaper: "God not only does play dice, but sometimes throws them where they cannot be seen".

Q: Very nice! But that's a lie.

P: Beauty and truth must be put side by side. Okay? Well, it doesn't matter, the fact is that I started to read what that kindly man had wanted to note down in my list. I was amazed by a variety of authors' professional origins I was not accustomed to: some of them were grammarians, some other ones philosophers, there were a few psychologists and mathematicians and many other people. I had thought that the study of languages was a matter of grammarians' concern, but, however, was seeing that it was not the case, that it was not a matter only of their concern.

Q: Perhaps it was that man's modern own oddity.

P: No, time and reading were convincing me that he was not on the wrong road. Because, moreover, my anonymous guide stressed very much that I should read the classics.

Q: Plato, for sure!

P: Among others.

Q: Plato said everything – don't forget it.

P: Well, I don't know. I read a good handful of works from previous centuries and noticed that the study of language is largely accumulative, unlike what we normally are used to see in the sciences.

Q: In other words, the study of language is not a science like others.

P: No, I'm not saying that, wait. I'm just telling you about my perplexity. Little by little, I was getting excited and learnt to put my concept of progress at issue. These people didn't seem to make progress, but to give different explanations to the same problem each time: how is it that we speak so well, so much and so early. They didn't raise it in such a way, but I was intrigued by this, because my experience as a teacher had showed me that it didn't happen the same way with the learning of mathematics.

Q: You should take note of that as an additional piece of information of a difference that you cannot deny.

P: I don't know whether or not I can deny it. But while reading some treatises written in the nineteenth century about the evolution of languages in history, I tended to pay especial attention to resemblances rather than to differences. Maybe it was the deformed view of a person with mathematical training, but the fact is that it was so and that I didn't believe I was the only one in the world to think so. Well, but some further detail awakened my curiosity: people who had worked on languages had been intrigued by the relationship that words have with things. Since my years in the faculty, I had retained my interest in discovering what it was I was learning had to do with real life, not only with machines and the world of development, but also with human beings and their world of passions and feelings. It turned out to be very comfortable to me that other people were concerned before with the relationship that languages have with reality.

Q: A merely instrumental relationship?

P: Not at all. Both the mathematical language and any language, which are of course structures composed of elements that perform certain functions, the same as cars which you like so much, just the same as cars, have their history, like rivers, which look for their bed in the land.

Q: Wait, I'm wondering whether this whole trip was really of some use to you.

P: It gave pleasure to me.

Q: That's all?

P: That's all.

Q: Just?

P: Just. Is it not so much?

Q: Frankly speaking, no.

P: Please let me continue, since that's not all. I tried to take a historical order, as that was that man's first tip should I wish to acquire a certain linguistic culture. And in this way I reached a book that I was about to not read, because it seemed to me that it was written in Latin: by a German – I believe – with a bit of a hectic life. His small book made me see that language was the key for clearing up something that had ever appeared to me as a huge bundle: what still continues to be called philosophy. Moreover, its style impressed me.

Q: Why have you never written a book, instead of reading so many others?

P: Well, look, because I've seen much redundancy circulating. That teutonic man... well, on second thoughts, I don't know whether he was born in Vienna... that man taught me that there are just a very few things to say that have not already been said.

Q: Okay, but perhaps it is worth saying them in another way.

P: Perhaps it is. I don't know. Many styles are possible, but the best is just one.

Q: The best! The best! What do you mean?

P: No, please, forget it, it doesn't matter, I'm in no mood now for profound philosophies with this so intense heat. I just know that such a book, which looks to have been written unwillingly and is the only one by him that I know, impressed me.

Q: It is evident that you need strong excitement.

P: Of course, otherwise how do you stir your imagination up? Those thrills are driving you, until one ends up living a life that is spinning with the remnants which often one inadvertently runs into.

Q: And which was, if I may ask, your next indescribable experience?

P: I'll tell it to you provided that you allow me to take a bunch of grapes out of the bag that is between your legs.

Q: All of them for you. You know that I've never been alcoholic.

P: I am, and many ancestors that precede me in that book of history where I'm sure I'll appear, join me in this hobby. I like to be out of place, to make me conspicuous...

Q: You don't need to remind me of it, I know it. You discuss the price of the ticket with the same clerk every day.

P: It's never with the same ticket clerk. They are different every day.

Q: Bah, please continue!

P: Don't get excited. So be it. I had heard that a certain Chomsky had been a specialist in politely not fitting in with the ideas that his colleagues supported in his time.

Q: What century are you talking about now?

P: About this one, the century when you and I will surely die. I had heard his work qualified as revolutionary. Not only in grammar, but also in philosophy, psychology, logic, politics and more fields. That is why I was quite favourably predisposed when I started to read him. The most attractive feature of his setting out was that he was opening a specialized territory to everybody, no matter their origin, which had something to say about language. That is, he invested even me with a very small authority. I shared with him his insistence on the necessity both to devote oneself to theory and to fearless go through the borders that academic institutions had set up, in their vain aspiration for drawing some limits on a field that is much larger than this one which you and I are kindly sharing.

Q: Pretty nice!

P: That's how it is. After Chomsky, one talks about language sciences, in plural. Among them, grammar, or linguistics as it is now usually called – I like grammar much more – occupies a relatively privileged position, because of the tradition that it has accumulated along the way. Moreover, that man was farsighted in summarizing which are the great areas in the study of languages in quite a simple way: their structure, acquisition and use. After him, there have been contributions to those areas by linguists as well as mathematicians, psychologists, philosophers and many others, since he made it categorically clear that a language is an immensely polyhedric object, and that any well aimed path we take in approaching it is perfectly legitimate at the time as it will enrich the other ones inevitably. That is why you'll see today that one talks about language sciences and no sensible person is wondering who is the boss there. Everyone uses the methods he learnt, and learns to listen the others and to understand what they are doing.

Q: Don't you think that all this is celestial music, in view of what is happening with the overspecialization of sciences?

P: They are not incompatible trends. Maybe these people are at a certain stage when they need to sit down and to interchange their results and tools, in order to get up later equipped with a greater strength and a more detailed map of the land that they will have to continue to explore. It has occurred in the history of science several times and is not bad. It is a kind of spiritual retreat...

Q: I observe once again that you don't take seriously even what is of interest to you. Poor Marilin!

P: I do. However, just a little bit of irony doesn't harm at all. Not even wise people have always learnt to take what they are doing with a bit of distance and a sense of humour.

Q: Nor politicians.

P: Those even less, engaged as they are to organize their fellow citizens' lives. Let them get on with it!

Q: And have your readings arrived up to there?

P: Yes, not much further. Those problems of language structure, acquisition and use seem big enough for me to daringly go forward: how are languages built, how is it that we are able to learn them in such a short time, and which abilities do we develop that allow us to use them so appropriately as you and I have been doing for years. Moreover, everything I'm reading on language has brought before my eyes some problems that interest me for the first time in my life. They interest me, they excite me, they intrigue me. Probably because they are the modern version of some ancient problems of the damned philosophy that I never succeeded in understanding and that I now finally grasp. While I was never a humanist, I'll end up being converted to the cause. I'm not interested in mathematical problems: they are like riddles or crosswords. I'm interested in bigger and pastier problems. But at the same time they overwhelm me, because they bring me from one matter to another even more obscure, and I'll finish by asphyxiating. Do you see why I refuse to open the book that my ex-pupil was so kind to give to me that day when she knew that she will not meet me again?

Q: Don't get sentimental.

P: No, you should understand that the sentimental state is the natural state, and any other one is a degeneration of it. That, thus, the feeling is the primary fact, the organ precedes the function, and the earth goes before the sky.

Q: I already live in my heaven well.

P: But you will fall down, and you have not become accustomed to crashes.

Q: Well, as you like. After all, you are the boss. And then you hit me.

P: It's not that, disagreements concerning principles can only be solved by force. And you will agree with me that it's not worth it.

Q: Certainly, and I'd like you to remember it every time you get angry.

P: I'll remember it. I promise you.

Q: Look, you'll die soon and we'll give up coming here every day. What will happen afterwards?

P: I don't know, that I'll finally go to your heaven – I guess. I don't know, I don't have qualities of a fortune-teller.

Q: All things you thought while reading, should you not put them down in writing?

P: What for? No.

Q: Not for others to follow your path.

P: No, it's better for them to go through it step by step. The only thing that it would not matter to me to write down in a few pages, a very few pages, is my feeling that the reflection on language, which is one of the few useless jobs that I find interesting in life, is becoming more and more immersed in another more general interdisciplinary environment: the worlds of computation, cognition and communication. I believe that these three words can well inherit in the future the attention paid to structure, acquisition and use in recent times. The attention and the money, as wherever there are dollars there is really some seed of utility. I've noticed recently that what are all these concerns on language for is starting to interest. And it seems that they are useful for something: for contributing to developing the computer industry, for preventing and correcting the problems some people encounter with their mother tongue, for making the social conflicts arising from the cohabitation of peoples with different languages and cultures more bearable, or for improving – and about time too! – the methods for teaching foreign languages.

Q: And is that bad?

P: No, I don't think so. If there is no money, these people will not be able to carry on. I'd like to still be here in order to see how they accommodate to this new situation. It has always been said that the money rots good intentions and deviates fair purposes. It seems to me that it remains to be proved. Anyway, there don't exist some intentions or purposes that can be so, abstractly, qualified as good.

Q: However, on that track, everything is right in the end.

P: Yes, but everything might be right. We tend to constantly value everything. One becomes really old when one says that things are unfortunately no longer like before. Of course, today is different from before. And it's all right to be so. Do you see?, it's all right. People working on artificial intelligence, which are also concerned with language – after all, they are intelligent and know what is worth spending in the time – say that the very characteristic of human beings is their capacity to learn and to adapt themselves to new situations. I don't know. Anyway, one must always be attentive to what is happening around one. And what is happening is what is happening: there is no alternative.

Q: Yes, master. So be it.

P: What a shame I have only a little time!

Q: No!

P: I'm hungry.

Q: What a change, now when I was already caught on the tone of your dissertation!

P: Let's go, my colleague, as Sister Trinidad must be waiting for us. And let me leash you, since the ticket inspector clearly doesn't understand you.

[After a few weeks, one morning P was discovered dead on the floor of his room. Within the drawer of his bedside table, the police could find, among many small, yellowish and muddled cigarette papers, three sheets with this text, which we literally transcribe for philological and psychiatric reasons. Apparently, he didn't sleep for some time and spent the nights, among the general silence, by giving up his mind through the recesses of so big problems that he was unable even to assign them to some concrete professionals. It appeared that his linguistic interests roused his concern with the subject you'll read about, that he then revived what his mathematical training had provided him with, and that, finally, he was carried away with his almost senile liking for philosophical speculation. This attitude was surely the last remnant of a life that was zigzagging and eluding answers. Or maybe it was all just about a matter of style.]

THE INFINITY, THE PARADOX AND THE HORROR: SOME QUESTIONS FOR TWO NIGHTS REFLECTING

(1) Are the infinite, the nonfinite, the indefinite and the unlimited all different things?

(2) Is the infinite something negative?

(3) According to the Apollonian view, Greeks show horror to the infinite, since they consider the reason as being unable to understand it; opposite to it, the Faustic view emphasizes the passionate tendency towards the infinite: is the reason moving in the emptiness when dealing with the infinite?

(4) How to reconcile that Greek art rejects the infinite whereas Greek philosophy admits it as a problem?

(5) Are the notion of the infinite, the belief that there exists some infinite reality, the feeling of the infinite, the expression of the infinite, the imagination of the infinite, etc., all quite different things?

(6) Sometimes one of the features showed by the infinity is the cyclicity or the eternal return. If there is repetition, could one claim infinity not to exist?

(7) Does the infinity always show the divisibility of the continuum as a feature?

(8) Does the infinity always show the eternity as a feature?

(9) Are there things that are unlimited in themselves, for example pleasure?

(10) Does the infinity emerge – as it looks to be – from the concept of the continuum and its associated one of limit?

(11) The infinity can be: by division (nondenumerable) or by addition (denumerable). Does our belief in the infinity result from the infinity of time and the divisibility of magnitudes?

(12) Is the infinity either a substance, an attribute or an extension/quantity?

(13) Is the infinity such that either it can not be gone through or, even being able to be, it is not actually gone through?

(14) Is it true that the infinity is such that there is something further on (not that there is nothing further on)?

(15) Does the (actual) infinity exist while not being enumerated?

(16) What has the infinity to do with the infinitive?

(17) Are infinity and perfection, as definability, attached to each other always?

(18) The most resolute infinity is God. Is the divine infinity different from either the numerical infinity or the infinite divisibility of a line?

(19) Could a point be regarded as infinite insofar as it has no divisible extension?

(20) Does the modern (post-medieval) age involve a certain movement from finitism towards infinitism, or from a closed world to an open one?

(21) To what extent have the scientific revolution and the progress of mathematical thinking (especially the concept of function and the infinitesimal calculus) contributed to the infinitization of our idea of the world, and to what extent should one attribute it to philosophy, literature and art?

(22) Is it legitimate to interpret infinitist trends in Baroque art?

(23) Does Descartes' ontological argument get it right when stating the idea that one must move from the infinite (as the most objective thing) to the finite, and not backwards?

(24) Is infinitism necessarily linked to pluralism?

(25) Is it true that the infinity can not be the mere feeling of something incommensurable, but is measurable and can be calculated?

(26) Is an ascending hierarchy of infinities like the following acceptable: numerical infinity first, then infinity of the world, and lastly infinity of God?

(27) Is the notion of the infinity inseparable from the feeling of the infinity, and in particular from the paradoxes and sublime truths it gives rise to and from the admiration it inspires, insofar as it makes the imagination burn and asks for both this latter as well as the thought?

(28) Is it true that the contemplation of the eternal silence of the infinite universe frightens and terrifies humans and makes them feel lost in the middle of that amazing marvel?

(29) How do we acquire the idea of the infinity and how to justify that we accept the infinity in duration better than in extension?

(30) Does the idea of the infinity arise always from the additive concatenation of finite segments?

(31) Might one say that the infinity should only be predicated about space, time (both quantitative notions, as they are related to numbers) and God (a qualitative idea, which is connected with perfection)?

(32) Why does the feeling of the infinite occupy a central position just in Romanticism?

(33) There are those who think that every passing from the finite to the infinite is rationally illegitimate. Some others believe that the infinite is just a way of speaking. Is it true that accepting the infinite generates unsolvable paradoxes, and that from the finite to the infinite there is a jump rather than a step?

(34) Is it of some use here to refer to Jorge Luis Borges' total library?

(35) "God made the integers; everything else is a human work" (L. Kronecker). Are there infinite sets?

(36) As the idea of the infinity i) cannot come from experience, because every object from experience is finite, and ii) cannot come from the imagination, because this latter is able only to reproduce data from the senses, and this would in any case give rise to the indefinite, might it be an 'a priori' idea?

(37) From the viewpoint of generalized relativity theory, the universe is finite but unlimited, like a sphere, which is finite but can be gone through in all directions without reaching any limit. Is being infinite clearly distinct from being unlimited?

(38) What would become of mathematical analysis without both the infinitely large and the infinitely small magnitudes?

(39) It was believed for some time that finite magnitudes consisted of infinitely many infinitely small indivisible parts, which were conceived as constants smaller than any finite magnitude. Should one rather think today that finite magnitudes are composed of an unlimited increasing number of components that decrease unlimited?

(40) Both infinitely small and infinitely large magnitudes are the basis for the definition of derivative (as the ratio of infinitely small magnitudes) and integral (as the addition of infinitely many infinitely small magnitudes). Does that coincide with some primary intuition?

(41) Can we calmly accept the numbers $|\infty$ and $-\infty$ in the theory of functions of one real variable?

(42) To say that a variable is, for instance, infinitely small makes sense only if the nature of the variation is described as a function of some other variable. What is the nature of the infinite sets of mathematical objects?

(43) Is enumeratively characterizing infinite sets of mathematical objects unimaginable?

(44) An infinite game is a noncooperative game, in particular a two-player game amounting to zero and with infinite sets of strategies at the players' disposal. Is it all an entelechy?

(45) The existence of infinite objects in a formal theory is guaranteed by an infinity axiom. Dedekind's infinity axiom, specifically, assures an one-to-one correspondence to exist between the set and some of its proper subsets. Is it an entelechy again?

(46) For intuitionism, a philosophy close to nominalism and idealism, there are no other infinite sets than the denumerable ones: abstract objects are creations of the human mind. In opposition to it, stands realism or platonism. Should we stop being platonists?

(47) May one talk about God's infinite love for human beings or *your infinite look*?

(48) Is the infinity a fiction? Why (what) the horror (of it)?

Finitc and Infinite iu Geography

Georges Nicolas

15 Rue Alfred de Musset
25300 Pontarlier, France
E-mail: nicorad@fc-net.fr

Abstract. Geography regards macroscopic entities perceived and represented at different scales. Microscopic or macrocosmic entities are not in the field of geography. Two conceptions of space coexist in geography. As empirical "reality", relations between the geographical locuses-objects observed by a geographer generate a geographical space. This space is "full" and constituted by a finite set of locuses-objects. To represent these geographical spaces with "geomaps" or "maps", geographers use a "void" cartographic space formed by an infinity of points. The transition between these two types of spaces needs a definition of the geographical entity as a Cartesian product between a locus and an object, which are differentiated on the surface of the Earth.

1 Introduction

The oldest reflections of Greek thought concerned nature and the origin of the Cosmos. They included an enquiry on the shape of the Earth considered as a Whole and on the *pattern of the inhabited world* to determine the situation of its Parts, the ones in comparison with the others.

Eratosthene (-275, -193) is the inventor of the word *geographiká* (from *gê* = the Earth, and *gráphein* = to make an incision in order to write or to draw). His *Geography* included two books. The first one was a polemic about the geographical value of descriptive poetry. It also contained a project for a scientific description of the Earth. The second book included a calculation of the dimensions of the Earth (considered as a sphere) built up with angular astronomical measures and evaluations of earthly distances. Erastosthene proposed the first geo-metrical picture of the world that will later be called *chart* or *map* (*carte*) from the 16th century in the Occident. This use will definitely triumph only at the end of the 19th century: in French, the word *cartographe* (*cartographer*) appears only in 1877.

The history of geography is inseparable from history of cartography. Reciprocal assimilation of geographical space and cartographic space is a fact in the middle of the 20th century. In this conception, geography begins with the location of objects with coordinates projected on axes drawn on a plane. This know-how offers the advantage of allowing the use of an Euclidean definition of the Earth in two or three dimensions. But the Whole/Parts logic applied to the surface of the Earth, with or without the help of cartography, is also present in the non-linear continuity of the history of occidental geography.

This very long story unfolded in a context where the relation between the words *finite* and *infinite* inverted, [4]. Until the Renaissance, cartography evolve chaotically and not always so rigorously as its founders wished. It used implicitly infinite sets to represent points, lines and areas. Geography for its part remained dependant on an Aristotelian world made of spherical universes fitted together. The Earth was the most central of these universes and the most external sphere was the finite limit. From the Renaissance onward, cartography assimilates the revolution of Copernicus and Newton, which inverts the relation between the finite and infinite. Closely submitted to political powers, geography puts a central humanist vision in place of its central cosmic vision of the Earth. It has not yet completely got free of this vision nowadays.

At the middle of the 20th century, classical cartography (principally topographical) has become completely mathematical. Geographers use intensively these maps in the field and in their laboratory. They also make graphical representations, paradoxically called "geographic maps". But this activity is not accompanied by a mathematical elaboration of their objects and of their methods. The making of their maps borrows techniques of classical cartography. However scientific support of geographic cartography disintegrates rapidly in the years 50–60 when geographers try to quantify invisible facts on topographical maps. Quantitative geographers then borrow their procedures to statistics without elaborating a properly geographical object on which measures can be made.

The absence of properly geographical formalization today generates difficulties in geographical computer science (geomatics). Methods and techniques being the same, data processing images of the surface of the Earth made by statisticians, infographists and geographers are at the present time similar. Assuming that, progress has been realized by mathematical research since the beginning of the 20th century concerning the finite and infinite. To what degree do they allow one to devise an object for proper geographical measuring?

2 Spatial Geographical Differentiation

Geography concerns macroscopic entities perceived at different scales. Microscopic or macrocosmic entities are not in the field of geography.

Definition 2.1 *An entity is spatial if it is formed by the association of a locus and an object.*

Definition 2.2 *An information is geographical if it differentiates either the locus or the object, or the locus and the object together, of a spatial entity situated or localized on the surface of the Earth.*

If Λ is a finite set of locuses and O is a finite set of objects, then the Cartesian product: $P = \Lambda \times O$ is the set of ordered pairs $p = \langle \lambda, o \rangle$, where λ belongs to Λ and o belongs to O. Two pairs $p_1 = \langle \lambda_1, o_1 \rangle$ and $p_2 = \langle \lambda_2, o_2 \rangle$ are *distinct*, and we write $p_1 \neq p_2$, if there is a *differentiation* (written with a t) of at least one of their components, the locus or the object. There are then four possibilities.

a) Locus and object differentiation: $\lambda_1 \neq \lambda_2$ and $o_1 \neq o_2$.
   Assuming that the relation $\neq$ (negation of $=$) is anti-reflexive, symmetrical and not transitive, strong differentiation would be anti-reflexive, symmetrical and not transitive.

b) Locus differentiation: $\lambda_1 \neq \lambda_2$ with $o_1 = o_2$.
   Assuming the properties of the relations $=$ and $\neq$, weak locus differentiation would also be anti-reflexive, symmetrical and not transitive.

c) Object differentiation: $\lambda_1 = \lambda_2$ with $o_1 \neq o_2$.
   For the same reasons as at point b), weak object differentiation is anti-reflexive, symmetrical and not transitive.

d) Indifferentiation: $\lambda_1 = \lambda_2$ and $o_1 = o_2$.
   Assuming the properties of equality $=$, indifferentiation (or equivalence) is reflexive, symmetrical and transitive.

Placing the four possibilities from top to bottom, that is from the strongest to the weakest, we have:

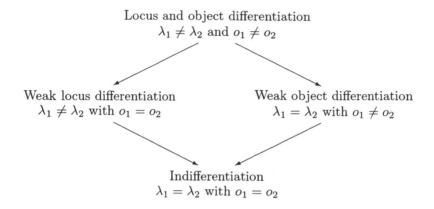

Locus and object differentiation
$\lambda_1 \neq \lambda_2$ and $o_1 \neq o_2$

Weak locus differentiation
$\lambda_1 \neq \lambda_2$ with $o_1 = o_2$

Weak object differentiation
$\lambda_1 = \lambda_2$ with $o_1 \neq o_2$

Indifferentiation
$\lambda_1 = \lambda_2$ with $o_1 = o_2$

3   Geomaps and Maps

Given the *situation S*, which is the relative position of certain geographical objects, expressed with order or non-metrical structures, a geomap M could be worked out and processed with S.

Given the localization L of geographical objects with numeric coordinates and the graphic representation C of these objects, a map C could be worked out and processed with L.

There is a duality between the pair (M, S) and the pair (C, L). Each relation between M and S involves automatically a similar relation where M can be replaced by C and S by L. This situation is analogous to the one of the De Morgan laws. On the other hand, analogy between (M, S) and (C, L) involves that M is to C what L is to S, that allows writing: $S \mid L \leftrightarrow M \mid C$. In other words, *geomappography* is to *cartography* what *situation* is to *localization*.

Scale expresses the relation between objects and their representations on a map or a geomap. The scale of a map is a connection between the dimensions of the object and the dimensions of its representation. This connection is opposite: the bigger the represented object the smaller its cartographic scale. The scale of a geomap is given by the ratio of the dimensions between represented geographical objects. This ratio is direct: the dimension and the scale of a geomap vary in the same way. The geomap scale of big geographical objects is a big scale. Inversely, the geomap scale of little geographical objects is a little scale.

Geomap represents relations between locuces-objects. It does not necessarily use localization. It can, for commodity reasons, be established on a map to evoke in the observer's mind the cartographic representation of the geographical space. But, since a geomap is not supposed to be used in the field, its design does not need to be as precise as a map. A geomap allows us to represent locus and object differentiation together ($\lambda_1 \neq \lambda_2$ and $o_1 \neq o_2$), that is, the identity of the geographical locus-object.

A map, built on localization, can express only one element of the entity locus-object. If the differentiation is generated by the locus (weak differentiation: $\lambda_1 \neq \lambda_2$), then only one sort of object ($o_1 \neq o_2$) corresponds to all localizations and for every object a map must be worked out and processed (analytical map). If, in contrast, differentiation is generated by the object (weak differentiation: $o_1 \neq o_2$), then all locuses are equivalent ($\lambda_1 = \lambda_2$) and it is possible to represent on the same map many objects in every localization (synthetic map). Finally, if neither the locus nor the object are differentiated, it is impossible to make a geomap or a map.

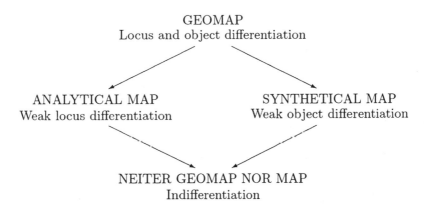

GEOMAP
Locus and object differentiation

ANALYTICAL MAP
Weak locus differentiation

SYNTHETICAL MAP
Weak object differentiation

NEITER GEOMAP NOR MAP
Indifferentiation

4 Finite and Infinite in Geography

The differentiate character of geographical entities is not a mathematical problem. It is however necessary to understand what its mathematical properties consist of. Difference indicates the subject's exteriority in comparison with reality. It also characterizes the subject's capacity to enter in the relations with reality and to recognize, to internalize, to incorporate and to transform it into an object. Difference generates identity because it allows us to distinguish what belongs to an object, to an individual, to a sensation or to an idea.

Historically, geographers use difference to know the Earth as a Whole. *Earth differentiation* is therefore the characteristic of geographical reality discovered by the geographer as a subject thanks to the cognitive mechanism of difference. But geographers cannot go beyond differentiation of the Earth as a Whole. The earthly Whole can be divided in Parts and every Part can be equal to a new Whole. But it is impossible to equal a Part to the Earth considered as a Whole. The earthly Whole is therefore at the start of comparison with other geographical Wholes.

Since there is no object without locus on the surface of the Earth, *geographical differentiation* regards all pairs formed by a locus and an object. Geographical differentiation as a process shows the two historical and logical ways followed to do geography. The first way is thinking on the entities formed by locuses and objects, which are simultaneously geographically differentiated on the surface of the Earth (strong differentiation). The second way is thinking on entities situated on the surface of the Earth whose locuses (but not objects) or objects (but not locuses) are geographically differentiated (weak differentiation). Finally, the last way, distinct from the two others, the way of undifferentiated entities on the surface of the Earth, concerns non-geographical spatial sciences (economy, geometry, psychology, topology, etc.).

Research on number theory by Georg Cantor (1845-1918) proved that there are many kinds of finites and infinites [1]. Cartographic space is evidently constituted by an infinity of points (points, lines and areas). This infinity allows two things. First, it makes possible the localization of locuses or objects with the help of coordinates projected on a system of axes on a plane, which is a blank space (without physical objects). Secondly, this non-enumerable infinite allows calculating localizations with precision. On the other hand, geographical space generated by locuses-objects relations has two limits: the empty set and the Earth considered as a Whole. The relation between the geographer subject and the locus-object "Earth" sets up the geographical objects. The finite number of geographical locuses-objects is an ordered set generated by this relation.

Consequently, there is not one single geographical space but as many spaces as there are possible ways of connecting locuses-objects. These geographical spaces are not "void" but "full". Their cardinal is finite and their limits are the empty set ($\emptyset$) and the Earth considered as a Whole.

GEOMAP
Locus and object differentiation
Space resulting from the relation between a finite number
of locuses-objects ordered by the geographers.
This space has two limits, ∅ and the Earth as a Whole.
The cardinality of this "full" space of locuses-objects is a finite set.

ANALYTICAL MAP
Weak locus differentiation

SYNTHETICAL MAP
Weak object differentiation

Representation of locuses or objects can be enumerated
and ordered in a "void" cartographic space.
This cartographic space is constituted of non-denumerable
sets of points (points, lines or areas)

NEITHER GEOMAP NOR MAP
Indifferentiation
"Void" space constituted by a non-denumerable set of points

References

[1] S. Baruk, *Dictionnaire de mathématiques élémentaires*, Paris, Seuil, 1992-1995.

[2] A. Bouvier, M. George, F. Le Lionnais, *Dictionnaire des mathématiques*, Paris, PUF, 1993.

[3] G. Nicolas, S. Marcus, Logique Tout/Partie, *Actes du Colloque IRI*, Sion, IKB-Eratosthéne, to be released in September 1999; diffusion: http://www.ikb.vsnet.ch/era3.html.

[4] J. P. Osier, Infini, *Encyclopédie philosophique universelle, Les notions philosophiques*, Paris, PUF, 1990, vol. 1, 1289–1291.

Ultrafilters, Dictators, and Gods[1]

Piergiorgio Odifreddi

Dipartimento di Informatica, Universita di Torino
Corso Svizzera 185, 10149 Torino, Italy
E-mail: piergior@di.unito.it

Abstract. We will show how a well-known and elementary fact about ultrafilters can be reinterpreted both politically and theologically, thus providing simple proofs of two central results in these areas. Mathematicians can consider such formulations as the true essence of such results. Outsiders can find comfort in a saying by Goethe, who once said (*Maxime und Reflectionen*, 1729): "Mathematicians are like Frenchmen: as soon as you tell them something, they translate it into their own language, and it immediately appears different".

1 Ultrafilters

In 1937 Henri Cartan introduced the following notions. Given a set A of elements, let $\mathcal{A}$ be a subset of the power set of A, i.e., a set of subsets of A. We use small letters such as x and y to refer to elements of A, and capital letters such as X and Y to refer to subsets of A. Let us consider the following possible properties of $\mathcal{A}$:

1. $\mathcal{A}$ is upward closed with respect to inclusion, i.e., if $X \in \mathcal{A}$ and $X \subseteq Y$, then $Y \in \mathcal{A}$;

---

[1]Solomon Marcus has often stressed the broad cultural interest of mathematics, writing extensively on its connections with apparently distant areas such as poetics and theology. A few years ago he published, together with Cristian Calude and Doru Ştefănescu, an interesting paper with an unusual title, The Creator versus its creation. From Scotus to Gödel, *Collegium Logicum, Annals of the Kurt-Gödel-Society*, Vol. 3, Institute of Computer Science, AS CR Prague, Vienna, 1999, 1-10, in which he was kind enough to quote a manuscript of ours on the mathematical modelling of God. His *Festschrift* seems a perfectly suited occasion not only to publish those observations, but also to dedicate them to him.

2. $\mathcal{A}$ is closed with respect to intersection, i.e., if $X \in \mathcal{A}$ and $Y \in \mathcal{A}$, then $X \cap Y \in \mathcal{A}$;

3. $\mathcal{A}$ does not contain all subsets of A, i.e., there exists at least an X which is not in $\mathcal{A}$;

4. given any subset X of $\mathcal{A}$, either it or its complement belongs to $\mathcal{A}$, i.e., $X \in \mathcal{A}$ or $\overline{X} \in \mathcal{A}$.

$\mathcal{A}$ is called a *filter* on A if it satisfies Properties 1 and 2. The idea is that $\mathcal{A}$ consists of "large" subsets of A, and Properties 1 and 2 are obviously necessary: if a set is larger than a large one, it is *a fortiori* large; and if two sets are large, their intersection should still be large. The following are examples of filters:

- The subsets of A containing a given element x. This is called the *principal filter* generated by x.

- The cofinite subsets of A, i.e., the subsets containing all elements of A with at most finitely many exceptions. Such a filter is not principal because, given any element x, the set $A - \{x\}$ obtained by taking x out of A is cofinite, and hence it belongs to the given filter, but it does not contain x, and hence it does not belong to the principal filter generated by x.

Properties 1 and 2 are not sufficient to forbid $\mathcal{A}$ to contain also "small" subsets of A. For example, nothing forbids $\mathcal{A}$ to contain *all* subsets of A. Property 3 is introduced to avoid this extreme possibility, and a filter that satisfies it is called a *proper filter*. Because of Property 1, *a filter is proper if and only if it does not contain the empty set $\emptyset$*, because the latter is contained in every subset of A.

Property 4 is, in a sense, opposite to 3: while the latter forbids the filter to be too large, the former forbids it to be too small, and a proper filter satisfying it is called an *ultrafilter*. Going back to the previous examples, we note that:

- A principal filter is an ultrafilter. Indeed, if it is generated by x, then either x is in X, or it is in $\overline{X}$.

- The cofinite subsets of A are not an ultrafilter. Indeed, if A is finite, then they form a filter which is not proper. And if A is infinite, there is a subset X of A such that both X and $\overline{X}$ are infinite, and thus neither X nor $\overline{X}$ is cofinite.

The fact we alluded at above is the following.

Theorem 1.1 *If A is finite, then every ultrafilter on A is a principal filter.*

Proof. Let $\mathcal{U}$ be an ultrafilter on a finite set A, and I be the intersection of all elements of $\mathcal{U}$.

Since A has only a finite number of elements, it also has a finite number of subsets, and $\mathcal{U}$ is finite. Thus I is obtained by a finite number of intersections of elements of $\mathcal{U}$, and it belongs to $\mathcal{U}$ by Property 2.

By Property 3, I cannot be empty. Thus, it contains an element x. By Property 4, either $\{x\}$ or $\overline{\{x\}}$ is in $\mathcal{U}$. But it is impossible that $\overline{\{x\}}$ be in $\mathcal{U}$, since it does not contain x, which is in I (and hence in all elements of $\mathcal{U}$). Then $\{x\}$ is in $\mathcal{U}$, and $I = \{x\}$.

It follows that $\mathcal{U}$ is the principal filter generated by x. Indeed, on the one hand, every element of $\mathcal{U}$ contains x, by definition of I. On the other hand, if x belongs to X as an element, then X contains $\{x\}$ as a subset, and thus it is in $\mathcal{U}$ by Property 1. □

Corollary 1.1 *If an ultrafilter on A contains the intersection of all its elements, then it is a principal filter.*

Proof. The finiteness assumption in the previous theorem is used only to deduce that the intersection of all elements of the ultrafilter is still in the ultrafilter. ⊔

We note in passing that the results just stated are not trivially true, since on every infinite set A there are ultrafilters that are not principal (for example, any ultrafilter containing the cofinite sets).

2   Dictators

In 1785 Jean Antoine Nicolas Marie de Caritat, better known as the Marquise of Condorcet, discovered the following paradox of the majority voting system. Consider three voters 1, 2 and 3, that have to choose among the alternatives A, B and C. Suppose they have the following cyclic orders of preferences:

$$
\begin{array}{cccc}
1: & A & B & C \\
2: & B & C & A \\
3: & C & A & B,
\end{array}
$$

to be read in the following way: 1 prefers A to B and B to C, 2 prefers B to C and C to A, and 3 prefers C to A and A to B. When the alternatives are put to a vote two by two, A wins over B by two votes (those of 1 and 3) to one (that of 2), and similarly B wins over C by two votes (1 and 2) to one (3). One could thus think that A should win over C, while the opposite happens, and C wins over A by two votes (2 and 3) to one (1).

The paradoxical aspect is that, while the individual preferences of the voters are linearly ordered, the social order that majority vote produces becomes circular.

In 1951 Kenneth Arrow asked himself whether one could find a voting system that would preserve the linear order of individual preferences. Given a set of voters, let us call a subset X of it *decisive* if, given two alternatives, one of them wins when all the elements of X vote for it, and all the elements of $\overline{X}$ vote for the other.[2]

The notion of a decisive set makes sense under the hypothesis of *vote dependence*, i.e., when the choice of the winner between two alternatives depends only on the votes they receive according to the individual preference orders, but not on other factors dependent on the alternatives.

Arrow's basic assumption is the following:

Axiom 2.1 *The decisive sets form an ultrafilter on the set of voters.*

This assumption can be justified as follows. Property 1 says that if a set of voters is decisive, then so is any set containing it. In other words, if an alternative wins when it receives a certain number of votes, it continues to win if it receives more votes (*monotonicity*).

Property 3 says that not every set is decisive, which is obvious if at most one alternative has to win.

Property 4 says that either a set or its complement are decisive, which is obvious if at least one of the two alternatives has to win, i.e., if there has to be no tie.

The following is thus the essential justification of Arrow's assumption.

Proposition 2.1 *Property 2, i.e., the fact that the intersection of two decisive sets is decisive, is equivalent to the fact that the social order is linear.*

Proof. Suppose X and Y are decisive, but $X \cap Y$ is not. Then $\overline{X \cap Y}$ is decisive by Property 4. Consider the following orders of preferences,

---

[2]In a majority vote, the decisive sets are those containing at least one half of the voters plus one. They satisfy Properties 1, 3 and 4, the latter under the hypothesis that the set of voters has an odd number of elements, but not Property 2, by Proposition 2.1 and Condorcet's Paradox.

and note the analogy with the system of preferences used in Condorcet's Paradox:

$$
\begin{array}{llll}
X \cap Y : & A & B & C \\
Y - X : & B & C & A \\
X - Y : & C & A & B \\
\overline{X \cup Y} : & C & B & A,
\end{array}
$$

to be read as follows: every voter on the left set has the order of preference indicated on the right. Then A wins over B because every element of X prefers A to B, every element of $\overline{X}$ prefers B to A, and X is decisive. Similarly, B wins over C because Y is decisive, and C wins over A because $\overline{X \cap Y}$ is decisive. Thus the social order is not linear.

Conversely, given any system of individual preference orders, consider the sets X, Y and Z of voters preferring, respectively, A to B, B to C and A to C. If A wins over B, then X must be decisive, otherwise $\overline{X}$ would be decisive by Property 4, and B would win over A. Similarly, if B wins over C, then Y must be decisive. If the intersection of two decisive sets is decisive, then $X \cap Y$ is decisive. Since every element of it prefers A to B (being in X) and B to C (being in Y), and hence A to C (because the linear orders are linear), it follows that $X \cap Y \subseteq Z$. By Property 4, then Z is decisive and A wins over C. Thus the social order is linear. $\square$

Note that the previous proof uses what Arrow calls the hypothesis of *individual freedom*: every possible combination of individual preference orders is admissible.

Now, since the set of voters is obviously finite in any application, we can use Theorem 1.1 and get a version of Arrow's result.

Theorem 2.1 *The decisive sets form a principal filter.*

In other words, there exists a voter whose vote alone determines the result of any election. Arrow calls such a voter a *dictator*, and deduces from this that it is impossible to find a democratic voting system satisfying the minimal conditions used above, in particular vote dependence, monotonicity and individual freedom. Thus every possible system produces a dictatorship, and democracy is impossible. For this result Arrow obtained the Nobel Prize for Economics in 1972.

3 Gods

In 1077 Anselm of Aosta discovered the following *ontological proof* of the existence of God: if we define God as a being with all perfections, then

God exists because existence is a perfection.

The unsatisfactory aspects of this proof are that, on the one hand, nothing ensures that the definition is not contradictory, and on the other hand, that existence cannot be considered as a property, and hence as a perfection.

In 1970 Kurt Gödel asked himself whether one could formalize the ontological proof, thus making it logically more acceptable. The idea is to replace perfections by *positive properties*, intended as particular subsets of the set of elements constituting the world.[3]

Gödel's assumption is the following:

Axiom 3.1 *The positive properties form an ultrafilter on the world.*

More explicitly, one assumes that a property that contains a positive property is positive, the intersection of two positive properties is positive, a positive property is non-empty, and the complement of a positive property is not positive.[4]

Such assumptions can be justified by analogy, i.e., by noticing that a number larger than a positive number is positive, the product of two positive number is positive, a positive number is not zero, and the opposite of a non-positive number is positive.

If we make the further assumption of *finiteness of the world*, we can apply Theorem 1.1 and get a version of Gödel's result.

Theorem 3.1 *The positive properties form a principal filter.*

In other words, there exists an object determined by the positive properties. Gödel calls such an object *God*, and deduces from this the existence of God, on the basis of the minimal conditions used in the justification of the axiom that the positive properties form an ultrafilter, *and* on the basis of the finiteness of the world.

To eliminate the latter unsatisfying assumption, Gödel replaces it by an additional axiom which, probably, Dostoevsky would not have accepted.

---

[3]A property can be identified with the set of elements satisfying it, and a set can be identified with the property of belonging to it. Then logical implication, conjunction and negation correspond to set theoretical inclusion, intersection and complement.

[4]Notice that the set of positive properties, being a filter, is closed with respect to finite intersection and arbitrary union: in particular, it forms a *divine topology* on the universe. Since the filter is proper, this topology is connected and not separable. More information on it can be found in the paper by Calude, Marcus and Ştefănescu quoted at the beginning.

Axiom 3.2 *Being God is a positive property.*

Since the property of being God is, by definition, the intersection of all positive properties, the effect of this axiom is to make such an intersection an element of the ultrafilter of all positive properties. By Corollary 1.1 such an ultrafilter is then principal, and the existence of God now follows as above.

4 Conclusion

Theorem 1.1 has shown that there is a connection between the existence of dictators and gods. We imagine that it would be difficult to make such a connection more explicit, if one excludes trivialities such as: God can be considered as a dictator, or dictators like to be considered as gods. However, as Umberto Eco has taught us, on what we cannot theorize, we must tell a story. We thus end with an anecdote.

In 1948 Gödel had to take the examination to become an American citizen. He studied the Constitution, and found out that in it there was a logical-legal possibility of turning the United States into a dictatorship. In the car trip to the ceremony, he communicated the discovery to Einstein, who was his witness. The latter tried to persuade Gödel that the ceremony would not have been the best time to talk about this, but by chance the judge noticed that Gödel was of Austrian citizenship, and referring to the *Anschluss* told him that in the United States he should not fear anymore the advent of a dictator. Einstein had to jump in, since Gödel had immediately seized the chance and started a communication of his most recent discovery.

We do not know which proof of the possibility of a dictatorship in the United States Gödel had found. But the developments reported in this note allow the reader to formulate an obvious conjecture.

On the Behaviour of Semiautomata

Sergiu Rudeanu

Faculty of Mathematics, University of Bucharest
Str. Academiei 14, 70109 Bucureşti, Romania
E-mail: rud@funinf.math.unibuc.ro

Abstract. It has been known for a long time that the Moore automaton constructed by Nerode is minimal, i.e., it has the smallest number of states among all the automata with the same behaviour. A theorem due to Goguen has revealed that the deep background of this property is the adjunction of two functors between the category of reachable Moore automata and the category of their behaviours. In this paper we introduce the category of F-automata and the category of F-behaviours and prove a Goguen-like theorem which includes as particular cases both the original Goguen theorem and a similar theorem for Mealy automata. We also characterize those functions which occur as behaviours of semiautomata and prove that the category of reachable semiautomata is isomorphic to the category of their behaviours.

1 Introduction

The starting point of this paper is a theorem due to Goguen [4], which establishes a pair of adjoint functors between the category of reachable Moore automata and the category of surjective behaviours $f : X^* \to Y$. Since semiautomata can be viewed as particular cases of Moore automata, it is natural to ask which behaviours can be realized by reachable semiautomata. Another natural question is whether the Goguen theorem can be extended to Mealy automata.

In Section 2 we prove that a surjection $f : X^* \to Y$ can be realized by a reachable semiautomata if and only if $\ker f$ is right invariant. In Section 3 we prove a theorem which includes as particular cases the Goguen theorem mentioned above and a quite similar result for Mealy automata. In Section 4 we deal with the particular case of semiautomata and prove the sharper result that the category of reachable semiautomata is isomorphic to the category of their behaviours.

All necessary automata-theoretical prerequisites will be recalled; cf. [1], [2], [3], [5].

2 Behaviours Realized by Semiautomata

In the following we use the following standard notation: A^* is the monoid of all the words over the alphabet A, λ denotes the empty word, $A^+ = A^* - \{\lambda\}$, while $\alpha^* : A^* \to A'^*$ stands for the monoid homomorphism which extends a map $\alpha : A \to A'$.

A *semiautomaton* is an algebra $\Sigma = (S, I; \delta, s_0)$, where $\delta : S \times I \to S$ and $s_0 \in S$ are known as the *transition function* or *next-state function* and the *initial state* or *central state* or *source*, respectively.

The same letter δ is used for the function $\delta : S \times I^* \to S$ which extends the transition function and satisfies

(1.1) $\delta(s, \lambda) = s, \ \forall s \in S$
(1.2) $\delta(s, w_1 w_2) = \delta(\delta(s, w_1), w_2), \ \forall s \in S, \ \forall w_1, w_2 \in I^*.$

The *reachability function* $\delta^0 : I^* \to S$ is defined by

(2) $\delta^0(w) = \delta(s_0, w), \ \forall w \in I^*,$

and the semiautomaton is said to be *reachable* provided δ^0 is surjective.

A *Moore automaton* (also called a *(sequential) machine* in [1], [3], [4]) is an algebra $A = (S, I, O; \delta, \mu, s_0)$ where $\Sigma(A) = (S, I; \delta, s_0)$ is a semiautomaton and $\mu : S \to O$ is known as the *output function*. The automaton is said to be *reachable* provided its *semiautomaton reduct* $\Sigma(A)$ is so.

The function $\beta : I^* \to O$ defined by

(3) $\beta(w) = \mu(\delta^0(w)), \ \forall w \in I^*$

is called the *behaviour* of A ; we also say that the automaton A *realizes* the function β.

This algebraic model represents a discrete-time technical device which has the set S of *internal states*, receives *input signals* from the set I and produces *output signals* belonging to the set O. The operation of the device obeys the law

(4) $s_{n+1} = \delta(s_n, i_n), \quad o_n = \mu(s_n),$

where x_k denotes the value of $x \in X$ at time $t := k \in \mathbf{N}$. So the internal state at time $t := 0$ is s_0 and if the device receives the sequence of inputs $i_0, i_1, \ldots, i_{n-1}$, then at time $t := n$ it will assume the state $\delta^0(w)$ and produce the output signal $\beta(w)$, where $w = i_0 i_1 \ldots i_{n-1} \in I^*$. Note that $\delta^0(\lambda) = s_0$ and $\beta(\lambda) = o_0$.

Now remark that a semiautomaton $\Sigma = (S, I; \delta, s_0)$ can be identified with the Moore automaton $A(\Sigma) = (S, I, S; \delta, 1_S, s_0)$, where 1_S is the identity map of the set S. The behaviour of $A(\Sigma)$ is simply δ^0. Having in view this remark it is natural to rename the reachability function δ^0 as the *behaviour* of the semiautomaton Σ and to say that Σ *realizes* the function δ^0. Proposition 2.1 below provides the answer to the following natural question: what are the functions $f : X^* \to Y$ that can be realized by reachable semiautomata?

Recall first that $\ker f$ is the equivalence defined on the domain of the function f by $x \ker f\, y \Leftrightarrow f(x) = f(y)$ and that an equivalence $\equiv$ on a monoid X^* is called *right invariant* provided $w_1 \equiv w_2 \Rightarrow w_1 w \equiv w_2 w$.

Proposition 2.1 *A function $f : X^* \to Y$ is realized by a reachable semiautomaton if and only if f is surjective and $\ker f$ is right invariant.*

Proof. If the function f satisfies the above conditions, then the semiautomaton

$$(5.1) \qquad SNf = (Y, X, \sigma_f, f(\lambda)),$$
$$(5.2) \qquad \sigma_f : Y \times X \to Y, \, \sigma_f(f(w), x) = f(wx), \, \forall w \in X^*, \forall x \in X,$$

is well defined.

Further we show that the extension $\sigma_f : Y \times X^* \to Y$ of σ_f is given by

$$(6) \qquad \sigma_f(f(w), w') = f(ww'), \, \forall w, w' \in X^*.$$

The proof is by induction on w'. For $w' := \lambda$ relation (6) holds by (1.1), while (6) implies, via (1.2) and (5.2), that for every $x \in X$,
$$\sigma_f(f(w), w'x) = \sigma_f(\sigma_f(f(w), w'), x) = \sigma_f(f(ww'), x) = f(ww'x).$$
Now (6) implies
$$(\sigma_f)^0(w) = \sigma_f(f(\lambda), w) = f(\lambda w) = f(w),$$
that is

$$(7) \qquad (\sigma_f)^0 = f.$$

Conversely, if Σ is a reachable semiautomaton, then δ^0 is surjective and $\ker \delta^0$ is known to be right invariant (the easy proof uses (2) and (1.2)). $\square$

The deep background of Proposition 2.1 will be pointed out in Section 4; cf. Section 5.

3   Mealy Automata and a Generalization of Goguen's Theorem

The definition of a *Mealy automaton* $A = (S, I, O; \delta, \mu, s_0)$ is similar to that of a Moore automaton, with the difference that we have now $\mu : S \times I \to O$.

The technical meaning of this concept is similar to that of a Moore automaton; in particular $o_n = \mu(s_n, i_n)$. So in this case it is natural to define the *behaviour* of the automaton as the function $\beta : I^+ \to O$ given by

$$(8) \qquad \beta(wi) = \mu(\delta^0(w), i), \ \forall w \in I^*, \ \forall i \in I,$$

whose meaning is the same as for the behaviour of a Moore automaton. Note that the initial output o_0 of a Mealy automaton is undetermined.

In this section we generalize the Goguen theorem mentioned in the Introduction in such a way as to include a similar result for Mealy automata; as a matter of fact, the proof itself will be quite similar to the original one.

Let $F : \textbf{Set} \to \textbf{Set}$ be a functor which preserves surjections. By an *F-automaton* we mean an algebra $A = (S, I, O; \delta, \mu, s_0)$, where $\Sigma A = (S, I; \delta, s_0)$ is a semiautomaton and $\mu : S \times FI \to O$ is called the *output function*. The automaton A is said to be *reachable* if ΣA is so. The *behaviour* of A is the function

$$(9) \qquad \beta : I^* \times FI \to O, \ \beta(w, j) = \mu(\delta^0(w), j), \ \forall w \in I^*, \ \forall j \in FI.$$

In the case when F is the constant functor $FI = \{\bullet\}$, where $\{\bullet\}$ is a fixed singleton, and $Ff = 1_{\{\bullet\}}$, we identify $I^* \times \{\bullet\}$ with I^* and the behaviour (9) with the behaviour (3), so that we obtain the concept of a Moore automaton. Taking $F := Id_{\textbf{Set}} =$ the identity functor of $\textbf{Set}$, we have $I^* \times FI = I^+$ and we identify the behaviour (9) with the behaviour (8), so that we obtain the concept of a Mealy automaton.

We define the category $F\textbf{Aut}$ of *F-automata* in the spirit of universal algebra: the objects of $F\textbf{Aut}$ are the F-automata, while the morphisms from $A = (S, I, O; \delta, \mu, s_0)$ to $A' = (S', I', O'; \delta', \mu', s_0')$ are the triples (a, b, c) where $a : S \to S'$, $b : I \to I'$ and $c : O \to O'$ satisfy the identities:

$$(10.1) \qquad a(\delta(s, i)) = \delta'(a(s), b(i)), \ \forall s \in S, \ \forall i \in I,$$
$$(10.2) \qquad c(\mu(s, j)) = \mu'(a(s), Fb(j)), \ \forall s \in S, \ \forall j \in FI,$$
$$(10.3) \qquad a(s_0) = s_0',$$

the composition of morphisms being defined componentwise.

It is easy to check, using the fact that F is a functor, that $F\textbf{Aut}$ is actually a category.

Lemma 3.1 *Properties* (10.1) *and* (10.3) *imply*

(11) $\qquad a(\delta(s, w)) = \delta'(a(s), b^*(w)), \forall s \in S, \forall w \in I^*,$

(12) $\qquad a(\delta^0(w)) = \delta'^0(b^*(w)), \forall w \in I^*,$

and if (10.2) *holds also, then*

(13) $\qquad c(\beta(w, j)) = \beta'(b^*(w), Fb(j)), \forall w \in I^*, \forall j \in FI.$

Proof. Identities (11) and (12) are well known: (11) is established by induction on w, via (1.1), (10.1) and (1.2), while (12) is obtained from (11) written for $s := s_0$ and taking into account (10.3). To prove (13) we use (9), (12) and (10.2):

$$
\begin{aligned}
\beta'(b^*(w), Fb(j)) &= \mu'(\delta'^0(b^*(w)), Fb(j)) \\
&= \mu'(a(\delta^0(w)), Fb(j)) \\
&= c(\mu(\delta^0(w), j)) \\
&= c(\beta(w, j)).
\end{aligned}
$$

$\qquad\qquad\qquad\qquad\qquad\qquad\qquad\qquad\qquad\qquad\qquad\qquad\qquad\qquad\qquad\qquad\square$

Let further $FBeh$ be the category of *F-behaviours*, whose objects are the functions of the form $f : X^* \times FX \to Y$, while the morphisms from f to $f' : X'^* \times FX' \to Y'$ are the pairs (b, c) where $b : X \to X'$ and $c : Y \to Y'$ satisfy the identity

(14) $\qquad c(f(w, z)) = f'(b^*(w), Fb(z)), \forall w \in X^*, \forall z \in FX,$

the composition of morphisms being again defined componentwise.

It is easy to check, using the fact that $*$ and F are functors, that $FBeh$ is actually a category.

Now let FA be the subcategory of $FAut$ which consists of those F-automata that are reachable and of those morphisms (a, b, c) for which b is surjective. Let also FB be the subcategory of $FBeh$ which consists of the same objects $f : X^* \times FX \to Y$ and of those morphisms (b, c) between them for which b is surjective.

The functor *external behaviour*

(15) $\qquad\qquad E : FA \to FB$

is defined by $EA = \beta = $ the behaviour (9) of A, and $E(a, b, c) = (b, c)$. (As a matter of fact the morphism sets of a category have to be disjoint, so that the morphisms of $FAut$ are in fact of the form $((a, b, c), A, A')$ and accordingly the latter definition should read $E((a, b, c), A, A') = ((b, c), EA, EA')$. However we conform to common practice and use a less sophisticated notation.)

Since the behaviour of an F-automaton is an F-behaviour by Lemma 3.1, (13), it is plain that E is actually a functor (as a matter of fact we could have defined $E : F\mathbf{Aut} \to F\mathbf{Beh}$). We are now going to construct a functor working in the opposite direction, from $F\mathbf{B}$ to $F\mathbf{A}$.

We associate with each function $f : X^* \times FX \to Y$ the equivalences $\ker_1 f$ and $\sim_f$ on X^* defined as follows: $\forall w_1, w_2 \in X^*$,

(16) $w_1 \ker_1 f \, w_2 \Leftrightarrow f(w_1, z) = f(w_2, z), \, \forall z \in FX,$
(17) $w_1 \sim_f w_2 \Leftrightarrow f(w_1 w, z) = f(w_2 w, z), \, \forall w \in X^*, \, \forall z \in FX.$

Lemma 3.2 $\sim_f$ *is the greatest right invariant equivalence relation included in* $\ker_1 f$.

Proof. It is routine to check that $\sim_f$ is a right invariant equivalence relation included in $\ker_1 f$. If $\equiv$ is another relation with these properties then $w_1 \equiv w_2 \Rightarrow w_1 w \equiv w_2, \, \forall w \Rightarrow w_1 w \ker_1 f \, w_2 w, \, \forall w \Leftrightarrow f(w_1 w, z) = f(w_2 w, z), \, \forall z, \, \forall w \Leftrightarrow f(w_1 w, z) = f(w_2 w, z), \, \forall w, \, \forall z \Leftrightarrow w_1 \sim_f w_2.$ $\quad\square$

Corollary 3.1 *For every F-automaton,* $\ker \delta^0 \subseteq \sim_\beta \subseteq \ker_1 \beta$.

Proof. Recall that $\ker \delta^0$ is right invariant, note that $\ker \delta^0 \subseteq \ker_1 \beta$ by (9) and apply Lemma 3.2. $\quad\square$

For every $w \in X^*$, let $[w]_f$ denote the coset of w modulo $\sim_f$.

Proposition 3.1 *Let $f : X^* \times FX \to Y$. Then:*
(i) *The following construction is a reachable F-automaton:*

(18) $Nf = (X^*/\sim_f, X, Y; \delta_f, \mu_f, [\lambda]_f),$
(19) $\delta_f : (X^*/\sim_f) \times X \to X^*/\sim_f, \, \delta_f([w]_f, x) = [wx]_f, \, \forall w \in X^*,$
         $\forall x \in X,$
(20) $\mu_f : (X^*/\sim_f) \times FX \to Y, \, \mu_f([w]_f, z) = f(w, z), \, \forall w \in X^*,$
         $\forall z \in FX;$
(ii) $\delta_f([w]_f, w') = [ww']_f, \, \forall w, w' \in X^*;$
(iii) $(\delta_f)^0 = nat \sim_f;$
(iv) $ENf = f.$

Comment. Nf will be called the *Nerode F-automaton* associated with f.

Proof. If $[w_1]_f = [w_2]_f$ then $w_1 x \sim_f w_2 x$ and $f(w_1, z) = f(w_2, z)$ by Lemma 3.2, therefore Nf is well defined.

Property (ii) is easily proved by induction on w'.

From (ii) we obtain $(\delta_f)^0(w) = \delta_f([\lambda]_f, w) = [w]_f$, that is (iii), which completes the proof of (i) as well.

Finally from (9)(iii) and (20) we infer

$$ENf(w,z) = \mu_f((\delta_f)^0(w), z) = \mu_f([w]_f, z) = f(w,z). \qquad \square$$

Lemma 3.3 *Let* $A = (S, I, O; \delta, \mu, s_0)$ *be a reachable F-automaton,* $f : X^* \times FX \to Y$ *and* $(b,c) \in FB(EA, f)$. *Then there is a unique function* $a : S \to X^*/\sim_f$ *such that* $(a, b, c) \in FA(A, Nf)$, *namely* $a(\delta^0(w)) = [b^*(w)]_f,$ $\forall w \in I^*$.

Proof. Since A is reachable, the elements of S are actually of the form $\delta^0(w)$ and in order to prove that a is well defined it remains to show that if $\delta^0(w_1) = \delta^0(w_2)$ then $b^*(w_1) \sim_f b^*(w_2)$, that is $f(b^*(w_1)w', z) = f(b^*(w_2)w', z)$ for every $w' \in X^*$ and every $z \in FX$. Recall that $EA = \beta : I^* \times FI \to O$. Since $b : I \to X$ is a surjection, so is b^* and since F preserves surjections it follows that Fb is a surjection. Therefore $w' = b^*(w)$ for some $w \in I^*$ and $z = Fb(j)$ for some $j \in FI$. Taking into account the homomorphism condition (14) written for $f := EA = \beta$ and $f' := f$, then (9) and the right invariance of $\ker\delta^0$, we obtain

$$\begin{aligned}
f(b^*(w_1)w', z) &= f(b^*(w_1)b^*(w), Fb(j)) \\
&= f(b^*(w_1 w), I'b(j)) = c(\beta(w_1 w, j)) \\
&= c(\mu(\delta^0(w_1 w), j)) = c(\mu(\delta^0(w_2 w), j)) = \cdots \\
&= f(b^*(w_2)w', z).
\end{aligned}$$

We prove similarly the homomorphism conditions (10) for $\delta' := \delta_f$, $\mu' := \mu_f$ and $s := \delta^0(w)$:

$$\begin{aligned}
a(\delta(\delta^0(w), i)) &= a(\delta(\delta(s_0, w), i)) = a(\delta(s_0, wi)) \\
&= a(\delta^0(wi)) = [b^*(wi)]_f \\
&= [b^*(w)b(i)]_f = \delta_f([b^*(w)]_f, b(i)) \\
&= \delta_f(a(\delta^0(w)), b(i)), \\
c(\mu(\delta^0(w), j)) &= c(\beta(w, j)) = f(b^*(w), Fb(j)) \\
&= \mu_f([b^*(w)]_f, Fb(j)) \\
&= \mu_f(a(\delta^0(w)), Fb(j)), \\
a(s_0) &= a(\delta(s_0, \lambda)) = a(\delta^0(\lambda)) = [b^*(\lambda)]_f = [\lambda]_f.
\end{aligned}$$

To prove uniqueness, suppose that $(a', b, c) \in FA(A, Nf)$ for some $a' : S \to X^*/ \sim_f$. Then Lemma 3.1 and Proposition 3.1 (iii) imply

$$a'(\delta^0(w)) = (\delta_f)^0(b^*(w)) = [b^*(w)]_f. \qquad \square$$

Proposition 3.2 *The following construction defines a functor $N : FB \to FA$. Take an object f of FB to the Nerode F-automaton Nf and if $(b, c) \in FB(f, f')$ define $N(b, c) = (a, b, c)$, where*

(21) $\qquad a : X^*/ \sim_f \to X'^*/ \sim_{f'},\ a([w]_f) = [b^*(w)]_{f'},\ \forall w \in X^*.$

Comment. N will be called the *Nerode functor.*

Proof. We have seen in Proposition 3.1 that Nf is an object of FA. Further let $(b, c) \in FB(f, f')$.

Since $ENf = f$ by Proposition 3.1 (iv), we have $(b, c) \in FB(ENf, f')$ and we can apply Lemma 3.3 with $A := Nf$ and $f := f'$, hence $S := X^*/ \sim_f$ and $X := X'$. So there is a unique function $\alpha : X^*/ \sim_f \to X'^*/ \sim_{f'}$ such that $(\alpha, b, c) \in FA(Nf, Nf')$, namely $\alpha((\delta_f)^0(w)) = [b^*(w)]_{f'}$. But $(\delta_f)^0(w) = [w]_f$ by Proposition 3.1 (iii), so that $\alpha = a$, therefore a is well defined and $(a, b, c) \in FA(Nf, Nf')$. Checking the functorial conditions for N is routine. $\qquad \square$

Theorem 3.1 (E, N) *is a pair of adjoint functors between the categories FA and FB.*

Proof. It follows from Proposition 3.1 (iv) and $EN(b, c) = E(a, b, c) = (b, c)$, that $EN = Id_{FB}$. This enables us to prove the desired conclusion in the following form: for every object f of FB, the pair $(ENf, 1_f)$, where $1_f : ENf \to f$ is the identity morphism, is a final object in the category E/f. To be specific, this claim means that for every object $u = (b, c) : EA \to f$ of E/f there is a unique morphism $\overline{u} : A \to Nf$ such that $u = 1_f \circ E\overline{u} = E\overline{u}$. Setting $\overline{u} = (a, b', c')$, the latter condition becomes $(b, c) = (b', c')$, hence the claim reduces to the existence of a unique a such that $(a, b, c) \in FA(A, Nf)$, which is true by Lemma 3.2. $\qquad \square$

4   An Isomorphism of Categories

Proposition 2.1 characterizes the behaviours of reachable semiautomata. In this section we apply Goguen's technique to obtain an isomorphism between the category of reachable semiautomata and the category of their behaviours.

Let $\boldsymbol{SA}$ be the category whose objects are the reachable semiautomata $\Sigma = (S, I; \delta, s_0)$ and whose morphisms from Σ to $\Sigma' = (S', I'; \delta', s_0')$ are the pairs (a, b) where $a : S \to S'$, $b : I \to I'$ is a surjection and the identities (10.1) and (10.3) hold.

Let $\boldsymbol{SB}$ be the category whose objects are the surjections of the form $f : X^* \to Y$ for which $\ker f$ is right invariant and whose morphisms from f to $f' : X'^* \to Y'$ are the pairs (b, a) where $b : X \to X'$ is surjective, $a : Y \to Y'$ and the following identity holds:

(22) $\qquad a(f(w)) = f'(b^*(w)), \ \forall w \in X^*.$

In both categories the composition of morphisms is defined componentwise.

Define a functor $SE : \boldsymbol{SA} \to \boldsymbol{SB}$ as follows. Using the same notation Σ, Σ' as before, set $SE\Sigma = \delta^0 : I^* \to S$ and for $(a, b) \in \boldsymbol{SA}(\Sigma, \Sigma')$ set $SE(a, b) = (b, a)$. Then δ^0 is an object of $\boldsymbol{SB}$ by Proposition 2.1. On the other hand, it follows from Lemma 3.1 that identity (12) holds, which is condition (22) for δ^0 and δ'^0; so $(b, a) \in \boldsymbol{SB}(\delta^0, \delta'^0) = \boldsymbol{SB}(SE\Sigma, SE\Sigma')$.

A functor $SN : \boldsymbol{SB} \to \boldsymbol{SA}$ is obtained which takes an object f of $\boldsymbol{SB}$ to the object SNf of $\boldsymbol{SA}$ defined by (5) in the proof of Proposition 2.1 and which takes a morphism $(b, a) \in \boldsymbol{SB}(f, f')$ to $SN(b, a) = (a, b)$. Then conditions (10.1) and (10.3) are fulfilled for the semiautomata SNf and SNf', because conditions (5) and (22) imply

$$a(\sigma_f(f(w), x)) = a(f(wx)) = f'(b^*(wx)) = f'(b^*(w)b(x))$$
$$= \sigma_{f'}(f'(b^*(w)), b(x)) = \sigma_{f'}(a(f(w)), b(x)),$$

$$a(f(\lambda)) = f'(b^*(\lambda)) = f'(\lambda).$$

Theorem 4.1 *The functors SE and SN establish an isomorphism between the categories $\boldsymbol{SA}$ and $\boldsymbol{SB}$.*

Proof. It follows from Proposition 2.1, (7) that $SE(SNf) = (\sigma_f)^0 = f$, while $SE(SN(b, a)) = SE(a, b) = (b, a)$. Therefore $SE \circ SN = Id_{\boldsymbol{SB}}$.

On the other hand, using (5) we obtain

(23) $\qquad SN(SE(\Sigma)) = SN(\delta^0) = (S, I; \sigma_{\delta^0}, \delta^0(\lambda)),$
$\qquad \sigma_{\delta^0}(\delta^0(w), i) = \delta^0(wi) = \delta(s_0, wi) = \delta(\delta(s_0, w), i) = \delta(\delta^0(w), i),$

that is, $\sigma_{\delta^0} = \delta$. Besides, $\delta^0(\lambda) = \delta(s_0, \lambda) = s_0$. Therefore property (21) becomes $SN(SE(\Sigma)) = \Sigma$. Since we have also $SN(SE(a, b)) = SN(b, a) = (a, b)$, it follows that $SN \circ SE = Id_{\boldsymbol{SA}}$. $\qquad \square$

5   Conclusions

As already noticed, taking the constant functor $FX := \{\bullet\}$, $Ff := 1_{\{\bullet\}}$, we obtain Moore automata and the behaviour (3). Moreover, the theory constructed in Section 3 is simplified by the disappearance of the factors FI and FX and of the second argument of the functions; in particular $\ker_1 f$ reduces to $\ker f$. Briefly, we obtain the original Goguen theorem and its proof. We have also noted that for $F := Id_{Set}$ we obtain Mealy automata and the behaviour (8). So, the corresponding theorem for Mealy automata and its proof are obtained from Section 2 by replacing FI, FX, $I^* \times FI$ and $X^* \times FX$ by I, X, I^+ and X^+, respectively.

It has been known for a long time that the Nerode automaton is minimal, i.e., it has the smallest number of states among all automata with the same behaviour. The Goguen theorem has revealed that the deep background of this property is the adjunction between his functors E and N. The Nerode F-automaton is also minimal, as shown by the same (easy) proof as for Moore automata.

Note also that Proposition 2.1 is in fact included in the isomorphism constructed in Theorem 4.1 between the category of reachable semiautomata and the category of their behaviours.

Acknowledgement. The author wishes to thank Professor Cristian Calude for his valuable remarks and in particular for the conjecture which led to Theorem 3.1.

References

[1] M. A. Arbib, E. G. Manes, *Arrows, Structures and Functors. The Categorical Imperative*, Academic Press, New York, 1957.

[2] P. Deussen, *Halbgruppen und Automaten*, Springer-Verlag, Berlin, 1971.

[3] A. Ginzburg, *Algebraic Theory of Automata*, Academic Press, New York, 1968.

[4] J. A. Goguen, Realization is universal, *Math. Systems Theory*, **6**, 4 (1973), 359–374.

[5] R. E. Miller, *Switching Theory*, Vol. 2, John Wiley & Sons, New York, 1965.

Infinity – An Outline of Conceptions in Mythology, Cosmology and Natural Philosophy

Jouko Seppänen

Helsinki University of Technology
Department of Computer Science
02150 Espoo, Finland
E-mail: jouko.seppanen@hut.fi

Abstract. The earliest conceptions of infinity and eternity apparently occurred in Oriental cosmological and cosmogonical myths and fables that have been passed down in popular tradition and epic texts. Some of these became codified in literary religions as holy scripts and were subsequently refined in interpretations and commentaries. The pre-Socratic Greek natural philosophers were the first to detach themselves from myths and religious authority and began to ponder questions of cosmology and cosmogony independently of religious tradition and offered a variety of original views, many of which involved notions of infinity and eternity in various senses and contexts. The Medieval patriarchs, theologians and scholastic philosophers digested and filtered the Oriental religious and the Greek philosophical thought, transferred the idea of infinity from cosmology to theology and carried speculations about the immensity of God and His virtues to the extreme. In the Renaissance the legacy of the Greek natural philosophy and Indian arithmetics were mediated to Europe by the Arabs which allowed the natural philosophers of the Renaissance to bring the notion back to physical cosmology and the astronomers and mathematicians to pave the way for the rise of science and new manifestations and interpretations of the notion of infinity.

In the history of classical and modern astronomy and cosmology infinity continued to play a variety of roles until it was finally banished by the relativistic cosmology in the large and by the quantum gravity in the small and became substituted for by a conception of a universe of finite age, size and scale but having unbounded space-time geometry and cosmology. Yet, the notion of infinity was to undergo a Renaissance of its own in mathematics – in geometry, number theory, analysis, set theory, classical and mathematical logic, discrete mathematics and the theory of computation. Moreover, the dimension of complexity of the real world would come to reveal new potential infinities in the dimensions of the evolution of life, mind, language, culture and human thought and civilization, including science, technology and arts – and the notion of infinity as reduced merely to a construction of the human mind.

1   Introduction

The notion of 'infinity' occupies a unique place in the mental and cultural history of human civilization. In implying something beyond any conceivable bound, endless, unlimited, boundless, eternal etc. in terms of space, time, quantity, quality or other dimension of meaning the notion has come to stand for the negation of anything actual and comprehensible. Its belonging to the world of imagination is witnessed by its occurrence typically in mythology, theology, philosophy and mathematics, all of which are expressly products of mental construction, rather than in everyday life in the real world. Yet, the idea has proved to be equally important to the early conceptions about the world and heavens, theological and philosophical speculations about the cosmos and gods as to scientific theories about the universe. The notion itself has been so central that it has become lexicalized as a variety of words on its different senses and aspects in all languages.

The idea of infinity has a long history of discovery, interpretation and refutation in different fields of knowledge. In discussing the cultural or conceptual history of some notion in a multidisciplinary setting many viewpoints and approaches can be taken. One can choose to consider aspects such as the words or terms used to refer to the notion in question in various languages. Lexical and semantic analyses can reveal important aspects of the way of thought and world view of different peoples and cultures and how they have arrived at the idea. Etymological analyses can disclose interesting deep connections between languages, myths and religions and between cultures. The philosophical and scientific definitions and interpretations of the notion as used today will then be seen against a broader cultural and historical context.

In historical perspective the use and meaning of infinity as well as its status in different fields of knowledge has undergone a complex evolution. In various phases infinity and infinite things have received diverse interpretations and raised a variety of problems and paradoxes, many of which continue to be discussed and remain by far not settled questions. In the following we will take an overview of the conceptual history of 'infinity' only as it has occurred in the mythologies of a few Oriental and Middle Eastern civilizations, in the ancient Greek and Roman natural philosophy, in the Mediaval patristic Christian theology and scholastic philosophy and in the physical cosmologies and natural philosophy of the Renaissance.

The earliest conceptions of infinity and eternity apparently occurred in Oriental cosmological and cosmogonical myths and fables that have been passed down in popular tradition and epic texts. Some of these became codified in literary religions as holy scripts and were subsequently refined in interpretations and commentaries. The pre-Socratic Greek natural philosophers were the first to detach themselves from myths and religious authority and began to ponder questions of cosmology and cosmogony independently of religious tradition and offered a variety of original views, many of which involved notions of infinity and eternity in various senses and contexts. The Medieval patriarchs, theologians and scholastic philosophers digested and filtered the Oriental religious and the Greek philosophical thought, transferred the idea of infinity from cosmology to theology and carried speculations about the immensity of God and His virtues to the extreme. In the Renaissance the legacy of the Greek natural philosophy and Indian arithmetics were mediated to Europe by the Arabs which allowed the natural philosophers of the Renaissance to bring the notion back to physical cosmology and the astronomers and mathematicians to pave the way for the rise of science and new manifestations and interpretations of the notion of infinity.

2   Prehistory – Infinities in Mythology

Looking at the sky or even at the open sea or desert or realizing the unending flow of time and events from the future to the past must have struck already the early man with the idea of infinity. Observation and wondering on the immensity of the heavens and the world around us go back to the dawn of human thought and cognition. The awe of the stars and of the celestial sphere must have incited the idea of what lay beyond and of endlessness in terms of spatial extensions and cyclic time. The questions of the constitution, origin and purpose of everything began to intrigue the human curiosity, awoke admiration and inspired imagination, and all these received their expression in the emergence of religions.

Archaic cultures conceived the universe as having three levels – sky, earth and underworld – connected by a central axis and themselves as living in the centre. The Earth appeared as flat and finite though so large that only gods and perhaps heros could reach the boundary. The huge heavenly vault gave shelter to the Earth and its endless turning together with the heavenly bodies conceived as gods determined the life on Earth and marked the boundary between the human and the divine realms while the underworld was the world of the dead. The idea of the human habitat being located in the centre of the Earth and heavenly spheres became archetypal cosmological symbols which received expression and interpretation in the mythologies of the prehistoric peoples of the world. Cosmographical notions such as World Axis, World Pillar, Pole Star, Sky Nail, World Tree, Cosmic Mountain etc. implied as their complement if not infinite then at least immense dimensions of the world and stirred speculations about what possibly obtained beyond the visible world.

The idea of infinity may have occurred in the context of questions about the conditions before the coming into being of the world as reflected in the cosmogonical myths and stories of creation. The oldest conceptions of infinity are, indeed, found in the mythical cosmogonies as witnessed by notions, names and words denoting or connotating the idea of infinity in one sense or another. The abundance of ordinary words like endless, boundless, unlimited, unterminating, inexhaustible, eternal etc. in all natural languages are also witnesses of a long mental, linguistic and cultural history of the idea of infinity.

Explicit terms for infinity occur in all ancient Oriental and other literary cultures, as in the holy scripts and philosophical texts of Chinese Daoism, Indian Hinduism and Buddhism, Iranian, Babylonian, Semitic, Egyptian, Greek and others. Many of these notions have been carried over and survived in the western spiritual, philosophical and literary traditions and modern language. In the following subsections, infinity in the Indian, Greek and Jewish mythical traditions are briefly introduced.

Indian – Aditi, the Unlimited

An ancient Hindu myth deriving at the latest from the time 1500 B.C., when the Aryan tribes migrated from Iran to Hindustan, tell about the battle between the <u>Danavs</u>, sanskr. *danaw*, bounded, and the <u>Adityas</u> (Vilenkin 1995), sanskr. *aditya*, unbounded, also sun, and growth. In the Veda texts Aditias are identical at times with Savitri or Surya, the sun god, sanskr. *surya*, sun, cf. also Finn. *suuri*, great, the greatest of the heavenly bodies. With the help of Indra, the son of Heaven and Earth, the Adityas triumphed and, by a miracle, grew to frightful size terrifying the Heaven and Earth to break off into two opposite parts making up the world.

In another version of the legend (Barber 1979) <u>Aditi</u>, sanskr. *aditi*, unlimited, eternity, infinite consciousness, mother of Adityas, gave birth to originally three and in later versions of the text to eight or twelve gods, one for each month of the year, including Mitra, Aryman, Varuna, Ansa, Indra, Dhatri and others. She rejected one of them, Marttanda, who became the sun. Other Adityas were Aryaman, the god of truth and the light of divine consciousness, Mitra, the ruler of early things and Varuna, his opposite, the ruler of spiritual things, Bhaga, the god of wealth, cf. Slav. *bog, god, bogatyi*, rich, Indra, the god of weather, Varuna, cf. Slav. *vorona*, magpie, the god of heavens, later of the oceans and rivers, etc. (Schumacher, Woerner 1994). Cf. also Finn. *äiti*, mother, *Maa-äiti*, mother-earth, *maa*, earth, Gr. *gaia, geo*, mother-earth.

Similar myths are found in the ancient Iranian tradition according to which the God Ormazd created Earth, man and all creatures and things on Earth, and Heaven in the form of an egg reaching at its highest point the infinite world. The theme of creation from a primal egg occurs in the myths of many other peoples too, among them in Finnish Kalevala.

In the Hindu and Buddhist traditions infinite was manifested also in the temporal dimension as the idea of an <u>eternal cycle</u>, Skr. *rta*, cycle, *ratha*, carriage, cf. Finn. *rata*, rail, path, *ratas*, wheel, *rattaat*, carriage, an endless return in incarnations of the human soul.

Greek – Empyreon and Cosmos

The Greek inherited elements from the ancient Egyptian and Oriental cosmologies and theologies in which the Sun continued to occupy a central role and was worshipped as the main God. In the Greek mythology the realm of the Sun God became known as the <u>Empyreon</u>, Gr. *en*, in + *pur, py-*

ros, fire, *empurein*, set in fire, *empurion*, *empurios*, the sphere of fire, Lat. *empyreus, caelum*, heaven, cf. Finn. *ympyrä*, circle, *ympäri*, around. Note also Gr. *peri*, about, around, roundabout, Lat. *per*, through, via, around, period, Gr. *peri + hodos*, way, course, periphery, perimeter, perihelion, etc., Finn. *peri, perä*, end, *periaate*, principle, *perusta*, foundation.

According to the Greek cosmogony the world emerged from <u>chaos</u>, Gr. *khaos*, a primordial abyss, formless void, utter confusion, also <u>chasm</u>, Gr. *khasm*, deep cleft, gorge, and turned into an ordered and harmonic <u>cosmos</u>, Gr. *kosmos*, jewel, beauty, order, ornament. In some myths of this type the chaotic state is never completely overcome but continues as an eternal struggle of the opposites, a central principle in the Chinese taoist tradition. While order emerges and may continue to develop in space and time, vestiges of chaos remain and the created order is always in danger of slipping back into chaos, or else chaos appears as the destiny of the cosmos when it has exhausted all of its meaning and potential. It is worthwhile to notice that chaos, as a modern mathematical and complex dynamical notion, has the same attribute of infinity and disorder, something endlessly changing and never repeating itself or any part thereof in motion, form, order or scale.

Jewish – Sephirot and Ein Sof

In the ancient Jewish cosmology as transmitted down with the <u>Kabbalah</u>, Hebr. *kabbalah*, tradition, the world is conceived as created by God from <u>nothing</u>, Lat. *ex nihilo*, and consisted of a system of ten <u>spheres</u>, Hebr. *sephiras, sphere*, pl. *sephitot*, emanating from infinity, Hebr. *ein sof, ein*, non + *sof*, end, endless, the symbol of God, who never revealed himself directly but only through his virtues and deeds. God transcended all the spheres, the uppermost of which was called the crown. Below it came the spheres of wisdom, intelligence, mercy and lesser realms down to the lowest kingdom of earthly life. It was only through the sephirot that man could approach the holy spirit and himself pursue spiritual fulfillment. In the Old Testament one can find statements about the omnipotence and other properties of God without bound or <u>limit</u>, Hebr. *ein gebul*, bound, limit.

Over time the Kabbalah spread via Alexandria to Spain where its philosophy came to its culmination in the thinking of the Jewish philosopher and theologian Moses Maimonides (1135–1204) as documented in his "Moreh nebouchim". His "Guide for the perplexed" spread to Europe, gained wide popularity and exerted considerable influence on Medieval mystics.

In Catholic philosophy the Spanish Franciscan theologian and mystic

Raimundus Lullus (1235–1315) wrote, instructed by a vision, "Ars magna et ultima" (1273) in which he developed a diagrammatical method known as Lullian art or Lullian circles, with the aim of mechanically proving the infinite and absolute qualities of God. His circles comprised of concentric discs which quoted the qualities of God, in Latin "Gloria", "Bonitas", "Magnitudo", "Duratio", "Potestas", "Sapientia", "Voluntas", "Virtus", "Veritas" as "Bonitas". By turning the discs he could obtain new combinations of these and other notions on other discs and thereby supposedly be able to prove higher order properties and to derive new knowledge. The work was, however, condemned in 1376 by the church for its attempt to link faith with reason, but the idea was sympathetically received by some scholastic philosophers who continued to refine their speculations. The Lullian ideas influenced also the Germ. mathematician and philosopher Gottfried Leibniz (1646–1716) to dream of a universal language and machine that would allow one to mechanically derive all possible knowledge.

3   Antiquity – Infinities in Philosophy

The Greek philosophers were the first to detach themselves from myths and religious authority and began to ponder questions of cosmology and cosmogony independently. The pre-Socratic scholars were to found natural philosophy and to introduce a number of new and original conceptions about the constitution and origin of the world. Many of the natural cosmologies also involved notions of infinity in some senses.

Apeiron and Myriad

The early Greek cosmologies of the pre-Socratic philosophers were, for obvious reasons, based more on imagination than evidence. The first naturalistic conception about the origin and nature of the world is generally ascribed to the Greek philosopher Thales of Miletus (c.625–547), often quoted as the first scientist. He was, indeed, engaged with astronomy and geometry, and began to suspect the myths and role of gods in the events of nature. To him the Earth appeared as flat and surrounded by an ocean and an atmosphere of vapour. The heavenly bodies were floating in the atmosphere and all things had come about from water and vapour. Thales, yet, is not known to have given any opinion about infinity. With his attempts to find natural causes and explanations to the origin and order of the world and physical phenomena and in basing them on observation and original reasoning rather than belief he become the founder of naturalism.

His disciple Anaximander of Miletus (c.610–c.547), the author of the first philosophical text "On Nature", was the first to take up the notion of infinity explicitely. To him the world was cylindrical with the heavenly bodies moving each on its shell at different distances from the Earth. To account for the origin of the world he introduced the terms <u>arkhe</u>, Gr. *arkhe*, origin, first principle, beginning of all things, and <u>apeiron</u>, Gr. *a*, not + *peirein*, try, apeiron, boundless, infinite, indefinite, eternal, endless motion. The universe had come into being from apeiron which was an inexhaustible and eternal chasm. The varieties of things and their qualities were formed by differentiation and isolation from the primal chasm and the eternal struggle of its opposites. Moreover, there were infinitely many <u>other worlds</u> and new worlds could come into being from the apeiron until they perished back into apeiron.

The last of the great Milesian philosophers Anaximenes (c.585–527), a disciple of Anaximander, identified air as the primal element which by condensation and rarefaction gave birth to all things. He reasoned that since air is the breath of life for man it must also be the main principle of the universe. The <u>infinite air</u> was the principle from which all things that are, that are becoming and that shall be, including men and gods.

Later in the 5th century Greek began to consider counting with numbers and the problem of boundlessness in mathematical terms. The first Gr. mathematician Pythagoras (c.580–500), who became the founder of the field by introducing the word 'mathematics', Gr. *mathematike, manthe*, mind, thought that the world was made out of numbers and that all things could be explained by <u>arithmos</u>, Gr. *arithmos*, number, natural numbers and their properties, ratios and arrangements.

The Greek had the word <u>myriad</u>, Gr. *murios*, countless, innumerable, meaning an inconceivably large number. Yet, the Pythagoreans were shocked to discover the idea of genuine <u>infinity</u> as manifested in the <u>incommensurate ratio</u> of the side of a square to its diagonal. According to the legend, which goes in many versions, Pythagoras and his disciples became horrified and kept the discovery as a divine secret until it was revealed to outsiders by a former but expelled member Hippasus of Metapontum (5th c.), and was duly punished. One account has it that a tombstone was raised to him while still alive, while another version has it that he was left to drown in a shipwreck.

According to the Pythagorean cosmology what lay outside the heavens was an infinite void by which they understood an endless extent of air. The notion of apeiron occurs also in Pythagorean dualism where it is opposed to the principles of <u>peras</u>, Gr. *peras*, limit, boundary, and <u>arithmos</u>, number. According to Pythagoreans, apeiron was the boundless and amorphous which, together with its opposite, the limit and number were the basis of everything that exists.

Interesting enough, by chance or by true connection, the Greek words *apeiron*, and *peras* bear close morphological and semantic, possibly also etymological connection either as a cultural loan or even true genetical relationship, with the Finn. root *per, peri, perä, perus*, etc. and their derivatives, cf. Finn.

 – *peri*, destination, range,
 – *perä*, rear, tail, end, back, foundation, *perässä*, after, following, next,
 – *perin*, thoroughly, extremely, exceedingly, utterly,
 – *perinne*, tradition, *perintö*, heritage, legacy,
 – *periaate*, principle,
 – *perus*, base, basis, foundation, bottom, primary, *perukka*, bottom, end, *perusaine*, primary matter, *perusluku*, cardinal number, *perussyy*, primary cause, prime mover, *perusta*, foundation, *peruste*, base, ground, basis, reason, cause.

Note also Engl. *experience*, Lat. *experiri*, Engl. *empirical, empirism*, Lat. *empiricus*, Gr. *empirikos*, ancient physician, *empeireia*, experience, *empeiros*, skilled, *en + peirein*, try, *peira*, trial, attempt, attack, cf. also Engl. *peril*, Lat. *periculum*, experiment, risk, danger.

Eternity and Infinity of Particles

The materialist philosopher Heraclitus of Ephesus (c.544–c.483), the philosopher of <u>becoming</u>, held in his "On Nature" that the primal material of the world was fire for it was the most capable of change. According to him, the <u>world</u> was <u>eternal</u>, not created by any god nor by man. It had always been and would continue to exist for ever and everything was in a nonending flux of change. One could not step into the same river twice for he who did would be "washed by other waters".

The head of the Eleatic school Parmenides (c.540–480), the philosopher of 'Being', vigorously opposed the idea of apeiron and conceived the world as <u>eternal</u> but <u>finite</u>, immobile and having the form of a sphere filled with things and free from void. He developed the principle of <u>being</u> in opposition to the principle of becoming of Heraclitus.

Another materialist philosopher Anaxagoras of Clazomenae (c.500–c.428) conceived in his book "On Nature" the world as consisting of an <u>infinite variety</u> of qualitatively dissimilar primary elements of <u>imperishable</u>

and infinitely divisible particles of matter which he called homoeomeries, Gr. *homo*, same + *eomer*, particle. All things were made of these 'seeds' which occurred throughout the universe. Their coming together by chance or necessity gave rise to individual things, and their separation entailed the passing away of individual things. According to Anaxagoras "there is no smallest among the small and no largest of the large, but always something still smaller and something still larger". The particles retained their character on division and there was a portion of everything in everything, except Mind. The motive force that conditioned the union and division of particles was nous, Gr. *nous*, mind, cf. Finn. *nousta*, arise, *nousu*, ascent, which he understood to be the substance of the lightest and finest kind, eternal and imperishable and capable of motion and communication of its properties to things.

Infinite Cosmos and Infinity of Worlds

The idea of infinite divisibility of matter, in its turn, was rejected by a third breed of materialist philosopher Leucippus (c.500–440), the founder of the School of Abdera, who introduced the notions of absolute vacuum, causality and atom, Gr. *a*, not + *tomos*, part. He regarded space without matter as the essential ingredient of cosmos, an infinite void in which atoms moved and without which no change would be possible. "Nothing arises without cause but everything arises on some grounds and by the force of necessity". He also proposed the principle that all qualitative differences in nature may be reduced to quantitative ones, opposing the principle of homogeneous being advocated by Parmenides as an infinite variety of dissimilar kinds but homogeneous particles.

To Leucippus atoms were also of infinite variety but indivisible and occurred throughout the infinite space and were separated from one another by not-being, that is empty space. The coming into being of things was caused by atoms coming together by motion in the otherwise empty space. The cosmos appeared to him not only as infinitely large but with infinitely many other worlds in it. The ideas of Leucippus were developed further and applied to explain human perception, memory and mind by his disciple Democritus (c.460–360) who became the founder of atomism as a general doctrine.

In the Hellenic period the materialist and atomist philosophies and cosmologies were adopted and cultivated notably by the Greek atheist philosopher Epicurus (341–270) and in the Roman time by the Roman materialist philosopher and poet Carus Lucretius (c.99–55). They recognized eternity of matter and motion of atoms as central principles, developed atomism into a materialist philosophy of life and became founders of <u>sensationalism</u>.

Infinitely Small and Infinitely Large

The question of infinity arose also in attempts to understand the nature of motion which was taken up by Zeno of Elea (490–430), a disciple of Parmenides. Defending the doctrine of being of his master as opposed to the doctrines of becoming, change and motion, he presented logical <u>paradoxes</u>, Gr. *para*, side, alongside + *doxa*, opinion, about the nature of motion and developed the method of logical argument and proof. With the paradoxes known as "Dichotomy", "Achilles and the Tortoise", "The Flying Arrow" and "The Stade" Zeno argued against the possibility of motion. Under the assumption of the infinite subdivisibility of space and time, he showed that the idea of motion was self-contradictory, and hence, not a valid principle for explanation of natural phenomena as the atomists had taught. His reasoning also made Anaxagoras (c.500–c.428) to reconsider his thought and to convince himself that there was no smallest quantity of anything.

If Zeno raised the question about infinity in the small, the mathematician Archytas of Tarentum (c.428–347) from the Pythagorean school, who was said to have solved the problem of doubling the cube, became the first to reason logically about an infinite cosmos in asking: "If I am at the extremity of the heaven of the fixed stars, can I stretch outwards my hand or staff?" and answering that "It is absurd to suppose that I could not. And if I can, what is outside must be either body or space. We may then in the same way get to the outside of that again, and so on. And if there is always a new place to which the staff may be held out, this clearly involves extension without limit".

The first of the great philosophers of Athens, Socrates (469–399), initiated the turn from naturalism and materialism to <u>idealism</u>. To him the structure of the world and the physical nature of things were unknowable. One can know only oneself, as he expressed it, "know thyself". With this turn Socrates opened a new dimension for philosophy to explore, that of mind and knowledge, which were to reveal a multitude of new kinds of infinity, although Socrates himself did not specifically discuss the question of infinity himself.

His disciple Plato (428–348) adopted from his teacher the notion of <u>idea</u>, Gr. *eidos*, and developed in his more than 30 dialogues the doctrine of <u>objective idealism</u> to fight the materialist teachings of his time. To Plato ideas were unworldly, <u>eternal</u> and independent of time and space. Ideas could neither come into being nor perish because they already consisted of a perfect and absolute totality. The source of such knowledge was the immortal human soul, which was, however, defective and only reminiscent of the perfection of the divine world of ideas.

But, strangely enough, to Plato there was no infinity, not in this world nor in the world beyond, which makes his idealism defective if not self-contradictory. He believed that even his ultimate idea, form, the Good, must be finite and definite. This was in sharp contradistinction with almost all later metaphysicians, who assumed the absolute necessarily to be infinite.

The Gr. mathematician and astronomer Eudoxus of Cnidus (c.400 – c.347) conceived the first model to account for the motions of celestial bodies which he thought to be carried around the Earth each on their own sphere. In mathematics he removed the limitation imposed by Pythagoras of allowing only rational proportions accepting also incommensurable ratios and magnitudes. Based on the idea of continued divisions and proportions he developed in geometry the method of exhaustion which made it possible, for the first time, to calculate areas under curves and volumes under curved surfaces by dividing them into ever smaller sections. With these ideas, which were to be adopted by Euclid (c.330–c.260) and were included in his Book V, Eudoxus anticipated by nearly two thousand years the calculi of indivisibles and infinitesimals.

Actual and Potential Infinity

The cosmology of multiple spheres was refined by the great philosophers of Athens and subsequently by astronomers. The greatest of all philosophers, logician and scientist, Aristotle (384–322), perfected the system of spheres with the Sun, Moon, planets and stars circling the Earth each on their spheres. As described in his "Physike", the Aristotelian cosmology consisted of a series of nine spheres and later up to 56 all set in motion by the outermost permanent and divine crystal sphere, the one which he called the prime mover that lay beyond the sphere of the fixed stars.

Yet, Aristotle denied the idea of an infinite void space on the grounds of his conception of mechanics. In his "De caelo" he declared categorically that there was neither time nor space nor void outside of the heaven. There was nothing infinite nor void, since things, possessed of dimensions, such as space, would have been equivalent to body and therefore incapable of receiving other material bodies. But even if void space existed, finite and successive motion in it would be impossible. In a vacuum there could neither be resistance to motion, so bodies would fall with infinite velocity. Nothing could be thrown through empty space, and there would be no way to distinguish one point from another, nor possibility of determining place or direction.

Neither was there anything infinitely small, infinitely large or infinitely numerous. For him, the idea of something being actually infinite was a privation, not a perfection or totality but only absence of a definite limit. What was limited, was actual and real with respect to what surrounded it. Infinity could only be potential, not actual and complete, like the sequence of natural numbers among which there was no largest number. The sequence could always be continued but would never be completed nor exist as an actual and finished thing.

Aristotle's reasoning was rigorous and settled, for the time being, the question of infinity in the negative. Satisfied with this view, natural philosophers, mathematicians and mechanicians could turn their attention to more earthly questions regarding astronomy, mechanics and mathematics. It also brought about what became known as the horror of infinity, which perhaps had also a negative effect especially to the development of mathematics.

The Greek and the earlier Oriental civilizations which developed geometry and arithmetics conceived also the possibility of unimaginably large and small in terms of measure and number. The idea of infinitely large and small must have occurred also early on, as witnessed by special names and attempts to give explicit designations to numbers as large as one could think of, speak about or write

In the third century the Greek mechanician and mathematician Archimedes (278–212), the "father of physics", wondered how many sand grains there might be on a beach and devised designations in his "Sand-reckoner" for huge numbers or myriads and demonstrated that even they can be counted. "Many people believe, King Gelon, that the grains of sand are without number. Others think that although their number is not without bound, no number can ever be named which will be greater than the number of grains of sand. But I shall try to prove to you that among the numbers which I have named there are those which exceed the number of grains in a heap of sand the size not only of the earth, but even of the universe".

The largest number ever having been given a name of its own apart from compound mathematical notations is the Buddhist number asankhyeya, Skr. *a*, non + *san*, together, number, *khyeya*, know, say, innumerable, uncountable, equal to 10^{140}. This is a fairly big number indeed, at least for practical purposes, if compared with for instance the number of atoms in the universe which is estimated to be of order 10^{85} "only".

There are also legends involving large quantities. According to a legend, the invention of the game of chess, Iran. *shah*, emperor, impressed and delighted the Shah of ancient Persia so much that he asked to see the inventor and to reward him. As introduced to the Shah the poor but learned inventor asked merely that one grain of wheat be put on the first square of his chess board, two on the second, four on the third and so on until the whole board was covered. The Shah, embarrashed with the modesty of the request, ordered the wish to be fulfilled, but was soon struck to realize that the wheat deposits of his empire would be depleted long before the completion of the request.

Stoics – Infinite Extracosmic Void

In the Hellenistic period conceptions of infinity from the oriental religions and cosmologies were assimilated with the Greek natural philosophy and the cosmologies were reinterpreted in a new light. The Stoic philosophers reconsidered Aristotle's view of the world and heavens as a system of spheres accepting it in outline, agreeing that the intracosmic world was spherical and finite but disagreeing with his conception about extracosmos. The various Stoic philosophers were not, however, unanimous about the nature of extracosmos themselves.

The Stoic philosopher Cleomedes (1st c. A.D.) denied Aristotle's view of finite space, reasoning in his "De motu circulari corporum caelestium" rather in favour of Archytas who had held that no body could exist beyond the physical world because no material could limit void and because it was absurd to suppose that void could limit void and make it terminate and finite. The existence of an infinite extracosmic void beyond the heavens seemed thus inevitable. The infinite extracosmic void was seen as a receptacle for the finite intracosmos and that appeared to be its only purpose because interaction between cosmos and void was seen possible (Grant 1981).

Like Aristotle, the later Stoic philosopher Philoponus (490–570) denied the existence of infinite extracosmic void rejecting also the interpretation of spatial dimensions. Moreover, void could never be empty of matter. The prominent Medieval commentator of Aristotle, Simplicius (6th c.), instead, became one of the loudest proponents of Aristotle's view in support of the finite intracosmos but infinite extracosmos in his "Commentary on de Caelo", paving the way for theological interpretations of the infinite extracosmos as the abode of God. In their conceptions about astronomy and physics the Stoics followed, however, Aristotle, agreeing with his view of the world and heavens as a finite system of spheres. In the Middle Ages the Stoic conception of the universe and the idea of an infinite extracosmic void were conveyed to the Latin West from a number of sources.

In spite of the great strides made by the Greek cosmology, astronomy, natural philosophy, physics and mathematics, it is often claimed that the development of science was hampered by the refusal to accept the notion of apeiron and rather to adopt the notion of atom as the primary building block of nature. Especially in mathematics the horror of infinity became detrimental. Yet, there were ideas about the space and time being infinitely divisible, the space extending indefinitely, and time and geometrical line being continuous and containing an infinity of points. The study of number sequences and ratios and the method of exhaustion in the calculation of areas and volumes witness, however, the use of potential infinity in geometry, number theory and the theory of proportions, anticipating their revival in the Renaissance.

Roman Time – Heavenly Spheres

The Roman philosophers and commentators had less concern for natural philosophy and little to add to the Greek philosophical thought. The Miletan, Pythagorean, Eleatic, atomist, Aristotelian and Stoic conceptions of cosmology were digested and perpetuated but with little original contribution.

The Roman poet and materialist philosopher Carus Lucretius (c.99–55) rendered the work of Epicurus (341–270) into a poetic form in his work "De Rerum Natura" and argued for the unboundedness of space reasoning in the following way: "Suppose for a moment that the whole of space were bounded and that someone made his way to the uttermost boundary and threw a flying dart. Apparently the dart must either fly past the boundary, whereby it is not a boundary, or else be stopped by it, in which case there must be something beyond the boundary that hence cannot be the limit of the universe", rephrasing the argument of Archytas.

The idealist Roman philosopher Plotinus (205–270), the founder of neoplatonism, the last Greek 'pagan' philosophy flavoured with Oriental mysticism, taught that the world began with the incomprehensible divine "One", which was the eternal source of all being. Reviving Plato's objective idealism, he disagreed, however, with his master's denial of the infinite and thought that at least God was infinite. According to his "Enneads" the absolute 'One' has no measure, stands outside of number and has no limit in regard to anything.

The universe consisted of a <u>hierarchy of spheres</u> of being, the highest being the realm of the "One" and the lowest the earthly time and space that can be perceived by the senses. In between lay the spheres beyond space and time, each derived from the one above, of higher degree of unity and each being a lesser image of the one above. As one moved down the scale, greater degree of multiplicity, separateness and increasing limitation become evident and at the lowest level separation into atoms of the spatio-temporal world occurred. The highest sphere from which everything was derived, was itself derived from the ultimate "Good" that transcended everything, was beyond being, absolutely simple and devoid, could be imagined or described and could be known only when the mind arose into union with it.

The neoplatonist cosmologist exhibited spiritual elements from the ancient Oriental cosmologies and religions and became the prime philosophical source of ideas for the Christian theology in the Middle Ages developed by the patristic and Catholic theologians and scholastic philosophers in support of their speculations about the immensity, omnipresence and other infinite qualities of God.

4 Middle Ages – Infinities in Theology

In the Middle Ages, as the Christian belief spread and the early church became gradually established, natural philosophy and mathematics suffered set-backs and were cut off in the Greco-Latin world. The Biblical cosmogony became canonized by the patriarchs and neoplatonism was seen as a suitable source of philosophical ideology for the Christian theology while the Stoic conception of infinite extracosmos was adopted as a worthy abode for God.

The Aristotelian thought was selectively digested by the Catholic theologians and scholastic philosphers but his denial of the actual infinity could in no way be reconciled with the immensity of God and the infinite virtues ascribed to Him alone. The Aristotelian and other Greek traditions of natural philosophy, cosmology and science, including astronomy, physics and mathematics survived, however, and continued to flourish in the Arab world and allowed the Greek legacy of wisdom and knowledge to be mediated to Europe in the Renaissance.

The Arab centres of wisdom in the Middle East maintained connections through the Silk Road also to Oriental civilizations and mediated achievements in astronomy and mathematics, notably the decimal notation for numbers and arithmetics, from India to Europe.

Indian Astronomy and Arithmetics

The oldest Indian work on astronomy and <u>mathematics</u>, Skr. *ganita*, dates from the turn of the 5th century. The text of the Jaina school mathematician Aryabhata (c.476) known as "Aryabhatiya" (c.499) discusses arithmetics, <u>algebra</u>, Skr. *rasi*, quantity, <u>geometry</u>, Skr. *ksetra*, field, and trigonometry. The Ind. mathematician Bhaskara I (c.629) was the first to give a systematic treatment of decimal arithmetics, the rules for all the six arithmetical operations including negative and irrational numbers and zero. There was also a rule for division by zero '0/0 = 0', which today would, however, be considered as mistaken.

His contemporary Brahmagupta (c.598-c.666) wrote "Siddhanta" (628) which consisted of 25 mathematical texts and were translated into Persian and Arabic. He called the result of '*a*/0' <u>zero-divisor</u>, Skr. *khakkheda*, but did not give value to the result. In the late Middle Ages Bhaskara II (1115–1185) continued the work of Brahmagupta and took division by zero to give <u>infinite</u>, Skr. *ananta-rasi*, infinite quantity. For the Hindus the wisdom of arithmetics was of divine origin and developed in association with the determination of the proportions and construction of temples and altars as cosmological symbols, whereby the notion "ananta", infinity, was attributed to the Supreme God, Vishnu.

Arab Astronomy and Mathematics

In the 7th century the Syrian bishop Severus (465–538) wrote praisingly about Indian astronomy and arithmetics which used nine different signs for the digits, that is the Hindu <u>numerals</u> and not letters of the alphabet as had been customary in the Middle Eastern and Greek civilizations. This was significant, indeed, since the introduction into the Arab world of what came to be known as the <u>Hindu-Arabic numerals</u> also led to the adoption in Europe into mathematics and astronomy the positional decimal number system which made the development of mathematics, unthinkable in Roman numerals, possible.

The Arab mathematician, astronomer and geographer al-Khwarizmi (c.800–c.847), a member of the House of Wisdom founded by the Abbasidian caliph al-Mamun (d.833) in Baghdad, expanded on the work "Arith-

metica" of the Greek mathematician Diophantus of Alexandria (c.250) dealing with number theory and equations of rationals with rational solutions
and became with his book "Algebra" the founder of the field. Diophantos
had introduced symbols to denote quantities, operations and relationships
as previously and long after numbers had been represented by geometric
arrangements of dots, number symbols and words only which had became
an obstacle to the development of mathematics. The Arab. astronomer
and mathematician Tabit ibn Qurra (c.835–901) applied algebra and the
theory of proportions to Euclid's geometry and discussed his fith axiom
and the question of whether parallel lines meet in infinity or not.

The Arab astronomer and optician al-Haytham Lat. Alhazen (c.965–
1038) from Basra wrote about a hundred works, among them "Opticae
thesaurus" which was translated into Latin in the 12th century and finally
published (1572). It was the first authoritative work to reject Aristotle's
theory of light that the eye sends out rays to the object looked at and his
belief that the speed of light was infinite. Alhazen's work on geometrical optics became the foundation of analytical geometry developed by the
French mathematician and philosopher René Descartes (1596–1650)

The Baghdad astronomer and mathematician al-Battani (8th c.) gave
up the Greek system of chords of angles in angular geometry and adopted
the far more convenient proportion known as sine and its converse cosine,
laying foundation to trigonometry. He then applied trigonometry to projection of figures from the surface of a sphere on to a plane anticipating
projective and spherical geometry and the idea of point in infinity.

The astronomer Kamal al-Din (9th. c.) extended algebra to handling
irrational numbers such as square root and to higher degree equations while
the Islamic poet and mathematician al-Khayyami (d.1022) discussed methods for finding roots of fourth, fifth and higher powers. The Arab. physician
Yahya al-Samaw'al (12th c.) wrote a book "The Dazzling" when only 19
years old. In it he discussed algebraic theory of powers and introduced
multiplication and division of powers, series of powers and inverse powers.
He also adopted the unifying convention that 1 can be expressed as power
zero laying the foundation to a general theory of series. Moreover, he was
the first to use negative numbers as distinct entities three hundered years
before they were understood and accepted in Europe. The Samarkand
mathematician al-Kashi (d. c.1429) applied series to the calculation of π
obtaining its value up to 17 decimals.

The legacy of Arab astronomy and mathematics was transmitted to
Europe in the beginning of the 13th century by the Ital. mathematician

Fibonacci of Pisa (c.1170–c.1250), known for the infinite sequence of numbers in which each member is the sum of the two preceding numbers. Fibonacci obtained his series as the answer to the problem of the number of rabbits bred in pairs in successive generations. The Latin translation of al-Khwarizmi's text "De numero indorum" appeared as part of Fibonacci's "Libri abbaci" (1202). There he explained the use and significance of the number zero '0', Skr. *sunya*, Arab. *cifr*, empty, Lat. *zephirum*, zero, *zephyris*, cipher, and introduced the symbol '–' bar, Lat. *solidus*, for ratios. The work became epoch-making in making possible the rebirth of mathematics in the Renaissance and its rapid development in the subsequent centuries.

Patristic and Scholastic Philosophy

The Christian apostles and Medieval patristic theologians adapted the Jewish and Platonic tradition and the scholasticians began to speculate with the immensity, omnipotence and other infinite qualities of God. In the late Middle Ages the scholastic philosophers adopted some of Aristotle's concepts but had difficulty with others. Although medieval and scholastic philosophers would follow Aristotle in many respects, they preferred to follow the Stoics in assuming the existence of infinite extracosmos beyond the heavens that fitted well with the Catholic theology.

The early Christian bishop of Hippo, St. Augustine (354–430) held views close to neoplatonism in suggesting that God not only was infinite but also could think infinite thoughts. According to him, omniscient God must know each and every natural number and that he even knew infinity in the form of all the numbers taken at once, for otherwise they would exceed his powers. God himself must thus lie beyond all numbers.

To the Pope St. Gregory the Great (c.540–604), founder of the Catholic philosophy, the infinitude of God was insurmountable to human comprehension. No matter how far our mind may reach in the contemplation of God, it does not attain to what He is, but merely to what is beneath Him. God's immensity transcended the universe and the human imagination.

Among those who developed early scholastic thought about the physical space was the Arab. philosopher Ibn Rushd l. Averroes (1126–1198) who lived in Spain during the Muslim faith: he tried to prove the eternity and uncreatability of matter and motion and the mortality of soul. As a forerunner of the Renaissance and in upholding the philosophy of twofold truth, Aristotelian and theological, he fell victim to brutal persecution both by the Muslim and the Catholic churches.

Distantia Terminorum and the Possibility of Motion

The antique paradox about the impossibility of motion was revived by the scholastic philosophers under the notion of "distantia terminorum" which became an argument in defence of the possibility of motion in the void. The Span. mechanician and philosopher Ibn Bajja Lat. Avenpace (d.c.1138) considered the celestial bodies and noted that their "motions are not impeded by anything but yet they move at a definite speed in definite time". The idea that not only distance but also time and velocity were equally indefinitely divisible and proportional to one another provided the key to the solution of the antique paradoxes. Motion was not only physically, but also logically possible, a conclusion which gave way to further elaboration of the notions of celestial and earthly mechanics.

The next argument sought to explain how finite motion could occur in a physical vacuum or resistanceless medium. According to the Engl. Franciscan philosopher Roger Bacon (c.1214–1292), one of the first to argue on this matter, motion through void was possible because bodies have a distance between their boundaries which, too, are indefinitely divisible.

The principle of distantia terminorum was considered and accepted also by the Dominican scholastic philosopher Thomas Aquinas (1225–1274). According to him, any motion had a definite velocity which arose from the ratio of motive power to the mobile, even when there should be no resistance. To support his reasoning Aquinas leaned on the example given by Avenpace. The idea of motive power which caused the motions of earthly and heavenly bodies and the assumptions of finite world and infinite extracosmic void gave rise to further speculations about God and his powers. In his "Summa Theologiae" Aquinas tried to give a sort of Aristotelian proof that "although God's power is unlimited, he still cannot make an absolutely infinite thing, no more than he can make an unmade thing".

The Ital. poet Dante Alighieri (1265–1321) told in his epic poem "Divina Commedia" (1307) the story of a journey to Hell and Purgatory, guided by Virgil, and finally to Paradise (1300), guided by his lifelong beloved Beatrice Portinari (c.1265–1290) passing through each of the nine spheres. Beyond these lay the nine spheres of angels and beyond these the abode of God, Empyrion, the highest heaven.

Scholastic concern for the question about infinite culminated in the contradiction of Aristotle's rejection in his "De caelo" of the infinite extracosmic space and the plurality of worlds in declaring categorically that "neither place, nor void, nor time" can exist outside the heaven and of the Stoic view of the existence of the extracosmic view. The dispute was further aggravated by the Condemnation of 1277 issued by the bishop of Paris, Étienne Tempier (d.1279) which on theological grounds forbade 219 propositions in Aristotelian teachings and other conceptions, including the eternity of the world. The scholasticians were occupied also by the questions of conditions that had prevailed prior to creation, namely the existence of infinite void space before creation in which the world had to be placed and with the location of God himself.

Infinite Space and the Immensity of God

In the 14th century the idea of infinite void extracosmic space beyond the finite spherical intracosmos and heavens took hold and won widespread acceptance by the 16th and 17th centuries. Even astronomers and natural philosophers who otherwise were opposed to scholastic speculations were inclined to accept the idea. The nature of the extracosmic void space was further elaborated by the scholastics and became assimilated with the omnipresence of God. Although devoid of matter the imaginary void could be identified with God and His attributes. Its reality was, however, controversial and often characterized as a pure negation of the substantial world and space.

The Mediaval theological philosophy was unable to deal with other kinds of infinitude than that of God. No infinite thing, collection or quality apart from divine could ever arise or be thought of in the earthly world.

5   Renaissance – Infinities in Cosmology

The ancient Mesopotamian, Greek and Aristotelian cosmologies had maintained that the universe was finite, spherical and geocentric whereas the Stoics conceived an infinite extracosmic void beyond the skies. The latter view had been adopted by the Medieval theologians who associated it with the immensity of God to be further refined in the speculations of scholastic philosophy. In the Renaissance, instead, the idea of an infinite physical space took root, and not only space but an infinite plurality of other worlds became subjects of speculation. The formerly theological questions were redressed in terms of natural philosophy, astronomy and mathematics. Among the forerunners of the Copernican and scientific revolution and of the new world outlook were dissident scholastic theologians and early Renaissance natural philosophers who became the first to break off from the official doctrines of the church. The introduction of the decimal system and arithmetics to the Latin world made possible the development of astronomy, celestial and earthly mechanics and mathematics.

Infinite Cosmos and Infinitely Many Worlds

The Engl. Franciscan philosopher Roger Bacon (c.1214–1292), an early critic of Aristotle and advocate of the scientific method, defined the extracosmic void as a space in which there is absolutely nothing. By altering Aristotle's definition he propounded the view that allowed for the possibility of space beyond the world, still accepting that the absolute void was not able to receive any material body.

The French physicist and nominalist philosopher Jean Buridan (1300–1362) proposed the idea of divinely motive force or impetus that enabled the planets to move on their orbits forever in the heavenly ether that was assumed to offer no resistance to their motion. Moreover, he revived the idea of Anaximander about the existence of an infinite plurality of other worlds in the infinite cosmos.

The Germ. cardinal and natural philosopher Nicholas of Cusa (1401–1461), who was too, a dissident thinker of his time, came to mark a transition from scholasticism to humanism and to anticipate the transition from natural philosophy to the new science. Denying the central position of the Earth in the world order he expounded in his "De docta ignorantia" (1440) that the Earth and heavenly bodies were in motion in the cosmos and that the cosmos was infinite and contained infinitely many other worlds. The cosmos must be infinite, since if it had limits, then beyond those limits there would be something and the universe would not be the whole universe. And since it was infinite, it could have neither a centre nor a circumference. Any point could be taken as the centre like at sea and the horizon would always be equally distant in all directions. Cusa also refashioned the concept of God with his doctrine of concordance or unity of opposites, Lat. *coincidentia oppositorum.* According to it, all opposites, finite and infinite, the smallest and the greatest, single and multiple, etc., coincided in God. As a forerunner of the mathematical treatment of infinity, Cusa proposed the doctrine of maximum, the idea of which was the principle of not being able to be exceeded. For instance, a line was a circle of infinite radius which was the maximum of the quality of being circular. With these ideas Cusa anticipated the introduction of infinite quantities and proportions and infinitesimals into mathematics.

The Germ. astronomer and mathematician Georg von Peurbach (1423–1461), instead, carried on the Aristotelian and Ptolemaic system with its finite firmament as the prime mover of the heavens, while the Germ. astronomer and mathematician Petrus Apianus (1495–1552) consented with the Stoic and theological notion of infinite void as Empyrium, the abode of God and angles.

God as the Infinite Universe

The Ital. Platonic scholastic philosopher Francesco Patrizi (1529–1597) followed the Stoics in criticizing Aristotle's cosmology in his "Nova de universis philosophia" (1591). Instead, he expounded the idea that the cosmos itself was God, an infinite good, and that the world derives from Him as a decreasing hierarchy of perfection, from the highest wisdom to life, intellect, soul and lesser qualities and forms of things and finally the earthly matter.

The Ital. dominican astronomer, cosmologist and platonist mystic Giordano Bruno (1548–1600) also rejected Aristotelian cosmology, the Ptolemaic world order and the concept of the cosmos as a crystal sphere. Instead, he adopted the Copernican view that the Earth and planets circles the Sun, that in the infinite universe thousands of other worlds circle thousands of other Suns and that some of them might be inhabited. In his "De l'infinito universo et mondi" (1584), written in dialogue form, Elpino asks "How is it possible that the universe can be infinite?" to which Philoteo, Bruno in disguise, answers with a counter question "How is it possible that the universe can be finite?". "To a body of infinite size there can be neither centre nor boundary ... Just as we regard ourselves at the centre ... so doubtless the inhabitants of the Moon believe themselves to be at the centre ..."

Still, Bruno upheld the pantheistic theological conception of the unity of God with the infinite cosmos. In a poem "De immenso et innumerabilibus" he writes "The One Infinite is perfect, simply and of itself nothing can be created or better than it. This is the One Whole everywhere, God, universal nature. Nothing but the Infinite can be a perfect image and reflection thereof, for the finite is imperfect, every sensible world is imperfect, wherefore evil and good, matter and form, light and darkness, sadness and joy unite, and all things are everywhere in change and motion. But all things come in infinity to the order of Unity, Truth and Good, whereby it is named universum". In his writings "Della causa" and "Principio ed uno", Bruno developed his ideas about the dialectical unity of oppositions, universal motion, and the coincidence of contraries in the infinitely great and the infinitely small.

Bruno travelled widely in Europe advocating his views until he was betrayed by a wealthy Venetian Giovanni Mocenigo who had been commissioned by the church to persuade Bruno to come to teach him "the art of memory and inventio", and was trapped and brought in front of the Inquisition. For eight years he was interrogated, kept imprisoned and tortured but refused to revert his views until finally condemned of heresy and burnt at a stake in Rome.

Omnipresent God in Infinite Physical Cosmos

The French physicist and astronomer Pierre Gassendi (1592–1655) also adopted the Stoic view of cosmos as an infinite void but as a physical and geometrical 3-dimensional infinite space, the ultimate container of all things, capable of holding also physical matter, whereas Bruno had held that it contained only ether and Patrizi that it contained only light. Unlike the scholastics who maintained the view of a created non-dimensional cosmos and identified God with the immensity of space, Gassendi assumed non-created dimensional infinite physical cosmos coeternal with God independent of cosmos but capable of omnipresence. Thus, Gassendi's cosmology and theology consisted of two kinds of independent actual infinities.

Gassendi accepted atomism, not as an infinity of eternal non-created atoms and infinity of worlds, but as a finite number of atoms created by God and making up a single and finite world placed in the infinite 3-dimensional space. The dimensional spatial and atomistic doctrine of a created finite world in an un-created infinite space of Gassendi, despite his theological mysticism, paved the way for the cosmology of an infinite physical space, time and matter that would shortly become accepted as the framework of the scientific revolution and physical world outlook of the New Time.

This view raised, however, serious theological controversies that could not be satisfactorily resolved without revision of the official Catholic doctrine. The dimensionality of space and of God came into contradiction as the immensity of space and God had been identified as one. Either the dimensionality of space would have to be rejected or God would have to be thought as finite and 3-dimensional.

This radical move was taken by the English philosopher and theologian Henry More (1614–1684), a central figure of the 'Cambridge Platonists', in his "Enchridion metaphysicum" (1671). This meant rejection of the explanation of the omnipresence of God by the doctrine of 'whole in every part' that he called holenmerism, Gr. holos, whole, Engl. holy, sacred, divine + Gr. mere, measure, Slav. mera, measure, Germ. mehr, more, cf. Finn. *määrä*, measure. Instead, he insisted that everything, whether corporeal or spiritual, including God, had finite dimension and extension and occupied a place. Thus, also God, alike with space, albeit capable of omnipresence, had dimensions and extension.

The views of Gassendi and More were accepted by the Engl. empiricist philosopher John Locke (1632–1704), the founder of deism, Lat. *deus*, god, Gr. *theos*, Zeus, Sanskr. *deva*, shining one, *daiva*, divinity, *dagh*, day, Finn. *taivas*, sky, Germ. *dag*, Engl. *day*, the doctrine according to which God had created the world but did not interfere with its subsequent events ever since. This view allowed people to maintain their belief in God but allowed them also to pursue goals themselves to change their condition and even the society, giving rise to the ideology of progress.

The Engl. physicist and mathematician Isaac Newton (1642–1727) came to bring these ideas to fruition in science. As the chief architect of classical mechanics he would construct a new physics and cosmology within the framework of an infinite 3-dimensional space, time and matter, and the laws of motion in his "Philosophia naturalis principia mathematica" (1687). Newton reflected also on the philosophical and theological aspects of space and God, but left them unpublished. To him God was an infinite omnipotent spirit that he preferred to call aether, "who could move bodies without resistance", as he noted in his "General Scholium" (1715). In his cosmology the Solar system was set in infinite and absolute Euclidean space. God was eternal and infinite and by existing always and everywhere constituted duration and spatial extension. Infinite suns were spread throughout the infinite space and were held in dynamic equilibrium by their mutual attractions. The physical cosmology of Gassendi, the Copernican world order and the Newtonian mechanics were to overthrow the prehistoric and Biblical myths, natural philosophical cosmologies of Antiquity and the Mediaval theological speculations.

6 Conclusion

From the times of Antiquity it was known that the heavens contained five wandering planets, the Sun and the Moon. The other three planets Uranus, Neptune and Pluto were unknown as they were invisible to the naked eye and could be detected only after the invention of the telescope in the 17th century. The questions concerning the size of the universe and whether the world was finite or infinite and unique or indefinitely plural remained matters of speculation even after the telescope became available. In classical and modern astronomy and cosmology, infinity continued to play a variety of roles until it was finally banished by relativistic cosmology in the large and by quantum gravity in the small and became substituted for by a concept of a universe of finite cosmology.

Yet, the notion of infinity was to undergo a Renaissance of its own in mathematics – in geometry, number theory, analysis, set theory, classical and mathematical logic, discrete mathematics and the theory of computation. Moreover, the dimension of complexity of the real world would come to reveal potential infinities in the dimensions of the evolution of life, mind, language, culture and human thought and civilization, including science, technology and arts – and the notion of infinity was reduced merely to a construction of the human mind.

Literature

Cultural History

1. L. Zippin, *Uses of Infinity*, Random House, New York, 1962.

2. J. L. Borges, *Labyrinths*, New Directions, New York, 1962.

3. R. Péter, *Playing with Infinity*, Dower, New York, 1976.

4. E. Grant, *Much Ado About Nothing*, Cambridge University Press, Cambridge, 1981.

5. E. Maor, *To Infinity and Beyond. A Cultural History of the Infinite*, Birkhüser, Basel, 1987.

6. J. Seppänen, Empty words – Conceptions of nothing in language, mythology, philosophy and science, *Design for Architecture Symposium – Empty Spaces*, Alvar Aalto Museum, Jyväskylä, 14-16 Aug. 1998.

Cosmology

1. N. S. Hetherington, *Cosmology. Historical, Literary, Philosophical, Religious, and Scientific Perspectives*, Garland Publishing, Inc., New York, 1993.

2. N. Ya. Vilenkin, *In Search of Infinity*, Birkhäuser, Basel, 1995.

3. A. M. Wilson, *The Finite in the Infinite*, Oxford University Press, Oxford, 1996.

4. A. Koyré, *From the Closed World to the Infinite Universe*, Johns Hopkins University Press, Baltimore, Md., 1974.

5. M. Blay, *Les raisons de l'infini: Du monde clos à l'univers mathématique*, Gallimard, Paris, 1993.

Philosophy

1. S. Saarnio, *Mitä tiedämme äärettömyydestä?*, Wsoy, Porvoo, 1969.

2. J. Hintikka, *Time and Necessity. Studies in Aristotle's Theory of Modality*, Oxford University Press, Oxford, 1973.

3. R. Rucker, *Infinity and the Mind. The Science and Philosophy of the Infinite*, Birkhäuser, Basel, 1982.

Encyclopedia

1. R. Barber, *A Companion to Mythology*, Kestrel Books, Harmondsworth, 1979.

2. M. Eliade, *The Encyclopedia of Religion*, Macmillan, New York, 1983.

3. S. Schumacher, G. Eoerne, eds., *The Encyclopedia of Eastern Philosophy and Religion*, Shambala, Boston, 1994.

4. * * *, *The Encyclopedia of Eastern Philosophy and Religion*, Shambala, Boston, 1989.

Finite Versus Infinite Neural Computation

Hava T. Siegelmann

Information Systems Engineering
Faculty of Industrial Engineering and Management
Technion, Haifa 32000, Israel
E-mail: iehava@ie.technion.ac.il

1 Introduction

The Turing machine was suggested in 1935 as a model of a mathematician that solves problems by a finitely specifiable algorithm using unlimited time, energy, pencils, and paper. Although frequently used, there are other possible computational models and devices, not all of them finitely specifiable. The brain, for example, could be perceived as a powerful computer with its excellent ability for speech recognition, image recognition, and the development of new theories. The nervous system, constituting an intricately interconnected web of 10^{10}–10^{11} neurons whose synaptic connection strength changes in an adaptive and continuous manner, cannot be perceived as a static algorithm; the chemical and physical processes affecting the neuronal states are *not specifiable by finite means*.

In this work, we focus on the *Analog Recurrent Neural Networks* (ARNNs). These constitute a *finite* number of *continuous* valued neurons connected in a general fashion (not necessarily layered or symmetrical.) We interpret the dynamical behaviour of networks as a process of computation, and study the effects of various constraints and different parameters on their computational power. Altering the networks' constitutive parameters between finite and infinitely described numbers allows the model to coincide with previously considered models that are computationally and conceptually different, from simple automata up to a model that encompasses and transcends digital computation. In this sense, the neural model is a natural framework to test the connection between the level of infiniteness in a computational machine and the associated computational power.

The analog neural network model is different from the digital computer in some significant ways:

(1) The classical (von Neumann) model of the computer consists of a region of memory and a different region of a processing unit where the program is executed. In the neural network's model, memory and processing are not separable. Each neuron is part of the processing unit while the memory is encoded in the weights. Hence if the network has an infinite number of memory registers (weights) like in the Turing model, here it will also have infinite processing power (neurons); and vice versa – a reasonable model of neural computation that has bounded processing power is constrained to a finite number of registers. This differs from the Turing model in which the finite control has access to an unbounded tape of memory cells.

(2) Another inherent difference is the nature of the atomic data types in the memory cells. Digital computers handle bits of information; in analog neurons, the values are reals. If the neurons had finite precision only, the finite interconnections would fall into the computationally weak framework of a finite automaton.

(3) This difference has to do with continuity in the update function of each neuron. In computational terms, there are no flow commands which rely on infinite precision tests, such as

> "If z is greater than 0, compute one thing, and otherwise continue in another computation path."

This principle of continuous flow is definitely a restriction on our model.

(4) The last principle is the description length required to describe the machine model. The Turing model is describable by a finite string of bits. The neural network, having a finite number of neurons, but incorporating exact real weights, should be described by a finite string of real numbers. (This usage of weights with infinite bit description is useless for the actual construction of the system, but it is still appealing for the *mathematical modeling of analog computation* that occurs in nature.)

Given the above, how does the analog network compare with the digital computer in terms of computational capabilities? We show that although the network is continuous, cannot branch on values, does not have a memory unit in the classical sense, and utilizes only a fixed number of neurons, it still can compute anything that a digital machine does. Furthermore, due to its real-valued weights, it is inherently richer than the digital computational model and can compute functions which no digital computer can. The network, although transcending the Turing model, is still sensitive to resource constraints and still computes only a small subclass of all possible functions, thus constituting a well-defined super-Turing computational model.

It is beyond the scope of this work, but we note that when the precision in the weights and the neurons is not perfect or when noise is present, the computational power reduces to the class of regular or even to definite languages [2], [8], [7], [13].

The requirement of continuity in the computation, as specified in item (3), translates to the interesting feature that although real weight neural networks are defined with unbounded precision and although the neurons are able to store infinite precision values, for time-bounded computation, only bounded (linear) precision is required.

Adding discontinuities (e.g., some neurons that compute exact tests for 0) to the analog network changes the resulting dynamical properties. Such augmented networks were called "arithmetic networks" [3]. The first evidence of the arithmetic networks' computational superiority relies on the finding that arithmetic networks can recognize some recursive functions arbitrarily faster than Turing machines and ARNN: they recognize arbitrarily complex recursive functions in linear time. The second evidence concerns the amount of precision required to implement arithmetic networks on digital computers. Unlike the ARNN with their linear precision, no fixed precision function is enough to simulate all arithmetic networks running in linear time.

The work described here on ARNN can be found in full details in [12]; the work on arithmetic networks can be found in [3].

2   The Model

ARNN is a finite size network of neurons in which every neuron updates by the equation

$$x_i(t + 1) = \sigma(\Sigma_{j=1}^{N} a_{ij} x_j(t) + \Sigma_{j=1}^{M} b_{ij} I_j(t) + c_i) \tag{1}$$

where σ is the *piecewise linear* (*semilinear* or *saturated-linear*) function:

$$\sigma(x) = \begin{cases} 0, & \text{if } x < 0, \\ x, & \text{if } 0 \le x \le 1, \\ 1, & \text{if } x > 1, \end{cases} \tag{2}$$

where a, b, and c are the weights and I is the external input [12]. There is nothing particular about the σ function except that it is easy to treat mathematically. The network is uniform in the sense that its size, structure and weights are fixed for all computations and inputs. The only unfixed values are the activation values of the neurons that update in the computational steps.

To consider the network as a computational device, we have to define the input/output (I/O) map generated by the network. We can think of the input as an initial state of the network; output will be defined then similarly as the final state. An equivalent point of view which better fits models from Control Theory uses input and output channels. In this view, the input is a stream of letters which are transferred one at a time on a few channels. A similar convention is applied to the output, which is a stream of letters as well.

For simplicity, we constrain attention only to I/O maps which are defined on bits. This is not at odds with the analog internal computation, and by doing so, we separate the inherently different properties of the network itself from interface (I/O) differences, and we enable a totally fair comparison with the classical digital model. So the initial and final states will be constrained to finite precision, or equivalently, the channels will be digital, transferring bits. For the sake of clarity we next describe *formal networks* for which all results reviewed here are obtained [12]. These networks comply with the I/O protocol that follows. The input arrives on two binary input lines. The first of these is a *data line*, which is used to carry a binary input signal; when no signal is present, it defaults to zero. The second is the *validation line*, and it indicates when the data line is active. It takes the value "1" while the input is present and "0" thereafter. We use "D" and "V" to denote the contents of these two lines, respectively, so

$$I(t) = (D(t), V(t)) \in \{0, 1\}^2$$

for each t. Similarly, there are also two output processors that take the role of data and validation lines; they are denoted $G(t)$ and $H(t)$, respectively.

We now encode each input string

$$\omega = \omega_1 \cdots \omega_k \in \{0, 1\}^+$$

by

$$I_\omega(t) = (D_\omega(t), V_\omega(t)) , \quad t \in \mathbf{N},$$

where

$$D_\omega(t) = \begin{cases} \omega_k, & \text{if } t = 1, \dots, k, \\ 0, & \text{otherwise,} \end{cases}$$

and
$$V_\omega(t) = \begin{cases} 1, & \text{if } t = 1, \ldots, k \\ 0, & \text{otherwise.} \end{cases}$$

A word $\omega \in \{0,1\}^+$ is *classified in time r* by a formal net starting from the initial state $x(1) = 0$, if the input lines (D_ω, V_ω) take the values $D_\omega = \omega 0^\infty$ and $V_\omega = 1^{|\omega|} 0^\infty$, and the output line component $H_\omega(t) = 0$ for $t < r$ and $H_\omega(r) = 1$. If $G_\omega(r)$ is "1" then the word is accepted, and if $G_\omega(r)$ is "0" the word is rejected.

3 Preliminary: Computational Complexity

A thorough theory of computation has been developed to classify functions into classes of equally difficult functions. Each class is associated with a set of rules that describe it or, equivalently, with some type of theoretical machines that can compute exactly this class.

Let's see some common classes:

1. A finite state automaton is a machine that has a fixed number of different internal states. A light switch has two states: on and off. A counter *mod* 7 has 7 states, $0, 1, \ldots, 6$. All tasks executed by any finite automaton are named *regular*.

   One could naively think that, because the world is finite, all machines can be modeled by finite automata. But consider this task: "Given a binary sequence, detect whether the number of 1's in the sequence is greater than the number of 0's there." There is no machine that can memorize the difference between the occurrences of 0's and 1's of *any* binary string using a fixed predetermined finite memory only, so the class of regular functions is pretty weak.

2. Another model is the general purpose computer as suggested by von Neumann. There are different types of computer hardware, computer languages, and ways to represent programs to the computer; but all general digital computers have some common features. The programs executable on them form the class of *recursive functions*, and the model describing them is called the Turing machine. In a Turing machine, a finite automaton is used as a control or main computing unit, but this unit has access to potentially infinite storage space.

   Formally, a Turing machine consists of a finite control and a binary tape, infinite in one or two directions. The tape is accessed by a read-write head. At the beginning of the computation, an input sequence is written in binary code on the tape, surrounded by infinite sequences of blanks on both sides. The head is located at the leftmost symbol of the input string. At each step the machine reads the tape symbol ($a \in \{0, 1, \#\}$) under the head, checks the state of the control ($q \in \{1, 2, \ldots, |Q|\}$), and executes three operations:

(a) It writes a new binary symbol into the current cell under the head ($b \in \{0, 1, \#\}$).

(b) It moves the head one step to either the left or to the right ($m \in \{L, R\}$).

(c) It changes the state of the control ($q' \in \{1, 2, \ldots, |Q|\}$).

The transitions of the machine are described by a function $g(a, q) = (b, m, q')$. When the control reaches a special state, called the "halting state," the machine stops. The output of the computation is defined by the binary sequence written on the tape extending from the read/write head to the first $\#$ symbol on the right side of the head. Thus, the I/O map, or the function, computed by a Turing machine is defined in terms of the binary sequences on its tape before and after the computation.

Turing machines can compute many more functions than finite automata but they are still limited. Consider for example *the halting problem*; this is the decision problem defined as follows. Any Turing machine $\mathcal{M}$ can be encoded by a finite binary sequence $\tau(M)$. Given two words $\tau(M)$ and ω, decide whether the machine $\mathcal{M}$ would halt when starting with the input ω; or in other words, whether $\mathcal{M}$ would ever reach its halting state. It is easily proven that no computer can execute this halting problem.

The field of Computational Complexity looks not only for functions defined by computational models, but also for functions which are efficiently computable on them. In the digital model, a function is considered *efficient* if it requires only *polynomial time* in the length of the input; more formally, if there exists a Turing machine M and a polynomial p_1 such that the machine halts after $p_1(n)$ steps on every binary input string of length n. The class of efficient functions is called P, and it constitutes the basis of complexity theory. Another type of efficiency is defined in terms of space constraints. A function is considered *space efficient* if the number of tape cells which have been used during the computation is bounded to polynomial in the length of the binary input. The class of space efficient classes is called PSPACE. Due to the deterministic characteristic of the Turing machine, PSPACE functions will be computed in up to exponential time. The equality P = SPACE is one of the most interesting open questions in the field of computational complexity.

3. A new model was born in the 80's, when Karp and Lipton introduced a *nonuniform* model of computation that is computationally stronger than the Turing machine [4]. Nonuniform computers are a generalization of the concept of monotonically increasing (e.g., polynomial) response time, that allows an increase in available hardware with input length: in this model, there exists a polynomial $p_2(n)$ such that output for an input of length n is calculated by an acyclic interconnection of $p_2(n)$ digital components (e.g., McCullough and Pitts neurons or logical gates). Inputs of different length are computed by different (nonuniform) hardware. *Note that the hardware for calculating input of length n requires $p_2(n)$ bits of description, and the whole family together cannot be described finitely.* Lipton and Karp showed that such a family of nonuniform acyclic digital circuits exceeds the computational capabilities of the Turing machine. If exponential time is allowed, the nonuniform families compute all binary functions. Under polynomial computation time these families output the nonuniform class P/Poly. The class P/Poly strictly contains P. It also computes some nonrecursive functions, but is still a very small subset of arbitrary binary languages.

4   Summary of Results on ARNN

ARNN constitute a parametric model of computation. Their computational power depends on the type of numbers utilized as weights.

For mathematicians, natural choices of number sets are the integers, rationals, or reals. For computer scientists, a natural choice of function classes is the regular functions, the recursive function class, and the class of all (arbitrary) binary functions. The correspondence between networks with different number domains and computational classes is summarized in Table 1 [15].

The left columns of the table say that networks with integers, rationals, or real weights compute regular, recursive, and arbitrary functions respectively when the time is not constrained. The third column describes the resulting classes for the case that available computing time is constrained to be polynomial in the length of the input. This class is still the regular functions for integer weights, the class P for rationals, and when the weights are real, the resulting class is AnalogP. Although our model is uniform, AnalogP was found to be equivalent to the super-Turing class P/Poly.

Table 1: The computational power of recurrent neural networks

Weights	Arbitrary Time	Polynomial Time
Z	Regular	Regular
Q	Recursive	P
R	Arbitrary	AnalogP

Focusing on the two bottom lines of this table, one may wonder about the essence of the difference between rational numbers and real numbers that causes such a difference in the associated power. For rationals, the framework adheres to the classical realm of computer science of Turing machines and recursive functions while for reals, the networks belong to the analog regime. This question is addressed by Balcazar, Gavalda, and Siegelmann in [1]. They describe rationals, reals, and a whole hierarchy between them in an information theoretic manner, using resource-bounded Kolmogorov-type characterization. This characterization is sensitive to both the amount of input data and the time that is needed to construct the numbers. They prove that there is a proper hierarchy of complexity classes defined by networks, whose weights have increasing Kolmogorov complexity. P and AnalogP are just the two ends of the hierarchy for efficient computation; recursive and arbitrary are the ends for unbounded computation time.

4.1 Networks which are Turing Equivalent

The main theorem in this section is as follows [15], [12]. It is clear that an ARNN of finite size and rational weights can be described in finite terms and be fully simulated by a Turing machine (with no more than polynomial slowdown). We next state that the reverse inclusion holds as well.

Theorem 4.1 *For any function* Φ : $\{0,1\} \rightarrow \{0,1\}^+$ *that is computable by a Turing machine M, there exists a formal network $\mathcal{N}$ with rational weights that computes it. Furthermore,*

  1. *If Φ is computable on M in time $T(n)$ for inputs of length n, then $\mathcal{N}$ requires only $T(n) + O(n)$ time.*

2. *The structure and size of the simulating network is independent of the computation time. These depend only on the structure of the Turing machine, that is, on the number of its tapes and the size of its finite control.*

3. *The statement holds for nondeterministic computation as well.*

Corollaries:

1. There exists a universal neural network that computes all recursive functions.

2. A derivative of the halting problem is that there is no algorithm that decides whether a neuron ever gets the value 1, or a value in the vicinity of 1; similarly, no algorithm can decide whether a given network ever converges.

A secondary theorem asserts that Turing universality is a relatively common property of ARNNs [5]:

Theorem 4.2 *Let ς be any general sigmoidal function (fully described in [5], [12], e.g., $\varsigma(x) = \frac{1}{1+e^{-x}}$). For any ς there exists a universal network of ς-neurons that computes all recursive functions and suffers of up to exponential slowdown.*

4.2 Networks which are Beyond Turing

This subsection considers the full network model when real weights are allowed. We first note the interesting property that although real weight neural networks are defined with unbounded precision, they demonstrate the feature referred to as "linear precision suffices" [14]:

Lemma 4.1 *For up to q steps of computation, only the first $O(q)$ bits of both the weights and the activation values of the neurons influence the result; the less significant bits have no effect on the outcome.*

This property is used to properly formulate the time-dependent resistance ("weak resistance") of the networks to noise and implementation error [14].

The main theorem of this part states that real weight neural networks compute in polynomial time the class AnalogP which is equivalent to P/Poly.

Theorem 4.3 *There exists an integer N such that for every circuit family $\mathcal{C} = \{\mathcal{C}_n | n \in \mathbf{N}\}$ of size S that computes the language $\psi_\mathcal{C}$ there exists a neural network $\mathcal{N}$ with N neurons and one real weight computing $\psi_\mathcal{N}$ so that $\psi_\mathcal{N} = \psi_\mathcal{C}$ and $T_\mathcal{N}(n) \in O(n\,S^2(n))$.*

We next conclude three characteristics of real weight neural networks.

Corollaries:

1. Because of the polynomial correspondence with nonuniform circuits, we conclude that, if exponential computation time is allowed, then one can specify a network for each binary language, including non-computable ones. In polynomial time, the networks compute exactly the language class AnalogP which is equal to P/poly. This class computes all unary languages including the unary encoding of the Turing machine halting problem. But it includes a very small fraction of all binary languages. Although AnalogP strictly includes all P, it is not likely to contain functions of the classical set NP, since otherwise the polynomial hierarchy will collapse to Σ_2 [4].

2. A nondeterministic version can also be specified: AnalogNP is equivalent to nonuniform families of nondeterministic circuits.

3. This conclusion is a corollary of the property "linear precision suffices". The amount of information necessary for the neural network is identical to the precision required by chaotic systems. Therefore, neural networks may constitute a framework for the modeling of physical dynamics.

Remark. Unlike the Turing machine, the analog recurrent neural network is not universal for digital input but only for input in the form of a finite string of real numbers.

4.3 Analog Computation

Up to now, we concentrated on the very particular model of the Analog Recurrent Neural Network. We next shift viewpoints to see this network as only one example of a class.

As a first step we generalize the networks to *generalized analog networks*. In vector form, a generalized analog network D updates via the equation

$$x^+ = f(x, I),$$

where x is the current state vector of the network, I is an external input (possibly a vector), and f is a composition of functions:

$$f = \vartheta \circ \pi,$$

where

$$\pi : \quad \mathbf{R}^{N+M} \to \mathbf{R}^N$$

is some vector polynomial in $N + M$ variables with real coefficients, and

$$\vartheta : \mathbf{R}^N \to \mathbf{R}^N$$

is any vector function that has a *bounded range* and is *Lipschitz*.

That is, for every $\rho > 0$, there exists a constant C such that for all $x, \tilde{x} \in \text{Domain}(\vartheta)$, if $|x - \tilde{x}| < \rho$, then $|\vartheta(x, u) - \vartheta(\tilde{x}, u)| \le C|x - \tilde{x}|$ for any binary vector u. A similar property holds for $f = \vartheta \circ \pi$.

It is not hard to prove that this whole class of networks is not computationally stronger than the basic homogeneous neural network [14]

Our next step is to add stochasticity to the networks by means of a heads/tails coin, where the probability to fall on head or tail is a real number.

Definition 4.1 *A stochastic network has additional input lines, called stochastic lines, that carry independent identically distributed (IID) binary sequences, one bit per line at each tick of the clock. The distributions may be different on the different lines. That is, for all time $t \ge 0$, the stochastic line l_i has the value 1 with probability p_i ($0 \le p_i \le 1$), and 0 otherwise.*

We say that the language $L \subseteq \{0,1\}^+$ is ϵ-recognized in time T by a stochastic network $\mathcal{N}$ if every input string $\omega \in \{0,1\}^+$ is classified in time $T(|\omega|)$ by every computation path of $\mathcal{N}$ on ω, and the error probability in deciding ω relative to the language L is bounded: $e_{\mathcal{N}}(\omega) < \epsilon < \frac{1}{2}$.

A relatively standard lemma shows that for error-bounded stochastic networks, the error probability can be reduced to any desired value [9]. This indicates that the complexity class of functions recognized by stochastic networks is well-defined.

The stochastic model reveals intermediate interesting classes between P and AnalogP. For example, stochastic networks with rational weights and real probabilities form a similar model to the quantum Turing machine and computes BPP/log. The full model with stochasticity still computes AnalogP.

We next form a more general framework of analog computational models, which do not have to adhere to a network structure at all. This one will handle real atomic data items, use real constants, and it will update by simple continuous functions. The time update can be synchronous, or nonsynchronous. The analog shift map is a good example [11]. It is a generalization of the well-known shift map by Smale and the generalized shift map by Moore [6]. The strong analog shift map computes exactly AnalogP.

Noting that the basic ARNN encompasses many other models of Analog computation, Sontag and Siegelmann proposed an analogy to the Church-Turing thesis of computability, but to the realm of analog computation. Their thesis of time bounded analog computation suggests that all reasonable analog computational models can be described by the simple ARNN:

> "No possible abstract analog device can have more computational capabilities (up to polynomial time) than Analog Recurrent Neural Networks."

This can be interpreted as suggesting that the neural network be considered a standard model in the realm of analog computation, functioning in a role similar to that of the Turing machine in the realm of digital computation.

5   Networks with Infinite Precision Tests

The class of analog systems introduced in the previous section relies on the feature of continuous update. We next show that the introduction of even a single discontinuity changes the computational properties of these systems. The material of this section describes the work from [3].

Arithmetic networks are finite size interconnections with two types of neurons. Some are continuous and some are not. In the simple model, some neurons will compute $\sigma(\phi(x, \Phi))$ where Φ is any polynomial and σ is the saturated linear function of equation (2); other neurons will compute $\sigma_H(\phi(x, I))$, where ϕ_H is the threshold function

$$\mathcal{H}(x) = \begin{cases} 1, & x \geq 0, \\ 0, & x < 0, \end{cases}$$

or the zero test function

$$\mathcal{Z}(x) = \begin{cases} 1, & x = 0, \\ 0, & x \neq 0 \end{cases}$$

(or any other function with "gap continuities" [3]).

A real number r is called *polynomial-time computable* if there is a polynomial p and a Turing machine M such that on input n, M will produce the first n digits of the fractional part of r in time $p(n)$. All algebraic numbers, constants such as π and e, and many others are polynomial-time computable. To emphasize how small this class is, we note that there are no more polynomial-time computable real numbers than Turing machines; hence, there are countably many of them. Furthermore, when used as constants in ARNN, the networks still compute the class P only, just like in the case where all constants are rational numbers.

The next theorem contrasts the "linear precision suffices" property of ARNN.

Theorem 5.1 *There is no computable precision function $r(n)$ such that "precision $O(r(t(n)))$ suffices" to simulate all arithmetic networks running in time $t(n)$. This is true even if only polynomial-time computable weights are used.*

The next theorem states the unlimited speed-up of arithmetic networks; it contrasts ARNN because with polynomially computable real weights, they compute only functions in P.

Theorem 5.2 *There are arithmetic networks that run in polynomial time, have polynomial-time computable weights, and yet they accept recursive languages of arbitrarily high time complexity (in the Turing machine sense).*

Theorems 5.1 and 5.2 are both consequences of the following theorem.

Theorem 5.3 *For every time-constructible function $t(n)$ there is an arithmetic network $\mathcal{N}$ such that:*

1. *The weights in $\mathcal{N}$ are computable in time $O(n)$;*

2. *$\mathcal{N}$ runs in time $2n$;*

3. *The language T accepted by $\mathcal{N}$ is recursive, but not decidable in time $O(t(n))$ by any Turing machine.*

4. *Precision $O(t(n))$ does not suffice to simulate $\mathcal{N}$, that is, if $\mathcal{N}$ is simulated with precision $O(t(n))$, then a language different from T is accepted, even in the soft acceptance sense.*

Remark. Although ARNN in which ϕ is a linear combination or a high-order polynomial are computationally equivalent, the high order property is required for the theorems on arithmetic networks.

Acknowledgments

This work was partially funded by the Israeli Ministry of Art and Sciences, the Binational US/Israel Foundation, the fund for promotion of research at the Technion, and the VPR fund at the Technion.

References

[1] J. L. Balcázar, R. Gavaldà, H. T. Siegelmann, Computational power of neural networks: A characterization in terms of Kolmogorov complexity, *IEEE Transactions on Information Theory*, **43**, 4 (1997), 1175–1183.

[2] M. P. Casey, The dynamics of discrete-time computation with application to recurrent neural networks and finite state machine extraction, *Neural Computation*, **8**, 6 (1996), 1135–1178.

[3] R. Gavaldà, H. T. Siegelmann, Discontinuities in recurrent neural networks, *Neural Computation*, **11**, 3 (1999), 715–745.

[4] R. M. Karp, R. Lipton, Turing machines that take advice, *Enseignment Mathematique*, **28** (1982), 191–209.

[5] J. Kilian, H. T. Siegelmann, The dynamic universality of sigmoidal neural networks, *Information and Computation*, **128**, 1 (1996), 48–56.

[6] C. Moore, Unpredictability and undecidability in dynamical systems, *Physical Review Letters*, **64** (1990), 2354–2357.

[7] W. Maass, E. D. Sontag, Analog neural nets with Gaussian or other common noise distribution cannot recognize arbitrary regular languages, *Neural Computation*, **11**, 3 (1999), 771–782.

[8] P. Orponen, W. Maass, On the effect of analog noise on discrete time analog computations, *Neural Computation*, **10** (1998), 1071–1095.

[9] I. Parberry, *Circuit Complexity and Neural Networks*, MIT Press, Cambridge, MA, 1994.

[10] R. Penrose, *The Emperor's New Mind*, Oxford University Press, Oxford, England, 1989.

[11] H. T. Siegelmann, Computation beyond the Turing limit, *Science*, **268** (April 1995), 545–548.

[12] H. T. Siegelmann, *Neural Networks and Analog Computation: Beyond the Turing Limit*, Birkhauser, Boston, MA, 1999.

[13] H. T. Siegelmann, A. Roitershtein, Noisy computational systems and definite languages, *Discrete Applied Mathematics*, 1999, to appear.

[14] H. T. Siegelmann, E. D. Sontag, Analog computation via neural networks, *Theoretical Computer Science*, **131** (1994), 331–360.

[15] H. T. Siegelmann, E. D. Sontag, On computational power of neural networks, *Journal of Computer and System Sciences*, **50**, 1 (1995), 132–150.

On Information-Theoretical Aspects
of Relational Databases

Dan A. Simovici, Szymon Jaroszewicz

Department of Mathematics and Computer Science
University of Massachusetts at Boston
Boston, MA 02125, USA
E-mail: {dsim,sj}@cs.umb.edu

Abstract. We introduce the notion of entropy for a set of attributes of a table in a relational database starting from the notion of entropy for finite functions. We examine the connections that exist between conditional entropies of attribute sets and lossy decompositions of tables and explore the properties of the entropy of sets of attributes regarded as an outer measure on the set of subsets of a heading of a table. Finally, we suggest a generalization of functional dependencies based on conditional entropy.

1   Introduction

This paper examines some applications of information theory to relational database systems. The starting point is our axiomatization of entropies of finite functions based on the composition operation between functions that was presented in [5].

Unless we state otherwise, all sets considered in this paper are finite.

Let $\mathcal{F}$ be the class of all functions between finite, nonempty sets. Define the partial order $\sqsubseteq$ on $\mathcal{F}$ by $f \sqsubseteq g$ if $f : A \to B$, $g : A' \to B'$, $A \subseteq A'$, $B \subseteq B'$ and $f(a) = g(a)$ for $a \in A$. Note that if A, B, C, D are finite sets, such that $A \cap B = C \cap D = \emptyset$, $f : A \to B$, and $g : C \to D$, then there exists $\sup\{f, g\}$. The function $f \sqcup g = \sup\{f, g\}$ is given by

$$(f \sqcup g)(x) = \begin{cases} f(x), & \text{if } x \in A, \\ g(x), & \text{if } x \in B. \end{cases}$$

It is easily verifiable that, whenever it is defined, the "$\sqcup$" operation is commutative and associative.

The composition of functions $f : A \to B$ and $g : B \to C$ is a function $gf : A \to C$, such that $gf(x) = g(f(x))$, for every $x \in A$.

The Cartesian product of functions $f : A \to B$ and $g : C \to D$ is a function $f \times g : A \times C \to B \times D$ defined as $(f \times g)(x, y) = (f(x), g(y))$, for every $x \in A$ and $y \in C$.

Let $A_1, A_2, \ldots, A_n$ be n disjoint subsets of a finite set A. The *generalized characteristic function* of sets $A_1, A_2, \ldots, A_n$ is a function:

$$\chi_{A_1, A_2, \ldots, A_n} : A_1 \cup A_2 \cup \ldots \cup A_n \to \{1, 2, \ldots, n\},$$

such that $\chi_{A_1, A_2, \ldots, A_n}(x) = i$ if $x \in A_i$, for $1 \le i \le n$.

Another important notion for this paper is the notion of *multiset*. We present a few useful facts (see [3] for further details).

Definition 1.1 *A multiset or bag on a set S is a subset M of $S \times \mathbf{N}$ such that for every $a \in S$ there is an $n \in \mathbf{N}$ such that $(a, m) \in M$ if and only if $m < n$. The set of all multisets on S is denoted by $\mathcal{M}(S)$.*

If S is a set, M is a multiset on S, and $a \in S$, then the number n such that $(a, m) \in M$ if and only if $m < n$ is unique since it is the least number r such that $(a, r) \notin M$. We call this number n the *multiplicity* of a in M and we denote it by $M(a)$.

A multiset on a set S is determined by the multiplicities of its elements. Note that for any set S, $\emptyset \subseteq \mathcal{M}(S)$. The cardinality of a multiset M on S is $|M| = \sum_{a \in S} M(a)$.

Let S be a set and let M and N be multisets on S. It is easy to see that $M \cup N$ is a multiset on S and that if $a \in S$ has multiplicity m in M and n in N, then a has multiplicity $\max(m, n)$ in $M \cup N$. Similarly, $M \cap N$ is a multiset on S and if $a \in S$ has multiplicity m in M and n in N, then a has multiplicity $\min(m, n)$ in $M \cap N$. In other words, we have

$$(M \cup N)(a) = \max\{M(a), N(a)\},$$
$$(M \cap N)(a) = \min\{M(a), N(a)\},$$

for every $a \in S$.

The *sum* of M and N (denoted $M + N$) is the multiset on S such that for each $a \in S$, the multiplicity of a in $M + N$ is the sum of the multiplicities of a in M and N.

Let S be a set. For each subset T of S, we can define a multiset M_T on S by defining

$$M_T = \{(a, 0) \mid a \in T\}.$$

(So, if $a \in T$, then a has multiplicity 1 in M_T, and if $a \notin T$, then a has multiplicity 0 in M_T.)

For any two subsets U and T of S we have $M_{U \cup T} = M_U \cup M_T$ and $M_{U \cap T} = M_U \cap M_T$.

If M_1 and M_2 are multisets on S_1 and S_2 respectively, then the Cartesian product $M_1 \times M_2$ of M_1 and M_2 is defined as a multiset on $S_1 \times S_2$ such that

$$(M_1 \times M_2)(a, b) = M_1(a) \cdot M_2(b),$$

for every $a \in S_1, b \in S_2$.

2   Functional Entropy

If $f : A \to B$, $g : A \to C$ are finite functions, let $p_f(b)$ and $p_{f,g}(b, c)$ be given by

$$p_f(b) = P(f = b) = |f^{-1}(b)|/|A|,$$
$$p_{f,g}(b, c) = P(f = b \wedge g = c) = |f^{-1}(b) \cap g^{-1}(c)|/|A|,$$

for $b \in B$, and $c \in C$. Note that

$$\sum_{c \in C} p_{f,g}(b, c) = p_f(b) \quad \text{and} \quad \sum_{b \in B} p_{f,g}(b, c) = p_g(c).$$

Definition 2.1 *Let $f : A \to B$ be a function between finite sets, and let $B = \{b_1, b_2, \ldots, b_m\}$. The discrete random variable associated with f is denoted by X^f and defined as:*

$$X^f = \begin{pmatrix} b_0 & b_1 & \cdots & b_{n-1} \\ p_f(b_0) & p_f(b_1) & \cdots & p_f(b_{n-1}) \end{pmatrix}.$$

Let $f : A \to B$ and $g : A \to C$ be functions between finite sets. The functions f and g are said to be independent *if the random variables X^f and X^g are independent, that is, if $p_{f,g}(b, c) = p_f(b)p_g(c)$ for every $b \in B$ and $c \in C$.*

Let $H : \mathcal{F} \to \mathbf{R}$ be a function assigning a real number to every $f \in \mathcal{F}$ that satisfies the following properties:

(E1) $H(f\alpha) = H(f)$, for every function $f : A \rightarrow B$ and bijection $\alpha : A' \rightarrow A$.

(E2) $H(gf) \leq H(f)$, for every $f : A \rightarrow B$ and $g : B \rightarrow C$.

(E3) Let A, C be finite sets, such that $|A| \leq |C|$, let $\alpha : A \rightarrow B$ and $\beta : C \rightarrow D$ be bijections. Then $H(\alpha) \leq H(\beta)$.

(E4) If $f : A \rightarrow B$ and $f : C \rightarrow D$ are functions between finite sets and $A \cap C = B \cap D = \emptyset$, then

$$H(f \sqcup g) = \frac{|A|}{|A \cup C|} H(f) + \frac{|C|}{|A \cup C|} H(g) + H(\chi_{A,C}).$$

(E5) $H(f \times g) = H(f) + H(g)$ for all functions $f : A \rightarrow B$ and $g : C \rightarrow D$.

We proved in [5] that for any such function H we have

$$H(f) = -\sum_{i=0}^{m-1} \frac{|A_i|}{|A|} \log \frac{|A_i|}{|A|},$$

where $f : A \rightarrow B$, $B = \{b_0, \ldots, b_{m-1}\}$ and $A_i = f^{-1}(b_i)$ for $0 \leq i \leq m - 1$.

The conditional entropy of two functions was defined in [5] as a function $H : \mathcal{F} \times \mathcal{F} \rightarrow \mathbf{R}$ that satisfies the following conditions:

(CE1) $H(f|c) = H(f)$ for every constant function c.

(CE2) For every $f : A \rightarrow B$, $g_1 : A_1 \rightarrow C_1$, $g_2 : A_2 \rightarrow C_2$, where $A_1 \cap A_2 = C_1 \cap C_2 = \emptyset$ and $A_1 \cup A_2 = A$ we have

$$H(f|g_1 \sqcup g_2) = \frac{|A_1|}{|A|} H(f_{A_1}|g_1) + \frac{|A_2|}{|A|} H(f_{A_2}|g_2).$$

If $B = \{b_0, \ldots, b_{m-1}\}$, $C = \{c_0, \ldots, c_{n-1}\}$, then, as we have shown in [5], the axioms (CE1-2) imply that the conditional entropy between the functions $f : A \rightarrow B$ and $g : A \rightarrow C$ has the form:

$$H(f|g) = -\sum_{j=0}^{n-1} \sum_{i=0}^{m-1} \frac{|A_j^i|}{|A|} \log \frac{|A_j^i|}{|A_j|},$$

where $A_j^i = f^{-1}(b_i) \cap g^{-1}(c_j)$, $A_j = g^{-1}(c_j)$ for $0 \leq i \leq m - 1$ and $0 \leq j \leq n - 1$.

Theorem 2.1 *Let $f : A \rightarrow B, g : A \rightarrow C$ be two functions. There exists a function $h : C \rightarrow B$ such that $f = hg$ if and only if $H(f|g) = 0$.*

Proof. Let $B = \{b_0, \ldots, b_{m-1}\}$ and $C = \{a_0, \ldots, a_{n-1}\}$. Suppose that $H(f|g) = 0$. Using the previous notations, for every nonempty set A^i_j we have $|A^i_j| = |A_j|$ for every i, $0 \leq i \leq m-1$. Since $A^i_j \subseteq A_j$ it follows that $A^i_j \neq \emptyset$ implies $A^i_j = A_j$. Thus, if there exists $a \in A$ such that $g(a) = c_j$ then $f(a) = b_i$. This shows that the mapping $h : C \rightarrow B$ given by $g(c_j) = b_i$ is well defined and $f = hg$.

Conversely, suppose that $f = hg$. Then, if $A^i_j \neq \emptyset$ we have $A_j = A^i_j$. Indeed, suppose that $a \in A_j$, that is $g(a) = c_j$, which implies $f(a) = h(c_j)$, which proves that $A_j \subseteq A^i_j$, where $b_i = h(c_j)$. The converse inclusion is obvious, and this implies immediately $H(f|g) = 0$.   □

The *mutual information* between functions $f : A \rightarrow B$ and $g : A \rightarrow C$, denoted by $I(f; g)$, is defined as $I(f; g) = H(f) - H(f|g)$ and can be written as:

$$I(f; g)$$
$$= -\sum_{b \in B} p_f(b) \log p_f(b) + \sum_{b \in B} \sum_{c \in C} p_{f,g}(b, c) \log \frac{p_{f,g}(b, c)}{p_g(c)} \quad (1)$$
$$= -\sum_{b \in B} \sum_{c \in C} p_{f,g}(b, c) \log p_f(b) + \sum_{b \in B} \sum_{c \in C} p_{f,g}(b, c) \log \frac{p_{f,g}(b, c)}{p_g(c)}$$
$$= \sum_{b \subset B} \sum_{c \subset C} p_{f,g}(b, c) \log \frac{p_{f,g}(b, c)}{p_f(b) p_g(c)}.$$

Lemma 2.1 *We have $I(f; g) \geq 0$ for all functions between finite sets $f : A \rightarrow B$ and $g : A \rightarrow C$, with equality if and only if f and g are independent.*

Proof. The statement follows immediately from formula (1).   □

Below we give a few facts concerning entropies of functions between finite sets.

Theorem 2.2 *We have $H(f|hg) \geq H(f|g)$, for every $f : A \rightarrow B$, $g : A \rightarrow C$ and $h : C \rightarrow D$.*

Proof. Let $D = \{d_1, \ldots, d_m\}$ and $A_i = (hg)^{-1}(d_i)$. From (CE2) and (CE1) it follows that

$$H(f|hg) = \sum_{i=1}^{m} \frac{|A_i|}{|A|} H(f_{A_i}). \quad (2)$$

Since $g(A_i) \cap g(A_j) = \emptyset$, for $i \neq j$, we can decompose g into a union $g = g_{A_1} \sqcup \cdots \sqcup g_{A_m}$, where $g_{A_i} = g|_{A_i}$ is a restriction of g to the set A_i. Using (CE2) we obtain:

$$H(f|g) = \sum_{i=1}^{m} \frac{|A_i|}{|A|} H(f_{A_i}|g_{A_i}). \tag{3}$$

From Lemma 2.1 it follows that

$$H(f_{A_i}) \geq H(f_{A_i}|g_{A_i}), \tag{4}$$

for $1 \leq i \leq m$. From (2), (3) and (4) we obtain: $H(f|hg) \geq H(f|g)$.   □

Let $f : A \to B$ and $g : A \to C$ be two functions. Define the function $(f;g) : A \to B \times C$ as $(f;g)(a) = (f(a), g(a))$ for every $a \in A$. We proved in [5] that

$$H(f;g) = H(f|g) + H(g) = H(g|f) + H(f). \tag{5}$$

3   Database Tables and m-Tables

Let $U = \{A_0, A_1, \ldots\}$ be a set of symbols referred to as attributes; U itself is referred to as the *universal set*. We assume that for each attribute $A \in U$ there exists a finite set $\mathrm{Dom}(A)$, the *domain* of A, that contains at least two elements.

A finite set of attributes $\{A_0, \ldots, A_{n-1}\}$ is denoted by $A_0 \ldots A_{n-1}$, which is the standard notation for sets of database attributes. The union of sets of attributes A and B will be denoted by AB. At times however, for the sake of clarity, the standard mathematical notation $A \cup B$ will be used.

If $L = A_{i_0} \cdots A_{i_{n-1}} \subseteq U$, then we denote by $\mathrm{Dom}(L)$ the Cartesian product

$$\mathrm{Dom}(A_{i_0}) \times \cdots \times \mathrm{Dom}(A_{i_{n-1}}).$$

Uppercase letters denote sets of attributes, while lowercase letters denote values from the domain of the corresponding set of attributes, e.g., $x \in \mathrm{Dom}(X)$ or $y \in \mathrm{Dom}(Y)$.

If $u \in \mathrm{Dom}(L)$ and $K \subseteq L$, we denote by $u[K]$ the restriction of u to K; we refer to $u[K]$ as the *projection* of u on K.

Let L, M be two finite subsets of the universal attribute set U. The tuples $u \in \mathrm{Dom}(L)$ and $v \in \mathrm{Dom}(M)$ are *joinable* if $u[L \cap M] = v[L \cap M]$. Their *join* is the tuple $u \bowtie v \in \mathrm{Dom}(L \cup M)$ defined by $(u \bowtie v)[A] = u[A]$

if $A \in L$ and $(u \bowtie v)[A] = v[A]$ if $A \in M$. The joinability condition insures that the tuple $u \bowtie v$ is well-defined.

Definition 3.1 *A* table *is a triple* $\sigma = (T, L, \rho)$, *where* T *is a string of characters referred to as the* name *of the table,* $L = A_{i_0} \cdots A_{i_{n-1}}$ *is a finite subset of* U *called the* heading *of* σ*, and* $\rho \subseteq \mathrm{Dom}(L)$ *is an n-ary relation referred to as the* extension *of* σ*. The pair* (T, L) *is the* format *of* σ*. An element* $t \in \rho$ *is called a* tuple *of* σ*.*

Let $\sigma = (T, L, \rho)$ be a table and let $X \subseteq L$. The *projection* of σ on the set X is the table $\sigma[X] = (T[X], X, \rho[X])$, where $\rho[X] = \{t[X] \mid t \in \rho\}$. Observe that for every $X \subseteq L$ we have a projection function $\pi_{\sigma,X} : \rho \to \rho[X]$, where $\pi_{\sigma,X}(t) = t[X]$ for every $t \in \rho$.

If $\sigma = (T, L, \rho), \sigma' = (T', L', \rho')$ are tables, then their join is the table $\sigma \bowtie \sigma' = (T \bowtie T', L \cup L', \rho \bowtie \rho')$, where

$$\rho \bowtie \rho' = \{u \bowtie v \mid u \in \rho, v \in \rho' \text{ and } u, v \text{ are joinable}\}.$$

The notations and terminology introduced above can be found, for example, in [9].

To allow duplicate tuples in database tables, we introduce the notion of *m-table* as follows:

Definition 3.2 *An* m-table *is a triple* $\tau = (T, L, \mu)$*, where* T, L *are as in Definition 3.1, and* μ *is a finite multiset on* $\mathrm{Dom}(L)$*.*

An element $t \in \mu$ will be called a tuple of τ.

If $X \subseteq L$ is a set of attributes, the *projection* of an m-table τ on X is the m-table $\tau[X] = (T[X], X, \mu[X])$, where

$$\mu[X](x) = \sum \{\mu(t) \mid t \in \mathrm{Dom}(L) \text{ and } t[X] = x\}.$$

Let $t = (r, n) \in \mu$, denote $t[X] = r[X]$.

A table $\sigma = (S, L, \rho)$ can be regarded as a special case of m-table $\tau = (S, L, \mu)$, where

$$\mu(t) = \begin{cases} 1, & \text{if } t \in \rho, \\ 0, & \text{otherwise,} \end{cases}$$

for every $t \in \mathrm{Dom}(L)$.

Example 3.1 *Consider the m-table* $\tau = (T, ABC, \mu)$*, where*

$$\mathrm{Dom}(A) = \{a_0, a_1, a_2\},$$
$$\mathrm{Dom}(B) = \{b_0, b_1\},$$
$$\mathrm{Dom}(C) = \{c_0, c_1, c_2\}.$$

Suppose that $\mu(a_0, b_1, c_2) = 2$, $\mu(a_1, b_0, c_1) = 3$, $\mu(a_2, b_1, c_2) = 1$ and $\mu(t) = 0$ for every other $t \in \text{Dom}(ABC)$. Then, we visualize this m-table as

$$T$$

A	B	C
a_0	b_1	c_2
a_0	b_1	c_2
a_1	b_0	c_1
a_1	b_0	c_1
a_1	b_0	c_1
a_2	b_1	c_2

The projection function $\pi_{\tau,X}$ for m-tables has a definition that is similar to the one used for common tables.

Definition 3.3 *Let $\tau = (T, L, \mu)$ be an m-table and let $X \subseteq L$. The function $\pi_{\tau,X} : \mu \to \text{Dom}(X)$ is given by $\pi_{\tau,X}(t) = t[X]$.*

The relational algebra selection operation for m-tables is defined in a similar way as for ordinary tables.

Let $\tau_1 = (T_1, L_1, \mu_1)$, and $\tau_2 = (T_2, L_2, \mu_2)$ be two m-tables. The *join* of m-tables τ_1 and τ_2 is the m-table $\tau_1 \times \tau_2 = (T_1 \bowtie T_2, L_1 L_2, \mu_1 \bowtie \mu_2)$, such that

$$(\mu_1 \bowtie \mu_2)(r_1 \bowtie r_2) = \mu_1(r_1) \cdot \mu_2(r_2)$$

for any joinable $r_1 \in \text{Dom}(L_1)$ and $r_2 \in \text{Dom}(L_2)$.

The behaviour of m-tables with respect to the join operation differs from the behavior of regular tables, as the next example shows.

Example 3.2 *Consider m-table $\tau = (T, ABC, \mu)$ given below*

$$T$$

A	B	C
a_0	b_1	c_3
a_0	b_1	c_4

Let us form an m-table $\tau' = \tau[AC] \bowtie \tau[AB]$:

$$T$$

A	B	C
a_0	b_1	c_3
a_0	b_1	c_4
a_0	b_1	c_3
a_0	b_1	c_4

Thus, in the case of m-tables, we may have $|\mu'| > |\mu|$ even if the functional dependency $A \rightarrow B$ holds.

4   Entropy of Attribute Sets

Let $\tau = (T, L, \mu)$ be an m-table, and $X \subseteq L$ be a set of attributes. The *active domain* of a set of attributes X in the table τ is the set

$$\text{adom}_\tau(X) = \{x \in \text{Dom}(X) \mid t[X] = x \text{ for some } t \in \mu\}.$$

Since tables are special cases of m-tables, the above notation is valid also for tables.

Now we introduce the definition of entropy of sets of attributes of an m-table.

Definition 4.1 *Let $\tau = (T, L, \mu)$ be an m-table and let $X \subseteq L$. The entropy of X in τ is defined as $H(\pi_{\tau,X})$. We will denote this entropy as $H_\tau(X)$. We assume that $H_\tau(\emptyset) = 0$, for any m-table $\tau = (T, L, \mu)$.*

The conditional entropy between the sets of attributes X, Y in τ is defined as

$$H_\tau(Y|X) = H(\pi_{\tau,Y}|\pi_{\tau,X}).$$

We assume that $H_\tau(\emptyset|X) = 0$, for any m-table $\tau = (T, L, \mu)$ and any $X \subseteq L$.

The joint entropy of X and Y denoted by $H_\tau(X;Y)$ is defined as

$$H_\tau(X;Y) = H(\pi_{\tau,X}; \pi_{\tau,Y}).$$

These definitions and the equalities (5) imply

$$H(X;Y) = H(X|Y) + H(Y) = H(Y|X) + H(X). \tag{6}$$

Let $\tau = (T, L, \mu)$ be an m-table and let

$$x \in \text{Dom}(X), \ y \in \text{Dom}(Y), \ z \in \text{Dom}(Z),$$

where $X, Y, Z \subseteq L$. Define the probabilities:

$$p_\tau(x) = |\{t \in \mu : t[X] = x\}|/|\mu|,$$
$$p_\tau(y, z) = |\{t \in \mu : t[Y] = y \wedge t[Z] = z\}|/|\mu|,$$
$$p_\tau(y|x) = |\{t \in \mu : t[Y] = y \wedge t[X] = x\}|/|\{t \in \mu : t[X] = x\}|,$$
$$p_\tau(y, z|x) = |\{t \in \mu : t[Y] = y \wedge t[Z] = z \wedge t[X] = x\}|$$
$$/|\{t \in \mu : t[X] = x\}|.$$

If τ is clear from context, we will omit this subscript.

Definition 4.2 *Two sets of attributes X and Y are* independent *if the corresponding functions π_X and π_Y are independent.*

Theorem 4.1 *Let $\tau = (T, L, \mu)$ be an m-table and $X, Y \subseteq L$ be sets of attributes. Then $H_\tau(XY) = H_\tau(X; Y)$.*

Proof. Let $R_{XY} = \{(x, y) \mid x \in \text{Dom}(X), y \in \text{Dom}(Y), x[X \cap Y] = y[X \cap Y]\}$. Define the function $(\pi_X; \pi_Y) : \mu \to R_{XY}$ by $(\pi_X; \pi_Y)(t) = (t[X], t[Y])$ for $t \in \mu$.

If $f : \text{Dom}(XY) \to R_{XY}$ and $g : R_{XY} \to \text{Dom}(XY)$ are given by $f(v) = (v[X], v[Y])$ for $v \in \text{Dom}(XY)$ and $g(x, y) = (x \bowtie y)$ for $(x, y) \in R_{XY}$, then $(\pi_X; \pi_Y)(t) = f[\pi_{XY}(t)]$ and $\pi_{XY}(t) = g[(\pi_X; \pi_Y)(t)]$ for every $t \in \mu$. From (E2) it follows that $H(\pi_X; \pi_Y) \leq H(\pi_{XY})$ and $H(\pi_{XY}) \leq H(\pi_X; \pi_Y)$, so $H(\pi_X; \pi_Y) = H(\pi_{XY})$.   □

The functional dependency between sets of attributes X and Y is denoted by $X \to Y$. The definition of functional dependencies for m-tables is analogous to the corresponding definition for tables.

Theorem 4.2 *Let $\tau = (T, L, \mu)$ be an m-table and let $X, Y \subseteq L$ be sets of attributes. The following statements are equivalent:*

1. *τ satisfies the functional dependency $X \to Y$;*

2. *$H_\tau(Y|X) = 0$;*

3. *$H(X) = H(XY)$.*

Proof. The m-table τ satisfies the functional dependency $X \to Y$ if and only if there exists a mapping $f : \text{adom}_\tau(X) \to \text{adom}_\tau(Y)$ such that $\pi_Y = f\pi_X$. Indeed, if such a mapping exists then $u[X] = v[X]$ amounts to $\pi_X(u) = \pi_X(v)$, which implies $\pi_Y(u) = f(\pi_X(u)) = f(\pi_X(v)) = \pi_Y(v)$ for $u, v \in \mu$. Thus, $u[Y] = v[Y]$, which shows that τ satisfies $X \to Y$. Conversely, if τ satisfies $X \to Y$, then the mapping $f : \text{adom}_\tau(X) \to \text{adom}_\tau(Y)$ given by $f(x) = y$ if there is a tuple $t \in \mu$ such that $t[X] = x$ and $t[Y] = y$ is well defined and $\pi_Y = f\pi_X$. Thus, by Theorem 2.1, τ satisfies $X \to Y$ if and only if $H_\tau(Y|X) = 0$.

To prove the equivalence between the second and the third statement observe that the equalities (6) and Theorem 4.1 imply

$$H_\tau(XY) = H_\tau(X|Y) + H_\tau(Y) = H_\tau(Y|X) + H_\tau(X). \tag{7}$$

Thus, $H_\tau(X) = H_\tau(XY)$ if and only if $H_\tau(Y|X) = 0$.   □

Theorem 4.3 *Let $\tau = (T, L, \mu)$ be an m-table and let $X, Y, Z \subseteq L$. The following statements hold:*

1. *if $X \subseteq Y$, then $H_\tau(X) \leq H_\tau(Y)$; the equality $H_\tau(X) = H_\tau(Y)$ holds if and only if τ satisfies the functional dependency $X \to Y$;*

2. *$H_\tau(Y|X) \leq H_\tau(YZ|X)$, with equality if and only if τ satisfies the functional dependency $XY \to Z$;*

3. *if $Y \subseteq Z$, then $H_\tau(Y|X) \leq H_\tau(Z|X)$, with equality if and only if τ satisfies $XY \to Z$.*

Proof. For the first part note that the equalities (7) imply $H_\tau(X) \leq H_\tau(XY) = H_\tau(Y)$. Moreover, if $H_\tau(X) = H_\tau(Y) = H_\tau(XY)$, then $H(Y|X) = 0$, which implies that τ satisfies $X \to Y$.

For the second part, we can write

$$H_\tau(Y|X) = H_\tau(XY) - H_\tau(X) \leq H_\tau(XYZ) - H_\tau(X) = H_\tau(YZ|X).$$

Thus, that the equality holds if and only if $XY \to Z$.

The argument for the last part is similar to the argument for the first part and is omitted. $\qquad\square$

Theorem 4.4 *Let $\tau = (T, L, \mu)$ be an m-table and let $X, Y, Z \subseteq L$. The following statements hold:*

1. *$H_\tau(Y|XZ) \leq H_\tau(Y|X)$;*

2. *if $Z \subseteq X$, then $H_\tau(YZ|X) = H_\tau(Y|X)$;*

3. *$H_\tau(YZ|XZ) \leq H_\tau(Y|X)$.*

Proof. For Part 1 note that $\pi_X = f\pi_{XZ}$ for some function f. It follows from Theorem 2.2 that $H(\pi_Y|\pi_X) = H(\pi_Y|f\pi_{XZ}) \geq H(\pi_Y|\pi_{XZ})$, hence $H_\tau(Y|XZ) \leq H_\tau(Y|X)$.

For Part 2 we have

$$H_\tau(YZ|X) = H_\tau(XYZ) - H_\tau(X) = H_\tau(XY) - H_\tau(X) = H_\tau(Y|X).$$

Finally, for Part 3 observe that from the first two parts it follows that $H_\tau(Y|X) \geq H_\tau(Y|XZ) = H_\tau(YZ|XZ)$. $\qquad\square$

Theorem 4.5 *Let* $X, Y \subseteq L$. *We have* $H_\tau(XY) \leq H_\tau(X) + H_\tau(Y)$, *with equality if and only if* X *and* Y *are independent.*

Proof. By Theorem 4.1 and Lemma 2.1 we have:

$$H_\tau(XY) = H_\tau(X;Y) = H(\pi_X;\pi_Y) = H(\pi_Y|\pi_X) + H(\pi_X)$$
$$\leq H(\pi_Y) + H(\pi_X) = H_\tau(X) + H_\tau(Y).$$

From Lemma 2.1 it follows that $H(\pi_Y|\pi_X) = H(\pi_Y)$ if and only if π_Y and π_X are independent, which completes the proof.   □

The restriction $\pi_Y|_{(\mu \text{ where } X=x)}$ will be denoted by $\pi_Y|_{X=x}$ for any sets of attributes X and Y.

Theorem 4.6 *Let* $X, Y, Z \subseteq L$. *We have* $H_\tau(YZ|X) \leq H_\tau(Y|X) + H_\tau(Z|X)$, *and the equality holds if and only if for every* $x \in \text{adom}_\tau(X)$ *the functions* $\pi_Y|_{X=x}$ *and* $\pi_Z|_{X=x}$ *are independent.*

Proof. Let $A_x = \{t \in \mu : \pi_X(t) = x\}$, By Theorem 4.5 we have:

$$\begin{aligned}
H_\tau(YZ|X) &= \sum_x \frac{|A_x|}{|\mu|} H_{(\tau \text{ where } X=x)}(YZ) \\
&\leq \sum_x \frac{|A_x|}{|\mu|} H_{(\tau \text{ where } X=x)}(Y) + \\
&\quad \sum_x \frac{|A_x|}{|\mu|} H_{(\tau \text{ where } X=x)}(Z) \\
&= H_\tau(Y|X) + H_\tau(Z|X).
\end{aligned}$$

The condition on equality follows from the equality condition of Theorem 4.5.   □

5   Relational Algebra and Entropy

Relational algebra operations may generate m-tables even if they are applied to tables. Let $\sigma = (T, L, \rho)$ be a table and let $X \subseteq L$ be a set of attributes. The *projection without removal of duplicates* of σ on X is the m-table $\sigma\{X\} = (T\{X\}, X, \mu)$, where μ is the multiset on $\text{Dom}(X)$ given by

$$\mu(x) = |\{t \in \rho \mid t[X] = x\}|$$

for $x \in \text{Dom}(X)$. This operation models the effect of a standard SQL query

$$\textbf{select } A_1, \ldots, A_k \textbf{ from } \mathbf{T},$$

where $A_1 \cdots A_k \subseteq L$.

Theorem 5.1 *Let* $X \subseteq L$. *Then,* $H_{\sigma\{X\}}(X) = H_\sigma(X) \leq H_\sigma(L)$.

Proof. The proof of the equality is straightforward; the inequality is an immediate consequence of Part 1 of Theorem 4.3. □

The table obtained by removal of duplicate tuples from the m-table τ will be denoted by $\overline{\tau}$.

Theorem 5.2 *Let* $\tau = (T, L, \mu)$ *be an m-table. We have* $H_{\overline{\tau}}(L) \geq H_\tau(L)$.

Proof. Note that after removal of duplicate tuples, the number of distinct tuples will not change; all of them, however, will occur equally often, so the function $\pi_{\tau,L}$ becomes a bijection and the inequality follows. □

However, the entropy of a proper subset of L may either increase or decrease after the removal of duplicate tuples from τ.

Let $\tau_1 = (T_1, L_1, \mu_1)$ and $\tau_2 = (T_2, L_2, \mu_2)$ be m-tables and such that $L_1 \cap L_2 = \emptyset$. Their Cartesian product is defined as an m-table $\tau_1 \times \tau_2 = (T_1 \times T_2, L_1 L_2, \mu_1 \times \mu_2)$. The following theorem holds.

Theorem 5.3 *We have*

$$H_{\tau_1 \times \tau_2}(L_1 L_2) = H_{\tau_1}(L_1) + H_{\tau_2}(L_2).$$

Proof. It is easy to see that $\pi_{\tau_1 \times \tau_2, L} = \pi_{\tau_1, L_1} \times \pi_{\tau_2, L_2}$. It follows from (E5) that $H(\pi_{\tau_1 \times \tau_2, L}) = H(\pi_{\tau_1, L_1}) + H(\pi_{\tau_2, L_2})$. □

6   Entropy and Table Decompositions

Let $U, V \subseteq L$ be a pair of sets of attributes of a table $\tau = (T, L, \rho)$. If $U \cup V = L$, then we refer to (U, V) as a decomposition of τ. Let $\rho' = \rho[U] \bowtie \rho[V]$. In general, we have $\rho \subseteq \rho'$ for every decomposition (U, V) of ρ. The set of *spurious tuples* of the decomposition (U, V) is the set

$$\mathsf{spr}(U, V) = \rho' - \rho = (\rho[U] \bowtie \rho[V]) - \rho.$$

If $\mathsf{spr}(U, V) = \emptyset$, then we refer to (U, V) as a *lossless decomposition*; otherwise, (U, V) is a *lossy* decomposition.

Let $\tau = (T, L, \rho)$ be a table. Denote by $n_x(\rho) = |\{t \in \rho \mid t[X] = x\}|$ the number of tuples in ρ containing x for $x \in \text{Dom}(X)$. Observe that $n_x(\rho[XY])$ is the number of different values of $y \in \text{Dom}(Y)$ present in tuples of ρ containing x. If the functional dependency $X \to Y$ holds, then $n_x(\rho[XY]) = 1$ for all $x \in \text{adom}_\tau(X)$. It is easy to see that $|\rho| = \sum_{x \in \text{adom}_\tau(X)} n_x(\rho)$.

Note that $|\pi_{\tau,Y}^{-1}(y)| = |\{t \in \text{Dom}(L) \mid t[Y] = y\}| = n_y(\rho)$ and a similar equality holds for $\pi_{\tau,X}$. Also, $\pi_{\tau,Y}^{-1}(y) \cap \pi_{\tau,X}^{-1}(x) = \{t \in \text{Dom}(L) \mid t[XY] = xy\}$. Therefore, the conditional entropy $H_\tau(Y|X)$ can be written as

$$H_\tau(Y|X) = -\sum_{y \in \text{adom}_\tau(Y)} \sum_{x \in \text{adom}_\tau(X)} \frac{n_{xy}(\rho)}{|\rho|} \log_2 \frac{n_{xy}(\rho)}{n_x(\rho)}.$$

Let (U, V) be a decomposition of τ and let $X = U \cap V$. It is clear that

$$\text{spr}(U, V) = \sum_{x \in \text{adom}_\tau(X)} n_x(\rho[U]) n_x(\rho[V]) - |\rho|.$$

Moreover, we have the following result.

Theorem 6.1 *Let (U, V) be a decomposition of τ. Define $X = U \cap V, Y = U - X$, and $Z = V - X$. Then, we have*

$$\text{spr}(U, V) \geq |\rho| \left(\frac{2^{H_\tau(Y|X) + H_\tau(Z|X)}}{\max_{x \in \text{adom}_\tau(X)} n_x(\rho)} - 1 \right).$$

Proof. For the conditional entropy $H_\tau(Y|X)$ we can write

$$H_\tau(Y|X) = -\sum_{y \in \text{adom}(Y)} \sum_{x \in \text{adom}(X)} \frac{n_{yx}(\rho)}{|\rho|} \log_2 \frac{n_{yx}(\rho)}{n_x(\rho)}.$$

If $x \in \text{adom}_\tau(X)$, then $n_x(\rho[XY])$ is the number of elements of the active domain of Y that occur in the relation $\rho_x = (\rho \text{ where } X = x)$ under Y and $n_x(\rho[XY]) = \sum_{y \in \text{adom}_\tau(Y)} n_{xy}(\rho)$. Consequently, the value of the entropy of ρ_x will not exceed $\log_2 n_x(\rho[XY])$.

Since

$$H_\tau(Y|X) = \sum_{x \in \text{adom}_\tau(X)} \frac{n_x(\rho)}{|\rho|} H_\tau(Y|X = x),$$

$$H_\tau(Z|X) = \sum_{x \in \text{adom}_\tau(X)} \frac{n_x(\rho)}{|\rho|} H_\tau(Z|X = x),$$

we can write

$$2^{H_\tau(Y|X)+H_\tau(Z|X)} = 2^{\sum_{x\in\mathrm{adom}_\tau(X)} \frac{n_x(\rho)}{|\rho|}\left(H_\tau(Y|X=x)+H_\tau(Z|X=x)\right)}.$$

The convexity of the exponential function implies:

$$2^{H_\tau(Y|X)+H_\tau(Z|X)} \leq \sum_{x\in\mathrm{adom}_\tau(X)} \frac{n_x(\rho)}{|\rho|} 2^{H_\tau(Y|X=x)+H_\tau(Z|X=x)}$$

$$\leq \sum_{x\in\mathrm{adom}_\tau(X)} \frac{n_x(\rho)}{|\rho|} 2^{\log_2 n_x(\rho[XY])+\log_2 n_x(\rho[XZ])}$$

$$= \sum_{x\in\mathrm{adom}_\tau(X)} \frac{n_x(\rho)}{|\rho|} n_x(\rho[XY])n_x(\rho[XZ])$$

$$\leq \frac{\max_{x\in\mathrm{adom}_\tau(X)} n_x(\rho)}{|\rho|} \cdot |\rho'|.$$

Therefore, we have

$$|\rho'| \geq \frac{|\rho|}{\max_{x\in\mathrm{adom}_\tau(X)} n_x(\rho)} \cdot 2^{H_\tau(Y|X)+H_\tau(Z|X)},$$

which implies

$$\mathsf{spr}(U,V) \geq |\rho| \left(\frac{2^{H_\tau(Y|X)+H_\tau(Z|X)}}{\max_{x\in\mathrm{adom}_\tau(X)} n_x(\rho)} - 1 \right).$$

□

7   The Entropy as a Measure

Let $\tau = (T, L, \mu)$ be an m-table. We assume that $H_\tau(X|\emptyset) = H_\tau(X)$.

Theorem 7.1 *Let $\tau = (T, L, \mu)$ be an m-table. The entropy of the sets of database attributes of the m-table τ conditioned on some fixed set of attributes $X \subseteq L$ is an outer measure on the set of attributes L, i.e., it satisfies the following properties:*

1. $H_\tau(\emptyset|X) = 0$ and $H_\tau(Y|X) \geq 0$, for all $Y \subseteq L$;

2. if $Y_1 \subseteq Y_2 \subseteq L$, then $H_\tau(Y_1|X) \leq H_\tau(Y_2|X)$;

3. $H_\tau(\bigcup Y_i|X) \leq \sum H_\tau(Y_i|X)$, *for every countable collection of sets of attributes* $\{Y_i\}$, *such that* $Y_i \subseteq L$ *for all* i.

Proof. Property 1 follows from the definition of conditional entropy between sets of attributes and from the fact that the conditional entropy of functions between finite sets is always non-negative. Property 3 is an immediate consequence of Part 3 of Theorem 4.3.

Since L is a finite set, in the proof of Property 3 we only need to consider finite collections of sets of attributes $\{Y_i\}$. The argument is by induction. For collections consisting of only one set of attributes Property 3 holds. Suppose now that the property holds for all collections $\{Y_i\}, i = 1, \ldots, n$ of n sets of attributes. Then, for every collection $\{Y_i\}, i = 1, \ldots, n, n+1$, of $n+1$ sets of attributes we have (using Theorem 4.6):

$$
\begin{aligned}
H_\tau\left(\bigcup_{i=1}^{n+1} Y_i|X\right) &= H_\tau\left(\bigcup_{i=1}^{n} Y_i \cup Y_{n+1}|X\right) \\
&\leq H_\tau\left(\bigcup_{i=1}^{n} Y_i|X\right) + H_\tau(Y_{n+1}|X) \\
&\leq \sum_{i=1}^{n} H_\tau(Y_i|X) + H_\tau(Y_{n+1}|X) \\
&= \sum_{i=1}^{n+1} H_\tau(Y_i|X).
\end{aligned}
$$

□

Note that, as a special case, when $X = \emptyset$, unconditional entropy of attributes is also an outer measure on the set of database attributes L.

For a set of attributes $E \subseteq L$, denote $E' = L - E$.

Definition 7.1 *A set of attributes E is measurable under an outer measure $H_\tau(\cdot|X)$ if for every $Y \subseteq L$ we have $H_\tau(Y|X) = H_\tau(Y \cap E|X) + H_\tau(Y \cap E'|X)$.*

Theorem 7.2 *If either $H_\tau(E|X) = 0$ or $H_\tau(E'|X) = 0$ for some set of attributes X, then both E and E' are measurable under $H_\tau(\cdot|X)$.*

Proof. Suppose that $H_\tau(E|X) = 0$. It is a general property of measures that a set whose outer measure zero is measurable. Another general property of measures is that a complement of a measurable set is also measurable, which proves the measurability of E'. The proof when $H_\tau(E'|X) = 0$ is analogous. $\square$

Corollary 7.1 $\emptyset$ and L are measurable under $H_\tau(\cdot|X)$ for every $X \subseteq L$.

Theorem 7.3 A set of attributes $E \subset L$, $E \neq \emptyset$ is measurable under an outer measure $H_\tau(\cdot|X)$ if and only if for every $x \in \text{adom}_\tau(X)$ the functions $\pi_E|_{X=x}$ and $\pi_{E'}|_{X=x}$ are independent.

Proof. Take any $Y \subseteq L$ and denote $Y_1 = Y \cap E$, $Y_2 = Y \cap E'$.

By Definition 7.1 and Theorem 4.6, E is measurable if for every $Y \subseteq L$ and $x \in \text{adom}_\tau(X)$, $p(y_1, y_2|x) = p(y_1|x)p(y_2|x)$ for every $y_1 \in \text{Dom}(Y_1)$, $y_2 \in \text{Dom}(Y_2)$.

Suppose that condition holds. Then, for every $x \in \text{adom}_\tau(X)$ we have, $p(e, f|x) = p(e|x)p(f|x)$ for every $e \in \text{Dom}(E)$, $f \in \text{Dom}(E')$.

Denote $E_{y_1} = \{e \in \text{Dom}(E) \mid e[Y_1] = y_1\}$ and $E'_{y_2} = \{f \in \text{Dom}(E') \mid f[Y_2] = y_2\}$, where $y_1 \in \text{Dom}(Y_1)$, $y_2 \in \text{Dom}(Y_2)$. From $Y_1 \subseteq E$ and $Y_2 \subseteq E'$ it follows that $p(y_1|x) = \sum_{e \in E_{y_1}} p(e|x)$ and $p(y_2|x) - \sum_{f \in E'_{y_2}} p(f|x)$.

Choose arbitrarily $Y \subseteq L$ and $x \in \text{adom}_\tau(X)$. Then, for every $y_1 \in \text{Dom}(Y_1)$ and $y_2 \in \text{Dom}(Y_2)$, we have:

$$p(y_1, y_2|x) = \sum_{e \in E_{y_1}} \sum_{f \in E'_{y_2}} p(e, f|x) = \sum_{e \in E_{y_1}} \sum_{f \in E'_{y_2}} p(e|x)p(f|x)$$

$$= \sum_{e \in E_{y_1}} \left(p(e|x) \sum_{f \in E'_{y_2}} p(f|x) \right) = \sum_{e \in E_{y_1}} p(e|x)p(y_2|x)$$

$$= p(y_2|x) \sum_{e \in E_{y_1}} p(e|x) = p(y_2|x)p(y_1|x).$$

Since the choice of Y and x was arbitrary, the above equality holds for every $Y \subseteq L$ and $x \in \text{adom}_\tau(X)$. This proves, that the condition of the theorem implies the measurability of E.

Suppose now that the condition of the theorem does not hold. We have:

$$H_\tau(L|X) = H_\tau(E \cup E'|X).$$

From Theorem 4.6 and the fact that the condition of the theorem does not hold, it follows that:

$$H_\tau(E \cup E'|X) < H_\tau(E|X) + H_\tau(E'|X)$$
$$= H_\tau(L \cap E|X) + H_\tau(L \cap E'|X).$$

This shows that when the condition is violated, E is indeed not measurable, which completes the proof. □

As a special case, when $X = \emptyset$, a set of attributes $E \subset L$, $E \neq \emptyset$ is measurable under unconditional entropy if and only if π_E and $\pi_{E'}$ are independent.

Another characterization of measurability is given by the following theorem.

Theorem 7.4 $E \subseteq L$ *is measurable under unconditional entropy if and only if* $H_\tau(E) + H_\tau(E') = H_\tau(L)$.

Proof. If $E = L$ or $E = \emptyset$, then the result obviously holds. The proof when $E \neq L$ and $E \neq \emptyset$ is an immediate consequence of Theorems 7.3 and 4.5.

□

Theorem 7.5 *If $E \subseteq L$ is measurable under unconditional entropy, then:*

(i) *the functional dependency $X \rightarrow Y$ does not hold for any $X \subseteq E$, $Y \subseteq E'$, unless $H_\tau(Y) = 0$, i.e., π_Y is constant or $Y = \emptyset$;*

(ii) *the functional dependency $X \rightarrow Y$ does not hold for any $X \subseteq E'$, $Y \subseteq E$, unless $H_\tau(Y) = 0$, i.e., π_Y is constant or $Y = \emptyset$.*

Proof. Suppose that E is measurable. There is a functional dependency $X \rightarrow Y$, and $H_\tau(Y) \neq 0$. Suppose that $X \subseteq E$ and $Y \subseteq E'$. For the set of attributes $X \cup Y$ we can write:

$$H_\tau[(X \cup Y) \cap E] + H_\tau[(X \cup Y) \cap E']$$
$$= H_\tau(X) + H_\tau(Y) \neq H_\tau(X) = H_\tau(X \cup Y).$$

Thus, E is not measurable. This completes the proof of part (i). The proof of (ii) is similar. □

Note that when in the above theorem $H_\tau(X) = 0$ and the functional dependency $X \rightarrow Y$ holds, then $H_\tau(Y)$ must also be equal to zero, so such a dependency does not exclude the measurability of E.

Theorem 7.6 *If the functional dependency $X \rightarrow E$ holds, then E is measurable under $H_\tau(\cdot|X)$.*

Proof. If $X \rightarrow E$, then $H_\tau(E|X) = 0$, and since E has outer measure 0, it is measurable. □

Theorem 7.7 *If $K \subseteq L$ is a candidate key, then K is measurable under unconditional entropy if and only if $H_\tau(K') = 0$.*

Proof. The proof that K is measurable when $H_\tau(K') = 0$ is straightforward. To prove the converse, note that since K is a candidate key, there is a functional dependency $K \to K'$, so by Theorem 7.5, K is not measurable, unless $H_\tau(K') = 0$. $\qquad\Box$

Theorem 7.8 *Every $X \subseteq L$ is measurable under an outer measure $H_\tau(\cdot|K)$, where K is a candidate key.*

Proof. If $X = \emptyset$, the proof is trivial. Suppose that $X \neq \emptyset$. Note that all tuples t for which $\pi_K(t) = k$ for some $k \in \mathrm{Dom}(K)$ are identical. Denote by t_k any tuple from the set $\{t \in \rho \mid t[K] = k\}$, where $k \in \mathrm{Dom}(K)$. Then for every $X \subseteq L$, $x_1 \in \mathrm{Dom}(X)$, $x_2 \in \mathrm{Dom}(X')$ and $k \in \mathrm{Dom}(K)$, the following holds

$$p(x_1|k) = \begin{cases} 0, & \text{if } t_k[X] \neq x_1, \\ 1, & \text{if } t_k[X] = x_1, \end{cases}$$

$$p(x_2|k) = \begin{cases} 0, & \text{if } t_k[X'] \neq x_2, \\ 1, & \text{if } t_k[X'] = x_2. \end{cases}$$

Therefore,

$$p(x_1, x_2|k) = \begin{cases} 0, & \text{if } t_k[X] \neq x_1 \text{ or } t_k[X'] \neq x_2, \\ 1, & \text{if } t_k[X] = x_1 \text{ and } t_k[X'] = x_2, \end{cases} = p(x_1|k)p(x_2|k),$$

and π_X and $\pi_{X'}$ are independent. $\qquad\Box$

8 Entropy and Integrity Constraints

We believe that the study of integrity constraints can benefit from the application of information-theoretical techniques. For instance, entropic dependencies introduced below as generalizations of functional dependencies can offer greater flexibility to database designers.

Definition 8.1 *Let $\tau = (T, L, \mu)$ be an m-table. We say that τ satisfies an entropic dependency between the sets of attributes X and Y, denoted by $X \to_\epsilon Y$, if $H_\tau(Y|X) \leq \epsilon$.*

Theorem 8.1 *Let* $\tau = (T, L, \mu)$ *be an m-table and let* $X, Y, Z \subseteq L$ *be sets of attributes. We have:*

1. $X \rightarrow_\epsilon YZ$ *implies* $X \rightarrow_\epsilon Y$;

2. $X \rightarrow_\epsilon Y$ *implies* $XZ \rightarrow_\epsilon Y$;

3. $X \rightarrow_\epsilon Y$ *implies* $XZ \rightarrow_\epsilon YZ$;

4. $X \rightarrow Y$ *if and only if* $X \rightarrow_0 Y$;

5. $X \rightarrow Y$ *and* $Y \rightarrow_\epsilon Z$ *imply* $X \rightarrow_\epsilon Z$;

6. $X \rightarrow_\epsilon Y$ *and* $Y \rightarrow Z$ *imply* $X \rightarrow_\epsilon Z$;

7. $X \rightarrow_{\epsilon_1} Y$ *and* $X \rightarrow_{\epsilon_2} Z$ *imply* $X \rightarrow_{\epsilon_1+\epsilon_2} YZ$.

Proof. Statement 1 is a direct consequence of Part 2 of Theorem 4.3, Statement 2 follows from the first part of Theorem 4.4; also, the third statement is implied by the second part of Theorem 4.4, while Property 4 follows from Theorem 4.2.

To prove Statement 5 note that

$$H_\tau(Z|X) = H_\tau(XZ) - H_\tau(X) = H_\tau(XYZ) - H_\tau(XY)$$
$$= H_\tau(Z|XY) \le H_\tau(Z|Y) \le \epsilon.$$

The last part of the theorem follows from

$$H_\tau(Z|X) \le H_\tau(YZ|X) = H_\tau(XYZ) - H_\tau(X)$$
$$= H_\tau(XY) - H_\tau(X) = H_\tau(Y|X) \le \epsilon.$$

$\square$

Let $\sigma = (T, L, \rho)$ be a table, and let X, Y be two nonempty subsets of L. Denote by Z the set $Z = L - XY$. We say that there is a *multivalued dependency* $X \twoheadrightarrow Y$ between sets of attributes X and Y if for every tuples $t_1, t_2 \in \rho$, such that $t_1[X] = t_2[X]$, there is a tuple $t \in \rho$ such that $t[X] = t_1[X] = t_2[X]$, $t[Y] = t_1[y]$, and $t[Z] = t_2[Z]$.

It is easy to see that the multivalued dependency $X \twoheadrightarrow Y$ holds if and only if for every $x \in \mathbf{adom}_\tau(X)$ we have

$$(\sigma \text{ where } X = x)[YZ] = (\sigma \text{ where } X = x)[Y] \times (\sigma \text{ where } X = x)[Z] \tag{8}$$

Multivalued dependencies can be expressed in terms of function independence as follows.

Theorem 8.2 *Let $\sigma = (T, L, \rho)$ be a table. If $X, Y \subseteq L$, denote by Z the set of attributes $Z = L - (X \cup Y)$. The multivalued dependency $X \twoheadrightarrow Y$ holds if and only if for every $x \in \mathrm{adom}_\tau(X)$, the functions $\pi_Y|_{X=x}$ and $\pi_Z|_{X=x}$ are independent.*

Proof. Suppose $X \twoheadrightarrow Y$ holds. Then, the equality (8) holds for every $x \in \mathrm{adom}_\tau(X)$. Let $n_1 = |(\sigma \text{ where } X = x)[Y]|$ and $n_2 = |(\sigma \text{ where } X = x)[Z]|$. Then,

$$p(y, z|x) = \frac{1}{n_1 \cdot n_1} = \frac{1}{n_1} \cdot \frac{1}{n_2} = p(y|x)p(z|x),$$

for every $y \in \mathrm{Dom}(Y), z \in \mathrm{Dom}(Z)$.

To prove the converse take any $y \in \mathrm{Dom}(Y)$ and $z \in \mathrm{Dom}(Z)$ such that $p(y|x) > 0$ and $p(z|x) > 0$. From the independence property it follows that $p(y, z|x) = p(y|x)p(z|x)$ is also greater than zero. Thus regardless of the choice of x, and y above, there exists a tuple $t \in \rho$ such that $t[X] = x$, $t[Y] = y$, and $t[Z] = z$. $\qquad\square$

We believe that the notion of entropic dependency merits further attention. It is clear, from Theorem 6.1, that any decomposition (XY, XZ) of a table with few spurious tuples entails the existence of entropic dependencies of the form $X \to_\epsilon Y$ and $X \to_{\epsilon'} Z$ with low ϵ, ϵ'. The reverse implication remains as an open problem. Also, an axiomatization of entropic dependencies (in the style of Armstrong's axiomatization of functional dependencies) would be very useful.

Acknowledgment

Szymon Jaroszewicz wishes to acknowledge the support received from the Fulbright Scholarship Board as a visiting researcher at the University of Massachusetts at Boston (grant no. IIE#:15984114-ICB).

References

[1] K. T. Cheng, V. Agrawal, An entropy measure of the complexity of multi-output Boolean function, *Proc. of the 27th Design Automation Conference*, 1990, 302–305.

[2] V. Cheushev, V. Shmerko, D. A. Simovici, S. Yanushkevich, Functional entropy and decision trees, *Proc. ISMVL'98*, 1998, 257–262.

322   D. A. Simovici, S. Jaroszewicz

[3] P. A. Fejer, D. A. Simovici, *Mathematical Foundations of Computer Science*, vol. I, Springer-Verlag, New York, 1991.

[4] C. H. Hwang and A. C. H. Wu, An entropy measure for power estimation of Boolean function, *Proc. of Asia and South Pacific Design Automation Conference* (ASP-DAC), 1997, 101–106.

[5] S. Jaroszewicz, D. A. Simovici, On axiomatization of conditional entropy of functions between finite sets, *Proc. of the 29th International Symposium for Multiple-Valued Logic*, Freiburg, May 1999, to appear.

[6] A. Lloris-Ruiz, J. F. Gomez-Lopera, R. Roman-Roldan, Entropic minimization of multiple-valued logic functions, *Proc. ISMVL'93*, 1993, 24–28.

[7] A. M. Mathai, P. N. Rathie, *Basic Concepts in Information Theory and Statistics – Axiomatic Foundations and Applications*, Halsted Press, John Wiley & Sons, 1975.

[8] D. A. Simovici, C. Reischer, On functional entropy, *Proc. of the 23rd International Symposium for Multiple-Valued Logic*, 1993, 100–104.

[9] D. A. Simovici, R. L. Tenney, *Relational Database Systems*, Academic Press, New York, 1995.

A Version of Ω for which ZFC Cannot Predict a Single Bit

Robert M. Solovay[1]

Department of Mathematics
The University of California
Berkeley, CA 94720 – 3840, USA
E-mail: solovay@math.berkeley.edu

1 Introduction

In [2], Chaitin introduced the real Ω and proved the following:

Theorem 1.1 *Assume that ZFC is arithmetically sound. (That is, any theorem of arithmetic proved by ZFC is true.) Then ZFC can determine the value of only finitely many bits of Ω. In fact we can explicitly compute a bound on the number of bits of Ω which ZFC can determine.*

Chaitin's theorem is much more general than what we have stated. ZFC can be replaced by any recursively axiomatizable theory in which Peano arithmetic can be interpreted.

The real Ω depends on the choice of "universal Chaitin machine". It is natural to suspect that by tuning this choice one can improve Chaitin's result.

Here is the main theorem of this paper:

Theorem 1.2 *We can choose the universal Chaitin computer U so that ZFC (if arithmetically sound) cannot determine any bit of the Ω associated with U.*

The rest of this paper is organized as follows. Section 2 contains a review of the basic definitions of Chaitin's theory that we use. In section

---

[1]I wish to thank the Isaac Newton Institute for providing the environment where this paper was written.

3, we recall the notion of "1-consistent". (The hypothesis that ZFC is arithmetically sound can be sharpened, in both my theorem and Chaitin's, to merely asserting that ZFC is 1-consistent.) Section 4 gives a more detailed discussion of Chaitin's theorem. The remaining sections of the paper are devoted to a proof of our theorem.

I am grateful to Greg Chaitin for giving me a copy of his book [1] which started me thinking about this circle of ideas.

2   Preliminary Definitions and Notation

2.1   Bit Strings

ω is the set of non-negative integers. I follow von Neumann so that each integer n is equal to the set $\{m \in \omega \mid m < n\}$ of integers less than n.

Σ^* is the set of finite sequences of 0's and 1's. Thus an element of Σ^* is just a finite function whose domain is in ω and whose range is included in $\{0, 1\}$.

The concatenation of the two bit strings s and t will be denoted by $s \frown t$. If j is one of 0 or 1, the bit string of length 1 whose sole component is j will be denoted by $\langle j \rangle$. Of course $\emptyset$ is the unique string of length 0.

If s is a bit string, we write $|s|$ for the length of s.

The usual theory of partial recursive functions is done considering functions whose domain and range are subsets of ω. We want to import this theory over to functions whose domain and range are subsets of Σ^* and for that, it is convenient to fix a canonical bijection between Σ^* and ω. This is done as follows.

We linearly order Σ^* by putting $s < t$ if either:

1. $|s| < |t|$ or

2. $|s| = |t|$ and s lexicographically precedes t.

With this ordering, there is a unique order isomorphism of Σ^* with ω (which will serve as the "canonical bijection" between the two sets).

2.2   Prefix-free Codes

A subset A of Σ is a prefix-free code if whenever s and t are members of A such that $s \subseteq t$ then $s = t$.

Associated to any prefix-free code, A, is a real number Ω_A defined by

$$\Omega_A = \sum_{s \in A} 2^{-|s|}.$$

This has the following probabilistic interpretation. Pick a real x in $[0, 1]$ at random using the Lebesgue measure on $[0, 1]$. Then Ω_A is the probability that some initial prefix of the binary expansion of x lies in A.

2.3 Chaitin Machines

A *Chaitin machine* is a partial recursive function whose domain and range are subsets of Σ^* and whose domain is a prefix-free code.

Let U be a Chaitin machine with domain A. Then we set $\Omega_U = \Omega_A$.

A Chaitin machine U is universal if it can simulate any other Chaitin machine. More precisely, U is universal if for every Chaitin machine V there is a bit string π_V such that the equality

$$U(\pi_V \frown s) \simeq V(s)$$

holds for any bit string s.

Here, as usual $x \simeq y$ holds between two partially defined objects x and y if (a) x is defined iff y is defined and (b) if they are both defined, then they are equal.

It is proved in [2] that universal Chaitin machines exist. Moreover, if U is universal, then Ω_U has a strong randomness property. (It is now known that Ω_U is Martin-Löf random.) As a corollary, Ω_U is irrational and does not have a recursive binary expansion.

2.4 Gödel Numbering Chaitin Machines

We fix one of the usual Gödel numberings $\{\varphi_i \mid i \in \omega\}$ of all partial recursive functions from Σ^* to Σ^*. Then the function $\Phi : \omega \times \Sigma^* \mapsto \Sigma^*$ given by

$$\Phi(i, s) \simeq \varphi_i(s)$$

is partial recursive.

It follows that the domain of Φ is recursively enumerable. Since it is clearly infinite, we fix a recursive enumeration without repetitions of the domain of Φ: $\langle \langle n_i, s_i \rangle \mid i \in \omega \rangle$. We can certainly arrange that this enumeration is primitive recursive.

We are going to construct a new function $\Psi : \omega \times \Sigma^* \mapsto \Sigma^*$. Defining $\psi_i : \Sigma^* \mapsto \Sigma^*$ (for $i \in \omega$) by

$$\psi_i(s) \simeq \Psi(i, s)$$

will yield the desired Gödel numbering of Chaitin machines.

Ψ will be the restriction of Φ to a certain subset of its domain. We proceed by induction on i to determine if the pair $\langle n_i, s_i \rangle$ will be placed in the domain of Ψ.

Place $\langle n_i, s_i \rangle$ in the domain of Ψ iff for no $j < i$ for which $\langle n_j, s_j \rangle$ has been placed in the domain of Ψ do we have $n_j = n_i$ and s_j compatible with s_i. (That is, $s_j \subseteq s_i$ or $s_i \subseteq s_j$.)

This construction has the following properties (whose proof is left to the reader):

1. The domain of ψ_i is prefix-free.

2. If the domain of φ_i is prefix-free, then $\psi_i = \varphi_i$.

2.5   Timing

We let D_n be the domain of ψ_n. Let $\Omega_n = \Omega_{D_n}$.

Intuitively, $D_n[t]$ consists of those elements of D_n which have appeared by time t. More precisely,

$$D_n[t] = \{s \mid (\exists j \leq t)(n_j = n \text{ and } s_j = s \text{ and } \langle n_j, s_j \rangle$$
$$\text{was placed into Dom}(\Psi) \text{ at stage } j\}.$$

We put $\Omega_n[t] = \Omega_{D_n[t]}$. Intuitively, this is the approximation to Ω_n computable at time s.

The following facts are evident:

1. $\Omega_n[t]$ is a rational number with denominator a power of 2.

2. Given n and t we can compute (by a primitive recursive function) the finite set $D_n[t]$ and the rational number $\Omega_n[t]$.

3. As t increases to infinity, the $\Omega_n[t]$ increase monotonically to the limit Ω_n.

3   1-Consistency

Throughout this section T is a theory with a recursive set of axioms in which Peano Arithmetic (PA) is relatively interpretable. We fix a relative interpretation of PA in T. Of course, the basic example we have in mind is ZFC equipped with the usual relative interpretation of PA in ZFC.

For brevity in what follows, we say "interpretation" rather than "relative interpretation".

Our main theorem will use the hypothesis that "ZFC is 1-consistent". In this first part of this section, we review known results without proof that describe the relationship of the notion of 1-consistency to other notions of soundness.

In the second part of this section, we derive from the assumption that ZFC is 1-consistent that any determination that ZFC makes about one of the binary digits of Ω_U (for some *universal* Chaitin computer U) is *true*. This will be the only use we make of the 1-consistency hypothesis.

3.1   A Spectrum of Soundness Hypotheses

3.1.1 ω-Models

Since there is a fixed interpretation of PA in T, any model of T determines a model of PA. We say that a model M of T is an ω-model if the associated model of PA is isomorphic to the standard model of PA.

Our first soundness assumption is that T has an ω-model.

3.1.2   Arithmetic Soundness

Each sentence of the language of PA has a translation into a sentence of the language of T, determined by the interpretation of PA in T. We shall blur the distinction between a sentence of PA and its translation. We use the phrase "sentence of arithmetic" to indicate a sentence of the language of T that is the translation of some sentence of PA.

Our second soundness assumption is that T is arithmetically sound. That is, if ϑ is a sentence of arithmetic which is a theorem of T, then ϑ is true (in the standard model of PA).

Remark. Our metatheory is ZFC. So we know that PA itself is arithmetically sound.

3.1.3 ω-Consistency

The notion of ω-consistency was introduced by Gödel in connection with his incompleteness theorems.

It is easiest to define when a theory T is *not* ω-consistent (i.e., is ω-inconsistent.) This happens if there is a formula of the language of T, $\theta(x)$ (having only the indicated free variable x) such that the following happens:

1. T proves "There is an $x \in \omega$ such that $\theta(x)$."

2. For each natural number n, T proves "$\neg\theta(\mathbf{n})$".

In theories like PA which have a canonical term to denote each natural number, $\mathbf{n}$ is the canonical term that denotes the integer n. In theories like ZFC that lack such terms the explication of what the formula $\theta(\mathbf{n})$ is, is a little more subtle, but we presume the reader will be familiar with the details.

3.1.4 1-Consistency

A theory T is 1-consistent, if whenever it proves a Σ_1^0 sentence, θ, then θ is true.

There is a notion of when a formula of PA is "primitive recursive". Basically these are the formulas that arise in implementing Gödel's proof that every primitive recursive predicate is expressible in PA.

A sentence of PA is Σ_1^0 if it has the form $(\exists x)P(x)$ where P is primitive recursive.

A special case of Σ_1^0 sentences are the formalizations of assertions that a particular Turing machine halts if started (in its canonical "start state") on an empty tape. Indeed, every Σ_1^0 sentence is provably equivalent in PA to such a "halting statement".

3.1.5 Consistency

T is consistent if it does not prove the assertion "$0 = 1$". Equivalently, T is consistent if it does not prove every sentence in the language of T.

3.1.6 Positive Relations Between the Different Notions of Soundness

These claims are all trivial. Every theory T that has an ω model is arithmetically sound and ω-consistent. If T is arithmetically sound or ω-consistent, then T is 1-consistent. If T is 1-consistent, then T is consistent.

3.1.7 Negative Relations Between the Different Notions of Soundness

The claims that follow are not entirely trivial, but are all well-known. Details will not be given. The proof of our main theorem does not depend on these results.

There are theories T_1, T_2, T_3, and T_4, all in the language of ZFC and all extending ZC (Zermelo set theory with choice) such that:

1. T_1 is arithmetically sound but not ω-consistent.

2. T_2 is ω-consistent but not arithmetically sound.

3. T_3 is consistent, but not 1-consistent.

4. T_4 is not consistent.

From now on, when we say that a theory T is 1-consistent, it is implied that the theory has a recursive set of axioms and comes equipped with some definite interpretation of PA in T.

3.2 Proving Facts About Ω

3.2.1 Binary Expansions

A *dyadic rational* is a rational number of the form $r/2^s$ where r and s are integers and $s \geq 0$.

If x is a real number which is not a dyadic rational, then x has a unique binary expansion. If x is a dyadic rational, it has two distinct binary expansions. In this paper, we shall always pick the one that ends in an infinite sequence of 0's.

With this convention in place, the following can easily be formalized in PA: "The ith binary digit of Ω_j is k." (Here $k < 2$, of course, if the assertion is true.)

We start numbering the digits of the binary expansion of a real with the 0th digit. Thus the 0th digit of the binary expansion of $1/3$ is 0; the 1st digit is 1; the 2nd digit is 0, etc.

Lemma 3.1 *Let ψ_j be a Chaitin machine which PA can prove universal. Let T be 1-consistent, and let T prove the assertion "The ith binary digit of Ω_j is k". Then this assertion is true.*

Our proof of this lemma will proceed in two steps. We first show that any Π_2^0 sentence proved by T is true. We then show the sentence in question in the lemma is provably equivalent in PA to a Π_2^0 sentence.

3.2.2 Π_2^0 Sentences

A Π_2^0 sentence is a sentence of PA of the form $\forall x \exists y P(x, y)$, where P is primitive recursive.

Suppose then, towards a contradiction, that T proves the translation of such a Π_2^0 sentence, and that the sentence is false. Then for some particular integer n, the sentence $\exists y P(\mathbf{n}, y)$ is false and provable in T. But this latter sentence is Σ_1^0, and this contradicts the assumption that T is 1-consistent.

3.2.3 Proof of the Lemma

We work in PA. We know that Ω_j is irrational. Hence, we can express the fact that the ith digit of Ω_j is k as follows:

$$(\forall m)(\exists nm)\ [\text{the } i^{th} \text{ digit of } \Omega_j[n] \text{ is } k].$$

(The proof of this claim is easy and left to the reader.) But the assertion just displayed is visibly Π_2^0. The lemma is proved.

4 Chaitin's Results About Predicting Bits of Ω

The results of this section, which are stated without proof, are not needed for the proof of the main theorem.

Throughout this section we fix a universal Chaitin machine U. (In the later parts of this section, we even implicitly specialize to a particular such U.)

We fix a j such that $U = \psi_j$. Formal assertions about U refer to this j.

We write Ω for Ω_U. As discussed in Section 3.2.1, we can easily formalize in PA the assertion "The ith binary digit of Ω is k". (This formalization uses the Gödel number, j, of Ω.)

Now let T be a 1-consistent theory of the sort discussed in Section 3. Following Chaitin, we want to give an upper bound on the set of $i \in \omega$ such that T proves a theorem of the form "The ith bit of the binary expansion of Ω is k" for some $k \leq 1$. We refer to this cardinality as "the number of bits of Ω that T can determine". Of course, a priori, this cardinality might be infinite; however, it will turn out to be finite.

4.1 $H(T)$

We wish to give a definition of the number of bits it takes to describe the arithmetical theorems of T.

We fix a Gödel numbering of the sentences of PA.

Now consider a theory T of the sort described above. We proceed to associate an recursively enumerable set of strings W_T to T. Let $s \in \Sigma^*$. Then s corresponds to an integer n_s as discussed in Section 2.1. Then $s \in W_T$ iff n_s is the Gödel number of a sentence of PA whose translation is a theorem of T.

Now let $s \in \Sigma^*$. We say that s is a program for W_T if

1. $U(s)$ is defined and has the value t. Let n be the integer corresponding to t. (Cf. Section 2.1.)

2. The domain of φ_n is W_T.

Finally, let $H(T)$ be the length of the shortest program for W_T.
We can now state Chaitin's theorem (proved in [2]).

Theorem 4.1 *Let U be a universal Chaitin computer. Then there is a positive constant C (depending only on U) such that (for T a 1-consistent theory) T can determine at most $C + H(T)$ bits of Ω.*

In [1], Chaitin describes a particular universal computer (whose implementation is done in a dialect of Lisp that Chaitin devised.) For a definition of $H(T)$ which is similar in spirit to the one I have given above, Chaitin proves the following:

Theorem 4.2 *Let U be the particular universal Chaitin computer defined in [1]. Let T be a 1-consistent theory. Then T determines at most $H(T) + 15328$ bits of Ω_U.*

5 Precise Statement of the Main Theorem. Outline of the Proof

Theorem 5.1 *Let T be a 1-consistent theory. Then there is a universal Chaitin computer, U (equal to ψ_j), such that:*

1. *PA proves the fact that U is universal.*

2. *T cannot determine even a single bit of Ω_U.*

In particular, our theorem applies to ZFC provided that ZFC is 1-consistent.

Of course, the U provided by the theorem depends on T.

Here is a sketch of the proof. (Some technical details have been omitted from this sketch. They will be provided in the following sections where the proof will be presented in detail.)

We fix a standard Chaitin universal computer V such that the universality of V is provable in PA.

Our computer U will be undefined on the string $\emptyset$. For strings of the form $\langle 0 \rangle \frown s$, we will have:

$$U(\langle 0 \rangle \frown s) \simeq V(s)$$

This will ensure the universality claims made about U.

We are still free to define U on strings of the form $\langle 1 \rangle \frown s$ as we wish. We will use this freedom to prevent T from guessing a single bit of Ω_U.

Thanks to the magic of the recursion theorem, we can assume that when defining U we know the Gödel number of U. Our algorithm when given a string of the form $\langle 1 \rangle \frown s$ first begins an enumeration of the arithmetical theorems of T, looking for the first one of the form "The nth bit of Ω_U is k". This search may go on forever without finding such a sentence. If it does succeed, we note the particular values of n and k. If s does not have length n, then $U(\langle 1 \rangle \frown s)$ is undefined.

Let r be the dyadic rational whose dyadic expansion begins with $s \frown \langle k \rangle$ followed by an infinite string of 0's. Let $r' = r + 2^{-(n+1)}$. We search for a t such that $\Omega[t]$ lies in the interval (r, r'). If we find such, we make $U(\langle 1 \rangle \frown s)$ be defined with the value $\emptyset$.

It seems reasonable that the final value of Ω should be at least $\Omega[t] + 2^{-(n+1)}$ since we have just added a new string of length $n+1$ to the domain of Ω. Thus the action we have just taken prevents Ω from being in the interval (r, r').

But clearly, if T has correctly predicted the value of the nth bit of Ω then Ω will lie in an interval of the form (r, r') for some length n bit string s. Thus our assumption that T can predict a single bit of Ω has led to a contradiction.

There are two points where we have to amplify the sketch to turn it into a correct proof.

1. We must check that the self-reference in the sketch can indeed be handled by the recursion theorem. (This is routine, but we shall treat this carefully in the final proof.)

2. The phrase "It seems reasonable" probably could be turned into a rigorous argument. But the detailed proof will proceed differently at this point.

6 Description of the Construction

We will be defining a function $U : \Sigma^* \mapsto \Sigma^*$ that depends on an integer parameter j. (Intuitively, j is a guess at the Gödel number of U.) We will specify the value of j presently.

U can be viewed as coming from a function $U_1 : \omega \times \Sigma^* \mapsto \Sigma^*$. (So $U(s) \simeq U_1(j, s)$.) Our construction will be such that U_1 is partial recursive.

As discussed in the sketch, we fix a universal Chaitin computer V such that the universality of V is provable in PA.

We proceed to define $U(s)$ by cases:

Case 1: $s = \emptyset$.

Then $U(s)$ is undefined.

Case 2: $s = \langle 0 \rangle \frown t$ for some bit string t.

Then we set $U(s) \simeq V(t)$.

Case 3: This is the case where $s = \langle 1 \rangle \frown t$ for some bit string t.

Our construction begins with a preliminary calculation to determine certain constants n and k. This preliminary calculation may not converge. In that case U will be undefined at s for any s falling under Case 3.

The preliminary calculation lists the theorems of T in some definite order (not depending on t) searching for a theorem of the form "The nth binary digit of Ω_j is k". If it finds such a theorem, then the value of n and k for the rest of the construction are those given by the first such theorem.

We will only define $U(\langle 1 \rangle \frown t)$ if $|t| = n$.

Suppose then that $|t| = n$. We define dyadic rationals r and r' as follows. r is the unique dyadic rational (in $[0,1)$) whose binary expansion starts with $t \frown \langle k \rangle$ and whose digits (after the nth one) are all 0. $r' = r + 2^{-(n+1)}$.

We now proceed to search for the least integer m such that $\Omega_j[m]$ lies in the open interval (r, r'). Of course, this search might fail. If so, $U(s)$ is undefined.

Recall that $D_j[m]$ is the finite set of strings in the domain of Ω_j that have contributed to the computation of $\Omega_j[m]$. (Cf. Section 2.5.) If s appears in $D_j[m]$, then $U(s)$ is undefined. Otherwise, we set $U(s) = \emptyset$.

The recursion theorem assures us that there is a value of j such that $\varphi_j(s) \simeq U_1(j,s)$. We fix such a j and set $U = \varphi_j$. Thus in the definition of U just given, the value of the parameter j was the Gödel number of U.

7 Analysis of the Construction

U is a Chaitin Machine

Suppose that s_1 and s_2 are two elements of the domain of U such that $s_1 \subseteq s_2$. We have to see that $s_1 = s_2$.

Since U is undefined on the empty string, $|s_1| \geq 1$. Let $r = s_1(0)$. Let $s_i = \langle r \rangle \frown t_i$. Clearly $t_1 \subseteq t_2$. If $r = 0$, then t_1 and t_2 are in the domain of the Chaitin computer V. Hence $t_1 = t_2$. So $s_1 = s_2$.

If $r = 1$, then for $U(s_1)$ and $U(s_2)$ to be defined, we must have the integer n defined in the course of the construction. But then $|s_1| = |s_2| =$

$n + 1$. So $s_1 = s_2$ as desired.

It follows that $U = \psi_j$ and that the real Ω_j used in the course of the construction is Ω_U. (Cf. Section 2.4.)

U is Universal

This follows from the definition of U on strings beginning with a 0. It is also clear that U inherits from V the fact that its universality is provable in PA.

Note that it follows that Ω_U is irrational. (Cf. Section 2.3.)

Now, towards a contradiction, assume that T can determine some bit of Ω_U. Then in the course of the construction the integers n and k are defined.

Let r be a dyadic rational with denominator 2^{n+1} such that $r < \Omega_U < r + 2^{-(n+1)}$. (We use here the fact that Ω_U is irrational.) Let $r' = r + 2^{-(n+1)}$.

Since T is 1-consistent, the assertion "The nth binary bit of Ω_U is k" is *true*. Hence the first $n + 1$ bits of the binary expansion of r have the form $t \frown \langle k \rangle$ where t is a bit string of length n. For all sufficiently large m, $\Omega_j[m]$ will lie in the interval (r, r').

Let $s = \langle 1 \rangle \frown t$. Consider now the computation of $U(s)$. The r and the r' involved in that computation are the ones we have just defined. The search for an m such that $\Omega_j[m] \in (r.r')$ will succeed.

Could it be that $s \in D_j[m]$? No, for then $U(s)$ would not be defined. But $D_j[m] \subseteq D_j$, so we would have $s \in D_j$, i.e., s *would* be in the domain of U after all, a contradiction.

So $U(s)$ is defined, and D_j contains in addition to the members of $D_j[m]$ the string s of length $n+1$. It follows that $\Omega_U \geq r + 2^{-(n+1)} = r'$. But this contradicts the definition of r. The proof of the main theorem is complete.

References

[1] G. J. Chaitin, *The Limits of Mathematics*, Springer-Verlag, Singapore, 1998.

[2] G. J. Chaitin, A theory of program size formally identical to information theory, *ACM Journal*, **22** (1975), 329–340.

On the Power of Reading
the Whole Infinite Input Tape

Ludwig Staiger

Martin-Luther-Universität Halle-Wittenberg
Institut für Informatik
Kurt-Mothes-Str. 1
D-06120 Halle (Saale), Germany
E-mail: staiger@informatik.uni-halle.de

1 Introduction

Finite devices accepting infinite strings are the topic of several papers (see the recent surveys [4], [8] or [10], [11]). Most of these papers deal with finite automata. Thus finite automata as devices accepting infinite strings are wellunderstood. The situation is a little bit more involved if one considers more complicated accepting devices like, e.g., pushdown automata or Turing machines.

A thorough treatment of a general class of accepting devices, so-called X-automata, on infinite words has been given in [4]. If one compares the results on the acceptance of infinite strings by Turing machines obtained in the papers [2], [3], [4] and [12], [9] one observes slight differences in the manner how the machines under consideration take into account their behaviour on the input tape.

Type 1: The approach of [12], [9] (cf. also [7], [8]) does not take into consideration the behaviour of the Turing machine on its input tape. Acceptance is based solely on the infinite sequence of internal states the machine runs through during its infinite computation.

   Thus the machine may base its decision on a finite part of the whole infinite input.

Type 2: For X-automata Engelfriet and Hoogeboom [4] require that, in addition to the fulfillment of certain conditions on the infinite sequence of internal states in order to accept an input, the machine has to read the whole infinite input tape. Cohen and Gold [1] considered this same type by for pushdown automata.

Thus, besides blocking[1] as for Type 1, machines have a further possibility to reject inputs.

Type 3: The most complicated type of acceptance for Turing machines was introduced by Cohen and Gold [2], [3]. They require in addition to Type 2 that the machine scans every cell of the input tape only finitely many times. This behaviour is termed as having a complete non-oscillating run.

The aim of our paper is to explore those sets of infinite words accepted by deterministic Turing machines with respect to Type 2 and compare them to the ones accepted with respect to Type 1 which are widely investigated (cf. [2][2], [12], [9], [7], [8]). Following the lines of [12], [9], [7] we provide also characterizations of Type 2 Turing accepted ω-languages in terms of the arithmetical hierarchy and automaton-free descriptions using recursively enumerable languages and limit operations.

Utilizing these characterizations, we show that in the case of deterministic Turing machines Type 2-acceptance is strictly more powerful than Type 1-acceptance. The stronger acceptance power of Type 2, however, contrasts with two elegant properties of Type 1-acceptance:

First, the classes of accepted ω-languages coincide fully with classes of the arithmetical hierarchy [12], [9] (cf. also [7], [8]), and secondly, the hierarchy of classes accepted by non-deterministic Turing machines does not collapse.

2 Notation

We start with some necessary notation. By $\mathbb{N} = \{0, 1, 2, \ldots\}$ we denote the set of natural numbers. We consider the space X^ω of infinite strings (sequences, ω-words) on a finite alphabet of cardinality $r := \operatorname{card} X \geq 2$. By X^* we denote the set (monoid) of finite strings (words) on X, including the *empty* word e. For $w \in X^*$ and $b \in X^* \cup X^\omega$ let $w \cdot b$ be their *concatenation*. This concatenation product extends in an obvious way to

---

[1]Most of the six state acceptance conditions considered in Section 3.2 can simulate blocking.

[2]Although Cohen and Gold [2] started their consideration of Turing machines on ω-words by considering the more powerful Type 2-acceptance, when turning to the deterministic case they weakened their model stipulating the so-called Property C (continuity property) which requires that the Turing machine under consideration *a priori* has to read the whole input tape in a non-oscillating manner.

subsets $W \subseteq X^*$ and $B \subseteq X^* \cup X^\omega$. As usual we denote subsets of X^* as languages and subsets of X^ω as ω-languages.

Furthermore $|w|$ is the *length* of the word $w \in X^*$, hence $X^n = \{w : w \in X^* \wedge |w| = n\}$. By $b[m..n] := b(m) \cdots b(n)$ we denote the *infix* extending from the mth to the nth letter[3] of a string $b \in X^*$, $|b| \geq n$, or $b \in X^\omega$, and $\mathbf{A}(b) := \{b[1..n] : n \in \mathbb{N} \wedge n \leq |b|\}$ and $\mathbf{A}(B) := \bigcup_{b \in B} \mathbf{A}(b)$ are the sets of all finite prefixes of $b \in X^* \cup X^\omega$ and $B \subseteq X^* \cup X^\omega$, respectively.

As usual we define Π_1-*definable* ω-*languages* $E \subseteq X^\omega$ as

$$E = \{\xi : \forall n(n \in \mathbb{N} \to \xi/n \in W_E)\} \tag{1}$$

where $W_E \subseteq X^*$ is a recursive language, and we define Σ_2-*definable* ω-*languages* $F \subseteq X^\omega$ as

$$F = \{\xi : \exists i(i \in \mathbb{N} \wedge \forall n(n \in \mathbb{N} \to (i, \xi/n) \in M_F))\}, \tag{2}$$

where M_F is a recursive subset of $\mathbb{N} \times X^*$. Σ_1-definable and Π_2-definable ω-languages are defined accordingly. A language $W \subseteq X^*$ is called Π_2-*definable* iff

$$W = \{w : w \in X^* \wedge \forall i(i \in \mathbb{N} \to \exists n(n \in \mathbb{N} \wedge (i, n, w) \in M_W))\},$$

for some recursive subset $M_W \subseteq \mathbb{N} \times \mathbb{N} \times X^*$.

3 Turing Machines

In order to be in accordance with the X-automata of [4] we consider Turing machines $\mathcal{M} = (X, \Gamma, Z, z_0, R)$ with a separate input tape on which the read-only-head moves only to the right, n working tapes, X as its input alphabet, Γ as its worktape alphabet, Z the finite set of internal states, z_0 the initial state, and the relation

$$R \subseteq Z \times X \times \Gamma^n \times Z \times \{0, +1\} \times (\Gamma \times \{-1, 0, +1\})^n$$

defining the next configuration.

Here $(z, x_0, x_1, \ldots, x_n; z', y_0, y_1, \ldots, y_n) \in R$ means that when $\mathcal{M}$ is in state $z \in Z$, reads $x_0 \in X$ on its input tape and $x_i \in \Gamma$ on its worktapes ($i \in \{1, \ldots, n\}$), $\mathcal{M}$ changes its state to $z' \in Z$, moves its head on the input tape to the right if $y_0 = +1$ or if $y_0 = 0$ does not move the head, and for ($i \in \{1, \ldots, n\}$) and $y_i = (x_i', m_i)$ with $x_i' \in \Gamma$ and $m_i \in \{-1, 0, +1\}$ the

---

[3]If $m > n$ or $m > |b|$ then $b[m..n] := e$, if $n > |b|$ then $b[m..n] := b[m..|b|]$.

machine $\mathcal{M}$ writes x_i' instead of x_i in its i-th worktape and moves the head on this tape to the left, if $m_i = -1$, to the right, if $m_i = +1$, or does not move it, if $m_i = 0$.

Unless stated otherwise, in the following we shall assume that our accepting devices be fully defined, i.e., the transition relation R is to contain for every situation $(z, x_0, x_1, \ldots, x_n)$ in $Z \times X^{n+1}$ at least one (exactly one, if the device is deterministic) move $(z, x_0, x_1, \ldots, x_n; z, y_0, y_1, \ldots, y_n)$.

For a detailed description of configurations (instantaneous descriptions) and the behaviour (computing process) of Turing machines the reader is referred to the literature (e.g., [6]).

To avoid clumsy notation, we will describe the construction of machines and their behaviour only in an informal manner, leaving details to the reader.

Let the input of the Turing machine be some sequence $\xi \in X^\omega$. We call a sequence $\zeta \in Z^\omega$ of states a *run* of $\mathcal{M}$ on ξ if ζ is the sequence of states the Turing machine runs through in its (some of its, if the machine is non-deterministic) computation with input ξ.

3.1   Acceptance Without State Conditions

In this section we are interested in the question which ω-languages can be accepted by Turing machines when we disregard the internal behaviour (states) and consider only the reading behaviour of the head on the input tape.

We say that an input word $w \in X^*$ is *read in full length* by a Turing machine $\mathcal{T}$ provided w is the empty word e or if $w = x_1 \cdot \ldots \cdot x_\ell$ then $\mathcal{T}$ starting on the leftmost letter x_1 moves its input head at least once onto the last letter x_ℓ.

Let $\mathcal{T}$ be a Turing machine. Define the function $\varphi : X^* \times \mathbb{N} \to X^*$ in the following way

$$\varphi(w, n) := \begin{cases} w[1..n], & \text{if } |w| \geq n \text{ and } \mathcal{T} \text{ reads at least the first} \\ & n \text{ letters of } w, \text{ and} \\ \text{undefined}, & \text{otherwise.} \end{cases}$$

Obviously, φ is a partial recursive function. Hence $\varphi(X^* \times \mathbb{N})$ is a recursively enumerable language.

Thus we have the following relationship between prefix-closed, that is, $W = \mathbf{A}(W)$, recursively enumerable languages and Turing machines reading words in full length.

On the Power of Reading the Whole Infinite Input Tape 339

Lemma 3.1 *A language $W \subseteq X^*$ is prefix-closed and recursively enumerable iff there is a Turing machine $\mathcal{T}$ such that W is the set of all input words which $\mathcal{T}$ reads in full length.*

We say that an ω-word $\xi \in X^\omega$ is lim-*accepted* by a Turing machine $\mathcal{M}$ if and only if there is a run of $\mathcal{M}$ on input ξ such that $\mathcal{M}$ reads every symbol of ξ, and we refer to an ω-language $F \subseteq X^\omega$ as lim-*accepted* provided there is a deterministic Turing machine $\mathcal{M}$ such that $F = \{\xi : \xi \in X^\omega \wedge \xi$ is lim-accepted by $\mathcal{M}\}$.

We obtain the following characterization of the class of lim-accepted ω-languages.

Theorem 3.1 *An ω-language $F \subseteq X^\omega$ is lim-accepted iff there is a recursively enumerable language $W \subseteq X^*$ such that $F = \{\zeta : \mathbf{A}(\zeta) \subseteq \mathbf{A}(W)\}$.*

Proof. Let F be lim-accepted by a Turing machine $\mathcal{M}$. Then from the above lemma we know that the language $L_\mathcal{M}$ of all words $w \in X^*$ which are read in full length by $\mathcal{M}$ is recursively enumerable. Thus $F = \{\xi : \mathbf{A}(\xi) \subseteq \mathbf{A}(L_\mathcal{M})\}$.

Conversely, let $F = \{\xi : \mathbf{A}(\xi) \subseteq \mathbf{A}(W)\}$ for some recursively enumerable language W. Then $\mathbf{A}(W)$ is also recursively enumerable, and utilizing a Turing machine $\mathcal{T}$ which enumerates the language $\mathbf{A}(W)$ it is easy to construct a Turing machine $\mathcal{M}$ which for an input word $v \in X^*$ reads only the longest prefix $w \sqsubseteq v$, $w \in \mathbf{A}(W)$. From the behaviour of $\mathcal{M}$ it follows that $\mathcal{M}$ lim-accepts F. $\qquad\square$

In [7] it is shown that the condition derived in Theorem 3.1 has several equivalent formulations in terms of the arithmetical hierarchy and topology.

To this aim we consider X^ω as a topological space with the basis $(w \cdot X^\omega)_{w \in X^*}$. Since X is finite, this topological space is homeomorphic to the Cantor discontinuum, hence compact. We mention here still that a subset F in X^ω is closed if and only if $F = \{\xi : \mathbf{A}(\xi) \subseteq \mathbf{A}(F)\}$. For more details the interested reader is referred to [8], Section 2.

Lemma 3.2 ([7]) *Let $F \subseteq X^\omega$. Then the following conditions are equivalent.*

1. *There is a recursively enumerable language $W' \subseteq X^*$ such that $F = \{\zeta : \mathbf{A}(\zeta) \subseteq W'\}$.*

2. *There is a recursively enumerable language $W \subseteq X^*$ such that $F = \{\zeta : \mathbf{A}(\zeta) \subseteq \mathbf{A}(W)\}$.*

3. *There is a recursive language $V \subseteq X^*$ such that $F = \{\zeta : \mathbf{A}(\zeta) \subseteq \mathbf{A}(V)\}$.*

4. *There is a Π_2-definable language U such that $F = \{\zeta : \mathbf{A}(\zeta) \subseteq U\}$.*

5. *F is closed in X^ω and $\mathbf{A}(F)$ is Π_2-definable.*

In the following we will refer to the class of lim-accepted ω-languages as

$$\mathsf{P} := \{F : F \subseteq X^\omega \wedge F \text{ is closed } \wedge \mathbf{A}(F) \text{ is } \Pi_2\text{-definable}\}.$$

We mention still some closure properties of the class P. To this end we consider homomorphisms $h : X \to X$. We extend the homomorphism h in the usual way to words. For ω-words $\xi \in X^\omega$ we define the extension of h, $\overline{h}$, in the following way:
Let $\overline{X} := \{x : x \in X \wedge h(x) = e\}$ be the set of letters erased by the homomorphism h.

$$\overline{h}(\xi) := \begin{cases} \text{undefined}, & \text{if } \xi \in X^* \cdot \overline{X}^\omega; \\ \lim_{w \to \xi} h(w), & \text{otherwise.} \end{cases}$$

Lemma 3.3 ([7]) *The class P is closed under union, intersection, non-erasing homomorphisms and their inverse mappings.*

3.2   State-Acceptance Conditions for Turing Machines

In this part we define the six conditions imposed on the runs of Turing machines in order to specify a set of input ω-words (cf. [2][4], [4], [8], [9]).
   We say that an input sequence $\xi \in X^\omega$ is accepted by $\mathcal{M}$ according to condition (mode) Ξ if there is a run ζ of $\mathcal{M}$ on ξ such that ζ satisfies Ξ. In the following we shall consider the following conditions using the notation of [4].
   Let $\alpha : Z^\omega \to 2^Z$ be a mapping which assigns to every ω-word $\zeta \in Z^\omega$ a subset $Z' \subseteq Z$, and let $R \subseteq 2^Z \times 2^Z$ be a relation between subsets of Z. We say that a pair $(\mathcal{M}, \mathcal{Z})$ where $\mathcal{Z} \subseteq 2^Z$ accepts an ω-word $\xi \in X^\omega$ if and only if

$$\exists Z' \exists \zeta (Z' \in \mathcal{Z} \wedge \zeta \text{ is a run of } \mathcal{M} \text{ on } \xi \wedge (\alpha(\zeta), Z') \in R) .$$

Here we shall be concerned with the following mappings α and relations R. For an ω-word $\zeta \in Z^\omega$ let $ran(\zeta) := \{z : z \in Z \wedge \exists i(i \in \mathbb{N} \setminus \{0\} \wedge \zeta(i) =$

---
[4]Cohen and Gold consider only the five conditions introduced by Landweber [5].

$z)\}$ be the *range* of ζ (considered as a mapping $\zeta : \mathbb{N} \setminus \{0\} \to Z$), that is, the set of all letters occurring in ζ, and let $inf(\zeta) := \{z : z \in Z \wedge \zeta^{-1}(z)$ is infinite$\}$ be the *infinity set* of ζ, that is, the set of all letters occurring infinitely often in ζ.

As relations R we shall use $=$, $\subseteq$ and $\sqcap$ where $Z' \sqcap Z'' :\Leftrightarrow Z' \cap Z'' \neq \emptyset$.

We obtain the six types of acceptance presented in the following table. For the sake of completeness we add a simple description and their correspondence to the five types originally defined by Landweber [5] and used in [2], [3].

$(ran, \sqcap)$	1-acceptance	at least once
$(ran, \subseteq)$	$1'$-acceptance	everywhere
$(ran, =)$		
$(inf, \sqcap)$	2-acceptance	infinitely often
$(inf, \subseteq)$	$2'$-acceptance	almost everywhere
$(inf, =)$	3-acceptance	

4 Classes of ω-Languages Accepted by Deterministic Turing Machines

In this section we consider the ω-languages defined by Turing machines according to Types 1 and 2 defined above and the six accepting conditions introduced in the previous section. First we give a brief review on the results concerning the six classes accepted according to Type 1 obtained in [12], [9] or [7]. Then we derive a relationship between the Type 1- and Type 2-classes based on the results of Section 3.1. Here it turns out that the class of lim-accepted ω-languages plays a crucial role. Finally, we use this relationship to obtain automata-free characterizations of the classes of ω-languages accepted according to Type 2. These descriptions yield immediately that Type 2-acceptance is, in some cases, strictly more powerful than Type 1-acceptance, whereas the proper Type 2-classes neither do coincide with classes of the Arithmetical hierarchy, as the Type 1 classes do, nor do they enjoy the same closure properties.

4.1 ω-Languages Accepted According to Type 1

We start with some characterizations in terms of recursive or recursively enumerable languages of the low level classes of the Arithmetical hierarchy of ω-languages (cf. [7], [8]).

Lemma 4.1
  1. *A set $E \subseteq X^\omega$ is in Σ_1-definable iff there is a recursive (or, equivalently, a recursively enumerable) language $W \subseteq X^*$ such that $E = W \cdot X^\omega$.*

  2. *A set $F \subseteq X^\omega$ is in Π_1-definable if and only if F is closed in X^ω and $X^* \setminus \mathbf{A}(F)$ is recursively enumerable, that is, $\mathbf{A}(F) \subseteq X^*$ is Π_1-definable.*

  3. *A set $E \subseteq X^\omega$ is in Π_2-definable iff there is a recursive (or, equivalently, a recursively enumerable) language $W \subseteq X^*$ such that $E = \{\xi : \xi \in X^\omega \wedge \mathbf{A}(\xi) \cap W \text{ is infinite}\}$.*

We mention still the following well-known closure properties of the classes of the Arithmetical hierarchy, analogously to the ones in Lemma 3.3.

Lemma 4.2
  1. *Each one of the classes Σ_1, Π_1 and Σ_2 of Σ_1-definable, Π_1-definable, or Σ_2-definable subsets of X^ω, respectively, is closed under union, intersection, non-erasing homomorphisms and inverse non-erasing homomorphisms.*

  2. *The class Π_2 of Π_2-definable subsets of X^ω is closed under union, intersection, and inverse non-erasing homomorphisms, but not under non-erasing homomorphisms.*

The class of all ω-languages $F \subseteq X^\omega$ Type 1-accepted by deterministic (non-deterministic) Turing machines with respect to mode (α, R) is denoted by $\mathbf{DT}^{(X)}(\alpha, R)$ (or $\mathbf{NT}^{(X)}(\alpha, R)$, respectively). The following characterization of the six **DT**-classes is shown in [12], [9] (cf. also [7], [8]). We start with the simple modes $\mathbf{DT}^{(X)}(\alpha, \subseteq)$ and $\mathbf{DT}^{(X)}(\alpha, \sqcap)$.

Theorem 4.1

$$\mathbf{DT}^{(X)}(ran, \subseteq) = \Pi_1 = \{F : F \subseteq X^\omega \wedge F \text{ is } \Pi_1\text{-definable}\}$$
$$\mathbf{DT}^{(X)}(ran, \sqcap) = \Sigma_1 = \{F : F \subseteq X^\omega \wedge F \text{ is } \Sigma_1\text{-definable}\}$$
$$\mathbf{DT}^{(X)}(inf, \subseteq) = \Sigma_2 = \{F : F \subseteq X^\omega \wedge F \text{ is } \Sigma_2\text{-definable}\}$$
$$\mathbf{DT}^{(X)}(inf, \sqcap) = \Pi_2 = \{F : F \subseteq X^\omega \wedge F \text{ is } \Pi_2\text{-definable}\}$$

Since Turing machines allow for a parallel composition, we obtain a characterization for the classes $\mathbf{DT}^{(X)}(\alpha, =)$. Here $\mathcal{B}(\mathcal{S})$ denotes the closure of the set $\mathcal{S} \subseteq 2^{X^\omega}$ under Boolean operations.

Corollary 4.1

$$\mathbf{DT}^{(X)}(ran, =) = \mathcal{B}(\mathbf{\Pi}_1) = \mathcal{B}(\mathbf{\Sigma}_1),$$
$$\mathbf{DT}^{(X)}(inf, =) = \mathcal{B}(\mathbf{\Sigma}_2) = \mathcal{B}(\mathbf{\Pi}_2)$$

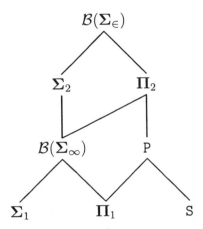

Figure 1: Inclusion relations between various classes of ω-languages accepted by Turing machines

Figure 1 shows the relations between the six classes of Theorem 4.1 and Corollary 4.1, the class P and its related subclass

$$\mathtt{S} := \{F : F \subseteq X^\omega \wedge F \text{ is closed } \wedge \mathbf{A}(F) \text{ is recursively enumerable}\} \subset \mathtt{P}.$$

Example 1.15 of [8] shows that S is incomparable to the classes of Π_1-definable, Σ_1-definable and Σ_2-definable subsets of X^ω. Thus, the inclusion properties presented in Figure 1 are proper and other inclusions than the ones represented do not exist.

4.2 Composition and Decomposition Theorem

In this section we analyze the behaviour of Turing machines having the additional possibility to reject by not reading the whole input utilizing the class of lim-accepted ω-languages and the classes accepted by Turing machines not regarding the moving of the input head.

We agree on the following notation. By $T(\mathcal{M}', \mathcal{Z})$ we denote the ω-language accepted by the pair $(\mathcal{M}, \mathcal{Z})$ according to Type 1, and by lim-$T(\mathcal{M}, \mathcal{Z})$ the ω-language accepted by the pair $(\mathcal{M}, \mathcal{Z})$ according to Type 2, and lim-$T(\mathcal{M})$ is the ω-language lim-accepted by $\mathcal{M}$.

We obtain the following easily verified decomposition theorem.

Theorem 4.2 (Decomposition Theorem) *Let $\mathcal{M} = (X, Z, \Gamma, z_0, f)$ be a deterministic Turing machine and let $\mathcal{Z} \subseteq 2^Z$. Then for $\alpha \in \{ran, inf\}$ and $R \in \{\subseteq, \sqcap, =\}$ we have lim-$T_R^\alpha(\mathcal{M}, \mathcal{Z}) = T_R^\alpha(\mathcal{M}, \mathcal{Z}) \cap$ lim-$T(\mathcal{M})$.*

In some sense we can also prove the converse of the Decomposition Theorem.

Theorem 4.3 (Composition Theorem) *Let $\mathcal{M} = (X, Z, \Gamma, z_0, f), \mathcal{Z} \subseteq 2^Z$, and $\mathcal{M}' = (X, S, \Gamma', s_0, g)$ be deterministic Turing machines. Then there is a deterministic Turing machine $\mathcal{M}''$ such that for $\alpha \in \{ran, inf\}$ and $R \in \{\subseteq, \sqcap, =\}$ it holds lim-$T_R^\alpha(\mathcal{M}'', \mathcal{Z}_{\alpha,R}) = T_R^\alpha(\mathcal{M}, \mathcal{Z}) \cap$ lim-$T(\mathcal{M}')$ for suitable $\mathcal{Z}_{\alpha,R}$.*

Proof. We give an informal description of how to construct an appropriate Turing machine $\mathcal{M}''$. $\mathcal{M}''$ is a parallel composition of $\mathcal{M}$ and $\mathcal{M}'$ in the following sense:

$\mathcal{M}''$ consists of one copy of $\mathcal{M}$ and $\mathcal{M}'$ each and an additional copying and step counting control.

$\mathcal{M}''$ starts working as $\mathcal{M}'$. As soon as $\mathcal{M}'$ reads a new letter from the input tape this letter is copied to an internal tape serving as the input tape of $\mathcal{M}$. Then $\mathcal{M}''$ switches to $\mathcal{M}$, resets the input head of $\mathcal{M}$ to the situation it had before copying and then simulates the work of $\mathcal{M}$ until $\mathcal{M}$ requires an input symbol not yet copied to its input tape or otherwise the number of steps given by the step counting control is exhausted. This step counting control ensures $\mathcal{M}''$ to switch to simulate $\mathcal{M}'$ even in the case $\mathcal{M}$ does not require further input letters. $\square$

As an immediate consequence of Theorems 4.2 and 4.3 we obtain that the classes $\overline{\mathbf{DT}}^{(X)}(\alpha, R)$ of ω-languages accepted by fulfilling the acceptance condition (α, R) *and* reading the whole input can be described as follows.

Theorem 4.4

$$\overline{\mathbf{DT}}^{(X)}(ran, \subseteq) = \mathrm{P} = \{F : F \text{ is closed} \wedge \mathbf{A}(F) \text{ is } \Pi_2\text{-definable}\},$$

$$\overline{\mathbf{DT}}^{(X)}(ran, \sqcap) = \{E \cap F : E \text{ is } \Sigma_1\text{-definable} \wedge F \in \mathrm{P}\},$$

$$\overline{\mathbf{DT}}^{(X)}(ran, =) = \{E \cap F : E \in \mathbf{DT}^{(X)}(ran, =) \wedge F \in \mathrm{P}\},$$

$$\overline{\mathbf{DT}}^{(X)}(inf, \subseteq) = \{E \cap F : E \text{ is } \Sigma_2\text{-definable} \wedge F \in \mathrm{P}\},$$

$$\overline{\mathbf{DT}}^{(X)}(inf, \sqcap) = \mathbf{DT}^{(X)}(inf, \sqcap) = \{E : E \text{ is } \Pi_2\text{-definable}\},$$
$$\overline{\mathbf{DT}}^{(X)}(inf, =) = \mathbf{DT}^{(X)}(inf, =).$$

Proof. The first four identities are obvious, and the last two follow from the fact that every ω-language $f \in P$ is Π_2-definable, and the class of Π_2-definable ω-languages is closed under intersection and union. $\qquad\square$

From the results presented in Figure 1 we obtain the following inclusion relations involving the new classes and the ones described above.

Lemma 4.3

$$P \subset \overline{\mathbf{DT}}^{(X)}(ran, \sqcap) \subset \overline{\mathbf{DT}}^{(X)}(ran, =) \subset \Pi_2,$$
$$\Sigma_2 \subset \overline{\mathbf{DT}}^{(X)}(inf, \subseteq) \subset \mathcal{B}(\Sigma_2).$$

In contrast to Lemma 4.2 the new classes $\overline{\mathbf{DT}}^{(X)}(ran, \sqcap)$, $\overline{\mathbf{DT}}^{(X)}(ran, =)$ and $\overline{\mathbf{DT}}^{(X)}(inf, \subseteq)$, however, are not closed under union, they are only closed under intersection. We provide an example.

Example 4.1 *Let $X := \{a, b, c\}$. Take an ω-language $F_1 \subset \{a, b\}^\omega$, $F_1 \in P$ which is not Σ_2-definable and let $F_2 := \{a, b\}^* \cdot c \cdot \{a, b, c\}^\omega$ which is Σ_2-definable, even Σ_1-definable, but not closed in X^ω. Then $F_1 \in P \subseteq \overline{\mathbf{DT}}^{(X)}(ran, \sqcap)$ and, since F_2 is Σ_1-definable, we have also $F_2 \in \overline{\mathbf{DT}}^{(X)}(ran, \sqcap)$. Assume now $F_1 \cup F_2 \in \overline{\mathbf{DT}}^{(X)}(inf, \subseteq)$.*

The smallest closed subset $F \subseteq \{a, b, c\}^\omega$ containing $F_1 \cup F_2$ is $\{a, b, c\}^\omega$ itself. Hence, $F_1 \cup F_2 = E \cap F$ with $E \in \Sigma_2$ and F closed in X^ω implies $F = \{a, b, c\}^\omega$, whence $F_1 \cup F_2 = E \in \Sigma_2$. But this contradicts the fact that $(F_1 \cup F_2) \cap \{a, b\}^\omega = F_1 \notin \Sigma_2$. $\qquad\square$

5 Non-deterministic Turing Machines

The classes of ω-languages accepted by non-deterministic Turing machines can be obtained from the deterministic classes via Projection lemmas for Type 1-acceptance and via the Homomorphism lemma [4], Lemma 3.11.

Let X, Y be finite alphabets, and let $\overline{pr}_1 : X \times Y \to X$ be a projection.[5]

Lemma 5.1 (Projection lemma for Turing machines [12], [9])
$F \in \mathbf{NT}^{(X \times Y)}(\alpha, R)$ *implies that* $\overline{pr}_1(F) \in \mathbf{NT}^{(X)}(\alpha, R)$.

---

[5]It is essential here that the mapping $\overline{pr}_1$ is a non-erasing homomorphism, whereas in Lemma 5.2 the homomorphism may erase letters.

Conversely, for every $E \in \mathbf{NT}^{(X)}(\alpha, R)$ *there is an* $F \in \mathbf{DT}^{(X \times Y)}(\alpha, R)$ *such that* $E = \overline{pr}_1(F)$.

This lemma yields in an elegant way an immediate characterization of the ω-languages Type 1 accepted by non-deterministic Turing machines. Because Lemma 4.2 shows that the classes $\mathbf{DT}^{(X)}(ran, \subseteq)$, $\mathbf{DT}^{(X)}(ran, \sqcap)$ and $\mathbf{DT}^{(X)}(inf, \subseteq)$ are closed under projections, according to the projection lemma they coincide with their non-deterministic counterparts.

The non-deterministic counterparts of the other classes are obtained as follows.

Theorem 5.1

$$\mathbf{NT}^{(X)}(ran, =) = \mathbf{NT}^{(X)}(inf, \subseteq) = \mathbf{DT}^{(X)}(inf, \subseteq),$$
$$\mathbf{NT}^{(X)}(inf, \sqcap) = \mathbf{NT}^{(X)}(inf, =) = \{F : F \subseteq X^\omega \wedge F \text{ is } \Sigma_1^1\text{-definable}\}.$$

Here, as usual, an ω-language is referred to as Σ_1^1-definable provided

$$F = \{\xi : \exists \eta (\eta \in X^\omega \wedge \forall n \exists m ((n, \eta[1..m], \xi[1..m]) \in M_F))\} \,,$$

for some recursive relation $M_F \subseteq \mathbb{N} \times X^* \times X^*$. Observe that the class of Σ_1^1-definable ω-languages equals the class of projections of Π_2-definable ω-languages.

Turning to Type 2 acceptance we have the following general relationship between the deterministic and non-deterministic classes.

Lemma 5.2 (Homomorphism lemma for Turing machines [4])
$F \in \overline{\mathbf{NT}}^{(X)}(\alpha, R)$ *implies that* $\overline{h}(F) \in \overline{\mathbf{NT}}^{(Y)}(\alpha, R)$.
Conversely, for every $E \in \overline{\mathbf{NT}}^{(Y)}(\alpha, R)$ *there is an* $F \in \overline{\mathbf{DT}}^{(X)}(\alpha, R)$ *such that* $E = \overline{h}(F)$.

Since a Turing machine can simulate even a counter, Theorem 3.11 of [4] proves that the hierarchy of classes $\overline{\mathbf{NT}}^{(X)}(\alpha, R)$ collapses. Using recursion-theoretic arguments one obtains:

Theorem 5.2 *For all* $\alpha \in \{ran, inf\}$ *and all* $R \in \{\subseteq, \sqcap, =\}$ *the class* $\overline{\mathbf{NT}}^{(X)}(\alpha, R)$ *coincides with the class of* Σ_1^1*-definable* ω*-languages over* X.

Thus, similar to the case of projections, where the class of Σ_1^1-definable ω-languages over X can be obtained from the class of Π_2-definable ω-languages over X via projections, all Σ_1^1-definable ω-languages over X are homomorphic images of (closed) ω-languages in P.

We can prove even more.

Lemma 5.3 *For every Σ_1^1-definable $F \subseteq X^\omega$ there are a Π_1-definable $E \subseteq Y^\omega$ and a homomorphism $h : Y \to X$ such that $F = \overline{h}(E)$.*

Proof. Since every Σ_1^1-definable ω-language is the projection (image under non-erasing homomorphism) of a Π_2-definable ω-language, it suffices to prove that every Π_2-definable $F \subseteq X^\omega$ is the image under arbitrary homomorphism of a Π_1-definable $E \subseteq Y^\omega$.

To this aim let $Y := X \cup \{d\}$ where d is a letter not in X. We define $h : X \cup \{d\} \to X \cup \{e\}$ to be the homomorphism erasing the letter d, that is,

$$h(y) = \begin{cases} y, & \text{if } y \in X, \\ e, & \text{if } y = d. \end{cases}$$

We use Lemma 4.1 to obtain a recursive language $W \subseteq X^*$ such that $F = \{\xi : \xi \in X^\omega \wedge \mathbf{A}(\xi) \cap W \text{ is infinite}\}$.

Next we pad the language W by inserting strings of ds in such a way that we can reconstruct from every padded word all (proper) prefixes of the original word.

Define for $w = w_1 \cdot w_2 \cdots w_n \in W$ where $\mathbf{A}(w) \cap W \backslash \{e\} = \{w_1 \cdot w_2 \cdots w_i : 1 \le i \le n\}$ its *padding* as follows:

$$\pi(w) := \{d^{|w_1|} \cdot w_1 \cdot d^{|w_2|} \cdot w_2 \cdots d^{|w_n|} \cdot w_n' : w_n' \sqsubseteq w_n\}, \text{ and let}$$
$$\pi(W) := \bigcup_{w \in W} \pi(w) \ .$$

Observe that $\pi(W)$ is a recursive language provided W is recursive, and, moreover, $\pi(W) = \mathbf{A}(\pi(W))$.

Let $E := \{\eta : \mathbf{A}(\eta) \subseteq \pi(W)\} = \{\eta : \mathbf{A}(\eta) \cap \pi(W) \text{ is infinite}\}$.

By definition $\overline{h}(\eta)$ is defined if η contains infinitely many prefixes ending with a letter of X. Thus, $\overline{h}(\eta)$, $\eta \in E$ is defined iff η has infinitely many prefixes of the form $d^{|w_1|} \cdot w_1 \cdot d^{|w_2|} \cdot w_2 \cdots d^{|w_n|} \cdot w_n'$ where $e \ne w_n' \in X^*$. From this we can conclude that $\overline{h}(\eta)$ has infinitely many prefixes in W, that is, $\overline{h}(\eta) \in F$.

Conversely, let $\xi \in F$. Then $\xi = w_1 \cdot w_2 \cdots w_i \cdots$ where $\mathbf{A}(\xi) \cap W \backslash \{e\} = \{w_1 \cdot w_2 \cdots w_i : i \in \mathbb{N} \backslash \{0\}\}$. Then $\eta := d^{|w_1|} \cdot w_1 \cdot d^{|w_2|} \cdot w_2 \cdots d^{|w_i|} \cdot w_i \cdots \in E$ and $\overline{h}(\eta) = \xi$. $\square$

Reading the previous proof more carefully, we observe that, since $\pi(W)$ is recursive, $E := \{\eta : \mathbf{A}(\eta) \subseteq \pi(W)\}$ is not only Π_1-definable, but also in S, and we obtain the following corollary showing the power of erasing.

Corollary 5.1 *For every Σ_1^1-definable $F \subseteq X^\omega$ there are an ω-language $E \subseteq Y^\omega$, $E \in \Pi_1 \cap S$ and a homomorphism $h : Y \to X$ such that $F = \overline{h}(E)$.*

References

[1] R. S. Cohen, A. Y. Gold, Theory of ω-languages I: Characterizations of ω-context-free languages, and II: A study of various models of ω-type generation and recognition, *J. Comput. System Sci.*, **15**, 2 (1977), 169–184 and 185–208.

[2] R. S. Cohen, A. Y. Gold, ω-computations on Turing machines, *Theoret. Comput. Sci.*, **6** (1978), 1–23.

[3] R. S. Cohen, A. Y. Gold, On the complexity of ω-type Turing acceptors, *Theoret. Comput. Sci.*, **10** (1980), 249–272.

[4] J. Engelfriet, H. J. Hoogeboom, X-automata on ω-words, *Theoret. Comput. Sci*, **110** (1993) 1, 1–51.

[5] L. H. Landweber, Decision problems for ω-automata, *Math. Syst. Theory*, **3**, 4 (1969), 376–384.

[6] H. Rogers, *Theory of Recursive Functions and Effective Computability*, McGraw Hill, New York 1967.

[7] L. Staiger, Hierarchies of recursive ω-languages, *J. Inform. Process. Cybernetics, EIK*, **22**, 5/6 (1986), 219–241.

[8] L. Staiger, ω-languages, in vol. 3 of *Handbook of Formal Languages*, (G. Rozenberg, A. Salomaa, eds.), Springer-Verlag, Berlin, 1997, 339–387.

[9] L. Staiger, K. Wagner, Rekursive Folgenmengen I, *Zeitschr. Math. Logik u. Grundl. Mathematik*, **24**, 6 (1978), 523–538.

[10] W. Thomas, Automata on infinite objects, in vol. B of *Handbook of Theoretical Computer Science* (J. Van Leeuwen, ed.), Elsevier, Amsterdam, 1990, 133–191.

[11] W. Thomas, Languages, automata, and logic, in vol. 3 of *Handbook of Formal Languages* (G. Rozenberg, A. Salomaa, eds.), Springer-Verlag, Berlin, 1997, 389–455.

[12] K. Wagner, L. Staiger, Recursive ω-languages, in *Fundamentals of Computation Theory*, (M. Karpiński, ed.), *Lecture Notes in Computer Science*, **56**, Springer-Verlag, Berlin 1977, 532–537.

Finite and Infinite in Logic

Victor Vianu

Department of Computer Science and Engineering
U. C. San Diego
La Jolla, CA 92093-0114, USA
E-mail: vianu@cs.ucsd.edu

1 Introduction

Classical logic generally focuses on infinite structures, such as the integers. However, in many applications, including computer science, finite structures are the primary focus. In this article we consider the following question: what happens to logic when structures are restricted to be finite? We survey several central results of *finite model theory* – the study of logic on finite structures. We start with the observation that many results that provide crucial tools in the general case, such as the Compactness theorem, fail in the finite case. We then present several key results of finite model theory that provide specific tools to replace some of the failed classical theorems. We focus on *Ehrenfeucht-Fraissé games* and *0-1 laws*.

These provide very effective tools for proving expressiveness results on finite structures, and some of the most elegant results in finite-model theory.

Some computer science applications involve structures that are conceptually infinite. For example, geographic information systems deal with regions in space, which in principle consist of infinitely many points. In such cases it is of interest to find finite summaries of the infinite structure that provide just sufficient information to answer the queries relevant to the application. We illustrate this approach with an example from geographic information systems, where the relevant set queries consists of *topological properties*, i.e., properties invariant under homeomorphisms of the space. We show that such properties can be summarized using a finite first-order structure called a *topological invariant*.

While conventional wisdom suggests that only finite representations are of interest in computer science, we conclude by presenting an alternative point of view: sometimes it makes more sense to model finite structures as infinite. As a concrete and somewhat amusing example, we consider the World Wide Web and show some advantages of modeling it as an infinite rather than finite structure.

2 Basics

We present an informal review of first-order logic. A formal presentation can be found in a standard logic textbook, such as [10].

2.1 Basic Logic Terminology

For simplicity, we only consider here relational structures, without explicit functions. A *relational vocabulary* **R** is a finite set of relation symbols with associated arities. We always assume that **R** contains the equality symbol. A *structure* over a relational vocabulary **R** (also called a *relational structure*) is a mapping **I** associating to each relation symbol R in **R** a relation of the same arity as R (the symbol $=$ is always interpreted as equality). A relational structure is *finite* if all of its relations are finite. The set of elements occurring in a structure **I** is called the *domain* of **I**, denoted $dom(\mathbf{I})$.

Relational first-order logic (FO) formulas are statements about relational structures. The statements are expressions formed using relation symbols $P, Q, R, \ldots$, variables $x, y, z, \ldots$, constants from some infinite domain **dom**, and the symbols $(,), \wedge, \vee, \rightarrow, \neg, \exists, \forall$. The simplest formula in FO, called an *atom*, is of the form $R(x_1, \ldots, x_k)$. It states that the tuple $x_1, \ldots, x_k$ of variables or constants belongs to relation R. Other formulas are obtained recursively from atoms by using the logical connectors $\wedge, \vee, \rightarrow, \neg$, and quantifiers $\exists, \forall$, with the usual semantics.

For example, consider a vocabulary consisting of a single binary relation G (the edges of a directed graph). The following sentence states that there exists a path in G of length three from node a to node b:

$$(\dagger) \quad \exists x \exists y (G(a, x) \wedge G(x, y) \wedge G(y, b)).$$

Classical logic has developed a powerful array of techniques for proving things about first-order logic. *Gödel's Completeness Theorem* provides a sound and complete set of axioms for proving *validity* of FO sentences (the property of being true in *all* structures, finite and infinite). As a consequence, the set of valid FO sentences is *recursively enumerable*, that is, can be enumerated by a Turing Machine.

Another consequence provides a technical tool which is a household name to all logicians: the *Compactness Theorem*, stating that a set Σ of FO sentences is satisfiable iff every finite subset of Σ is satisfiable. As an example of its effectiveness as a tool in understanding the expressive power of FO, consider the question of whether *graph connectivity* is FO definable. Here is a simple proof that it is not, using compactness. Suppose there exists an FO sentence σ defining graph connectivity. For each $i > 0$ consider an FO sentence σ_i stating that the distance between two nodes a, b in the graph is more than i (for example, σ_3 is the negation of the sentence (†) above). Let $\Sigma = \{\sigma_i \mid i > 0\} \cup \{\sigma\}$. Clearly, every finite subset of Σ is satisfiable. However, there is no graph satisfying Σ: in such a graph there would be no path of finite length from a to b, but a and b would be connected. The contradiction shows that no such σ can exist.

2.2   The Impact of Finiteness

Restricting structures to be finite has a dramatic impact on the landscape of classical logic. We mention two major examples.

First, Gödel's Completeness theorem fails in a very strong sense: not only is the set of validity axioms no longer complete, but there can be no sound and complete inference mechanism for proving validity on finite structures. This follows from another remarkable fact: the set of FO sentences valid on finite structures is not recursively enumerable. For suppose it is. Clearly, the set of invalid sentences is also recursively enumerable (an algorithm generating the invalid sentences would alternatingly generate sentences and finite structures, and would output a sentence when a finite structure which does not satisfy the sentence is found). But then the set of valid sentences has to be recursive. However, this is known to be false – it is undecidable if a sentence is valid on finite structures [8]. This shows that the set of FO sentences valid on finite structures is not recursively enumerable.

Perhaps as dramatically, the Compactness theorem also fails on finite structures. This is easily seen: consider a vocabulary consisting of a unary relation R (a set) and for each $i > 0$, let σ_i be an FO sentence stating that R has at least i elements. Let $\Sigma = \{\sigma_i \mid i > 0\}$. Clearly, every finite subset of Σ is satisfiable by a finite structure. However, there is no finite structure satisfying Σ.

The failure of the Compactness theorem deprives logicians of a crucial tool. For example, how might one prove that connectivity of finite graphs is not definable in FO? We next consider two important results that provide elegant, effective tools for answering such expressiveness questions about FO on finite structures.

3   Ehrenfeucht-Fraïssé Games

This section presents a technique based on a two-person "game", due to Ehrenfeucht [9] and Fraïssé [12], that can be used to prove that certain properties, including *evenness*[1] and connectivity, are not FO definable. While the game we describe is geared towards the first-order queries, games provide a general technique that is used in conjunction with many other languages.

The connection between FO sentences and games is, intuitively, the following. Consider as an example a FO sentence of the form

$$\forall x_1 \, \exists x_2 \, \forall x_3 \, \psi(x_1, x_2, x_3).$$

One can view the sentence as a statement about a game with two players, Spoiler and Duplicator, who alternate in picking values for x_1, x_2, x_3. The sentence says that Duplicator can always force a choice of values that makes $\psi(x_1, x_2, x_3)$ true. In other words, no matter which value Spoiler chooses for x_1, Duplicator can pick an x_2 such that, no matter which x_3 is chosen next by Spoiler, $\psi(x_1, x_2, x_3)$ is true.

The actual game we use, called the *Ehrenfeucht-Fraïssé* game, is slightly more involved, but is based on a similar intuition. It is played on two structures. Suppose that **R** is a relational vocabulary. Let **I** and **J** be structures over **R**, with disjoint sets of elements. Let r be a positive integer. The *game of length r associated with **I** and **J*** is played by two players, Spoiler and Duplicator, making r choices each. Spoiler starts by picking an element occurring in **I** or **J** and Duplicator picks an element in the opposite structure. This is repeated r times. At each move, Spoiler has the choice of the structure and an element in it, and Duplicator must respond in the opposite structure.

Let a_i be the ith elements picked in **I** (respectively, b_i in **J**). The set of pairs

$$\{(a_1, b_1), ..., (a_r, b_r)\}$$

is a *round* of the game. The *substructure* of **I** generated by $\{a_1, ..., a_r\}$, denoted $\mathbf{I}/\{a_1, ..., a_r\}$, consists of all tuples in **I** using only these elements, and similarly for **J**, $\{b_1, ..., b_r\}$ and $\mathbf{J}/\{b_1, ..., b_r\}$.

---

[1]Evenness is the property that a finite set has even cardinality.

Duplicator *wins the round* $\{(a_1, b_1), ..., (a_r, b_r)\}$ iff the mapping $a_i \to b_i$ is an isomorphism of the substructures $\mathbf{I}/\{a_1, ..., a_r\}$ and $\mathbf{J}/\{b_1, ..., b_r\}$.

Duplicator *wins the game of length r* associated with $\mathbf{I}$ and $\mathbf{J}$ if he has a winning strategy, i.e., Duplicator can always win any game of length r on $\mathbf{I}$ and $\mathbf{J}$, no matter how Spoiler plays. This is denoted by $\mathbf{I} \equiv_r \mathbf{J}$. Note that the relation $\equiv_r$ is an equivalence relation on structures over $\mathbf{R}$. Intuitively, the equivalence $\mathbf{I} \equiv_r \mathbf{J}$ says that $\mathbf{I}$ and $\mathbf{J}$ cannot be distinguished by looking at just r elements at a time in the two structures. We will need the notion of *quantifier depth* of a FO formula: the maximum number of quantifiers in a path from the root to a leaf in the representation of the sentence as a tree. The main result concerning Ehrenfeucht-Fraissé games states that the ability to distinguish among structures using games of length r is equivalent to the ability to distinguish among structures using some FO sentence of quantifier depth r.

Example　Consider the sentence $\forall x \ (\exists y \ R(x, y) \wedge \exists z \ P(x, z))$. Its syntax tree is represented in Figure 1. The sentence has quantifier depth 2. Note that, for a sentence in prenex normal form, the quantifier depth is simply the number of quantifiers in the formula.

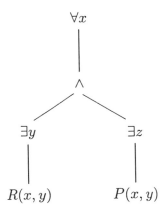

Figure 1: A syntax tree

The main result about Ehrenfeucht-Fraissé games, stated in Theorem 3.1 below, is that if $\mathbf{I}$ and $\mathbf{J}$ are two structures such that Duplicator has a winning strategy for the game of length r on the two structures, then $\mathbf{I}$ and $\mathbf{J}$ cannot be distinguished by any FO sentence of quantifier depth r. Before proving this theorem, we note that the converse of that result also holds. Thus, if two structures are indistinguishable using sentences of quantifier depth r, then they are equivalent with respect to $\equiv_r$. Although interesting, this is of less use as a tool for proving expressibility results, and we leave it as a (non-trivial!) exercise. The main idea is to show that each equivalence class of $\equiv_r$ is definable by a sentence of quantifier depth r.

Theorem 3.1 *Let* **I** *and* **J** *be two finite structures over relational vocabulary* **R***. If* **I** $\equiv_r$ **J** *then for each FO sentence* φ *over* **R** *with quantifier depth* r*,* **I** *and* **J** *both satisfy* φ *or neither does.*

Proof. Suppose that **I** $\models \varphi$ and **J** $\not\models \varphi$ for some φ of quantifier depth r. We prove that **I** $\not\equiv_r$ **J**. We provide here only a sketch of the proof on an example.

Let φ be the sentence: $\forall x_1 \exists x_2 \forall x_3 \ \psi(x_1, x_2, x_3)$, where ψ has no quantifiers, and let **I** and **J** be two structures such that: **I** $\models \varphi$, **J** $\not\models \varphi$. Then

$$\mathbf{I} \models \forall x_1 \exists x_2 \forall x_3 \ \psi(x_1, x_2, x_3) \quad \text{and} \quad \mathbf{J} \models \exists x_1 \forall x_2 \exists x_3 \ \neg\psi(x_1, x_2, x_3).$$

We will show that Spoiler can prevent Duplicator from winning by forcing the choice of elements a_1, a_2, a_3 in **I** and b_1, b_2, b_3 in **J** such that **I** $\models \psi(a_1, a_2, a_3)$ and **J** $\models \neg\psi(b_1, b_2, b_3)$. Then the mapping $a_i \to b_i$ cannot be an isomorphism of the substructures $\mathbf{I}/\{a_1, a_2, a_3\}$ and $\mathbf{J}/\{b_1, b_2, b_3\}$, contradicting the assumption that Duplicator has a winning strategy. To force this choice, Spoiler always picks "witnesses" corresponding to the existential quantifiers in φ and $\neg\varphi$ (note that the quantifier for each variable is either $\forall$ in φ and $\exists$ in $\neg\varphi$, or vice versa).

Spoiler starts by picking an element b_1 in **J** such that

$$\mathbf{J} \models \forall x_2 \exists x_3 \ \neg\psi(b_1, x_2, x_3).$$

Duplicator must respond by picking an element a_1 in **I**. Due to the universal quantification in φ,

$$\mathbf{I} \models \exists x_2 \forall x_3 \ \psi(a_1, x_2, x_3)$$

regardless of which a_1 was picked. Next, Spoiler picks an element a_2 in **I** such that

$$\mathbf{I} \models \forall x_3 \ \psi(a_1, a_2, x_3).$$

Regardless of which element b_2 in **J** Duplicator picks,

$$\mathbf{J} \models \exists x_3 \neg\psi(b_1, b_2, x_3).$$

Finally, Spoiler picks b_3 in $\mathbf{J}$ such that $\mathbf{J} \models \neg\psi(b_1, b_2, b_3)$; Duplicator picks some a_3 in $\mathbf{I}$, and $\mathbf{I} \models \psi(a_1, a_2, a_3)$. $\qquad\square$

Theorem 3.1 provides an important tool for proving that certain properties are not definable by FO. Indeed, it is sufficient to exhibit, for each r, two structures $\mathbf{I}_r$ and $\mathbf{J}_r$ such that $\mathbf{I}_r$ has the property, $\mathbf{J}_r$ does not, and $\mathbf{I}_r \equiv_r \mathbf{J}_r$. In the next proposition, we illustrate the use of this technique by showing that finite graph connectivity is not expressible in FO.

Proposition 3.1 *Let $\mathbf{R}$ be a relational vocabulary consisting of one binary relation. Then the query conn defined by:*

$$conn(\mathbf{I}) = true \ \textit{iff} \ \mathbf{I} \ \textit{is a connected graph},$$

is not definable in FO.

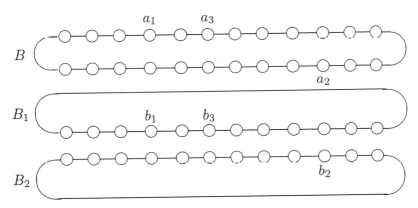

Figure 2: Two indistinguishable graphs

Proof. Suppose that there is a FO sentence φ defining graph connectivity. Let r be the quantifier depth of φ. We exhibit a connected graph $\mathbf{I}_r$ and a disconnected graph $\mathbf{J}_r$ such that $\mathbf{I}_r \equiv_r \mathbf{J}_r$. Then, by Theorem 3.1, the two structures satisfy φ or none does, a contradiction.

For a sufficiently large n (depending only on r) the graph $\mathbf{I}_r$ consists of a cycle B of $2n$ nodes and the graph $\mathbf{J}_r$ of two disjoint cycles B_1 and B_2 of n nodes each (see Figure 2). We outline the winning strategy for Duplicator. The main idea is simple: two nodes a, a' in $\mathbf{I}_r$ that are far apart "behave" in the same way as two nodes b, b' in $\mathbf{J}_r$ that belong to different cycles. In particular, Spoiler cannot take advantage of the fact that a, a' are connected but b, b' are not. To do so, Spoiler would have to exhibit a path connecting a to a', which Duplicator could not do for b and b'. However, Spoiler cannot construct such a path because it requires choosing more than r nodes.

To illustrate, if Spoiler picks an element a_1 in $\mathbf{I}_r$, then Duplicator picks an arbitrary element b_1, say in B_1. Now, if Spoiler picks an element b_2 in B_2, then Duplicator picks an element a_2 in $\mathbf{I}_r$ far from a_1. Next, if Spoiler picks a b_3 in B_1 close to b_1, then Duplicator picks an element a_3 in $\mathbf{I}_r$ close to a_1. The graphs are sufficiently large that this can proceed for r moves with the resulting subgraphs isomorphic. The full proof requires a complete case analysis on the moves that Spoiler can make. □

The above technique can be used to show that many other properties are not expressible in FO, for instance *evenness*, 2-colorability of graphs, or Eulerian graphs (i.e., graphs for which there is a cycle that passes through each edge exactly once).

4   0-1 Laws

We now present a powerful tool that provides a uniform approach to resolving in the negative a large spectrum of expressibility problems. It is based on the probability that a property is true in structures of a given size. This study, initiated by Fagin [11] and Glebskiĭ [13], shows a very surprising fact: all properties of finite structures definable in FO are "almost surely" true, or "almost surely" false.

Let σ be a sentence over some relational vocabulary $\mathbf{R}$. For each n, let $\mu_n(\sigma)$ denote the fraction of finite structures over $\mathbf{R}$ with entries in $\{1, ..., n\}$ that satisfy σ, i.e.,

$$\mu_n(\sigma) = \frac{|\{\mathbf{I} \mid \sigma(\mathbf{I}) = true \text{ and } dom(\mathbf{I}) = \{1, ..., n\}\}|}{|\{\mathbf{I} \mid dom(\mathbf{I}) = \{1, ..., n\}\}|}.$$

Definition 4.1 *A sentence σ is* almost surely true *(respectively, false) if* $\lim_{n \to \infty} \mu_n(\sigma)$ *exists and equals 1 (respectively, 0). If every sentence in a language L is almost surely true or almost surely false, the language L has a* 0-1 law.

To simplify the discussion of 0-1 laws, we continue to focus on constant-free sentences.

We will show that FO has a 0-1 law. This provides substantial insight into limitations in the expressive power of FO. It can be used to show that it cannot express a variety of properties. For example, it follows immediately that *evenness* is not expressible in FO. Indeed, $\mu_n(evenness)$ is 1 if n is even and 0 if n is odd. Thus, $\mu_n(evenness)$ does not converge, so *evenness* is not expressible in any language that has a 0-1 law.

While 0-1 laws provide an elegant and powerful tool, they require the development of some non-trivial machinery. Interestingly, this is one of the rare occasions when it is needed to consider *infinite* structures while proving something about finite structures!

For simplicity, we consider only the case of vocabularies consisting of a binary relation G (representing edges in a directed graph with no edges of the form $\langle a, a \rangle$). It is straightforward to generalize the development to arbitrary vocabularies.

We will use an infinite set $\mathcal{A}$ of FO sentences called *extension axioms*, that refer to graphs. They say, intuitively, that every subgraph can be extended by one node in all possible ways. More precisely, $\mathcal{A}$ contains, for each k, all sentences of the form

$$\forall x_1 ... \forall x_k ((\bigwedge_{i \neq j} (x_i \neq x_j)) \Rightarrow \exists y (\bigwedge_i (x_i \neq y) \wedge connections(x_1, ..., x_k; y))),$$

where $connections(x_1, ..., x_k; y)$ is some conjunction of literals containing, for each x_i, one of $G(x_i, y)$ or $\neg G(x_i, y)$, and one of $G(y, x_i)$ or $\neg G(y, x_i)$. For example, for $k = 3$, one of the 2^6 extension axioms is:

$$\forall x_1, x_2, x_3 \ (\ (x_1 \neq x_2 \wedge x_2 \neq x_3 \wedge x_3 \neq x_1) \Rightarrow$$
$$\exists y \ (x_1 \neq y \wedge x_2 \neq y \wedge x_3 \neq y \wedge$$
$$G(x_1, y) \wedge \neg G(y, x_1) \wedge \neg G(x_2, y) \wedge \neg G(y, x_2) \wedge G(x_3, y) \wedge G(y, x_3)))$$

specifying the pattern of connections represented in Figure 3.

A graph G satisfies this particular extension axiom if for each triple x_1, x_2, x_3 of distinct vertices in G, there exists a vertex y connected to x_1, x_2, x_3 as shown in Figure 3.

Note that $\mathcal{A}$ consists of an infinite set of sentences, and that each finite subset of $\mathcal{A}$ is satisfied by some infinite structure. (The structure is obtained by starting from one node and repeatedly adding nodes required by the extension axioms in the subset.) Then, by the Compactness theorem there is an infinite structure satisfying all of $\mathcal{A}$, and by the Löwenheim-Skolem theorem (see [10]) there is a countably infinite structure $\mathcal{R}$ satisfying $\mathcal{A}$.

The following lemma shows that $\mathcal{R}$ is unique up to isomorphism.

Lemma 4.1 *If $\mathcal{R}$ and $\mathcal{P}$ are two countably infinite structures over G satisfying all sentences in $\mathcal{A}$, then $\mathcal{R}$ and $\mathcal{P}$ are isomorphic.*

Proof. Suppose that $a_1 a_2 ...$ is an enumeration of all elements in $\mathcal{R}$, and $b_1 b_2 ...$ an enumeration of those in $\mathcal{P}$. We construct an isomorphism between $\mathcal{R}$ and $\mathcal{P}$ by alternatingly picking elements from $\mathcal{R}$ and from $\mathcal{P}$. We construct sequences $a_{i_1} ... a_{i_k} ...$ and $b_{i_1} ... b_{i_k} ...$ such that $a_{i_k} \rightarrow b_{i_k}$ is an isomorphism from $\mathcal{R}$ to $\mathcal{P}$. The procedure for picking the kth elements a_{i_k} and b_{i_k} in these sequences is defined inductively as follows. For the base case, let $a_{i_1} = a_1$ and $b_{i_1} = b_1$. Suppose that sequences $a_{i_1} ... a_{i_k}$ and $b_{i_1} ... b_{i_k}$ have been defined. If k is even, let $a_{i_{k+1}}$ be the first element in $a_1, a_2, ...$ which does not occur so far in the sequence. Let σ_k be the sentence in $\mathcal{A}$ describing the way $a_{i_{k+1}}$ extends the subgraph with nodes $a_{i_1} ... a_{i_k}$. Since $\mathcal{P}$ also satisfies σ_k, there exists an element b in $\mathcal{P}$ which extends the subgraph $b_{i_1} ... b_{i_k}$ in the same manner. Let $b_{i_{k+1}} = b$. If k is odd, the procedure is reversed, i.e., it starts by choosing first a new element from $b_1, b_2, ...$ This "back and forth" procedure ensures that (i) all elements from both $\mathcal{R}$ and $\mathcal{P}$ occur eventually among the chosen elements, and (ii) the mapping $a_{i_k} \rightarrow b_{i_k}$ is an isomorphism. $\qquad\square$

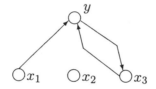

Figure 3: A connection pattern

Thus, the above shows that there exists a unique (up to isomorphism) countable graph $\mathcal{R}$ satisfying $\mathcal{A}$. This graph, studied extensively by Rado [21] and others, is usually referred to as the *Rado graph*. We can now prove the following crucial lemma. The key point is the equivalence between (a) and (c), called the *transfer property*: it relates satisfaction of a sentence by the Rado graph with the property of being almost surely true.

Lemma 4.2 *Let $\mathcal{R}$ be the Rado graph and σ a FO sentence. The following are equivalent:*

(a) *$\mathcal{R}$ satisfies σ;*

(b) *$\mathcal{A}$ implies σ; and*

(c) *σ is almost surely true.*

Proof. $(a) \Rightarrow (b)$: Suppose (a) holds but (b) does not. Then there exists some structure $\mathcal{P}$ satisfying $\mathcal{A}$ but not σ. Since $\mathcal{P}$ satisfies $\mathcal{A}$, $\mathcal{P}$ must be infinite. By the Lowënheim-Skolem theorem we can assume that $\mathcal{P}$ is countable. But then, by Lemma 4.1, $\mathcal{P}$ is isomorphic to $\mathcal{R}$. This is a contradiction, since $\mathcal{R}$ satisfies σ but $\mathcal{P}$ does not.

$(b) \Rightarrow (c)$: It is sufficient to show that each sentence in $\mathcal{A}$ is almost surely true. For suppose this is the case and $\mathcal{A}$ implies σ. By the Compactness theorem, σ is implied by some finite subset $\mathcal{A}'$ of $\mathcal{A}$. Since every sentence in $\mathcal{A}'$ is almost surely true, the conjunction $\bigwedge \mathcal{A}'$ of these sentences is almost surely true. Since σ is true in every structure where $\bigwedge \mathcal{A}'$ is true, $\mu_n(\sigma) \geq \mu_n(\bigwedge \mathcal{A}')$, so $\mu_n(\sigma)$ converges to 1 and σ is almost surely true.

It remains to show that each sentence in $\mathcal{A}$ is almost surely true. Consider the following sentence σ_k in $\mathcal{A}$:

$$\forall x_1 ... \forall x_k ((\bigwedge_{i \neq j}(x_i \neq x_j)) \rightarrow \exists y(\bigwedge_i (x_i \neq y) \wedge connections(x_1, ..., x_k; y))).$$

Then $\neg \sigma_k$ is the sentence

$$\exists x_1 ... \exists x_k ((\bigwedge_{i \neq j}(x_i \neq x_j)) \wedge$$
$$\forall y(\bigwedge_i(x_i \neq y) \rightarrow \neg connections(x_1, ..., x_k; y))).$$

We will show the following property on the probability that a structure with n elements *does not* satisfy σ_k:

$$(\dagger) \ \mu_n(\neg \sigma_k) \leq n \cdot (n-1) \cdot ... \cdot (n-k) \cdot (1 - \frac{1}{2^{2k}})^{(n-k)}.$$

Since $\lim_{n \to \infty}[n \cdot (n-1) \cdot ... \cdot (n-k) \cdot (1 - \frac{1}{2^{2k}})^{(n-k)}] = 0$, it follows that $\lim_{n \to \infty} \mu_n(\neg \sigma_k) = 0$, so $\neg \sigma_k$ is almost surely false, and σ_k is almost surely true.

Let N be the number of structures with elements in $\{1, ..., n\}$. To prove $(\dagger)$, observe the following:

1. For some fixed distinct $a_1, ..., a_k, b$ in $\{1, ..., n\}$, the number of $\mathbf{I}$ satisfying some fixed literal in $connections(a_1, ..., a_k; b)$ is $\frac{1}{2} \cdot N$.

2. For some fixed distinct $a_1, ..., a_k, b$ in $\{1, ..., n\}$, the number of $\mathbf{I}$ satisfying $connections(a_1, ..., a_k; b)$ is $\frac{1}{2^{2k}} \cdot N$ (since there are $2k$ literals in $connections$).

3. The number of $\mathbf{I}$ *not* satisfying $connections(a_1, ..., a_k; b)$ is therefore $N - \frac{1}{2^{2k}} \cdot N = (1 - \frac{1}{2^{2k}}) \cdot N$.

4. For some fixed $a_1, ..., a_k$ in $\{1, ..., n\}$, the number of **I** satisfying

$$\forall y(\bigwedge_i (a_i \neq y) \rightarrow \neg connections(a_1, ..., a_k; y))$$

is $(1 - \frac{1}{2^{2k}})^{n-k} \cdot N$ (since there are $(n-k)$ ways of picking b distinct from $a_1, ..., a_k$).

5. The number of **I** satisfying $\neg\sigma_k$ is thus at most

$$n \cdot (n-1) \cdot ... \cdot (n-k) \cdot (1 - \frac{1}{2^{2k}})^{(n-k)} \cdot N$$

(from the choices of $a_1, ..., a_k$). Hence (†) is proven.

$(c) \Rightarrow (a)$: Suppose that $\mathcal{R}$ does not satisfy σ, i.e., $\mathcal{R} \models \neg\sigma$. Since $(a) \Rightarrow (c)$, $\neg\sigma$ is almost surely true. Then σ cannot be almost surely true, a contradiction.   □

The 0-1 law for FO now follows immediately.

Theorem 4.1 *Each sentence in FO is almost surely true or almost surely false.*

Proof. Let σ be a FO sentence. The Rado graph $\mathcal{R}$ satisfies either σ or $\neg\sigma$. By the transfer property $((a) \Rightarrow (c)$ in Lemma 4.2), σ is almost surely true or $\neg\sigma$ is almost surely true. Thus, σ is almost surely true or almost surely false.   □

The 0-1 law for FO also holds for recursive extensions of FO such as *fixpoint* [5], [23] and *while* [17]. The proof uses once again the Rado graph and extends the transfer property to the *while* sentences. It follows immediately that many queries, including *evenness*, are not *while* sentences. The technique of 0-1 laws has been extended successfully to languages beyond *while*. Many languages which do not have 0-1 laws are also known, such as *existential second-order logic*. The precise "border" separating languages that have 0-1 laws from those that do not has yet to be determined, and remains an interesting and active area of research.

5 Turning the Infinite Into Finite

Most computer science applications involve structures which are finite. In some cases, however, the structures of interest are conceptually infinite.

For example, geographic information systems deal with regions in space, which in principle consist of infinitely many points. In such cases it is of interest to find finite summaries of the infinite structure that provide just sufficient information to answer the queries relevant to the application. We exemplify this approach with geographic information systems when the target set queries consists of *topological properties*, i.e., properties invariant under homeomorphisms of the space.

We use a spatial model that speaks about *regions* in the two-dimensional plane. Regions are specified by inequalities involving polynomials with rational coefficients (such regions are called *semi-algebraic*). Topological properties of regions are those that are invariant under homeomorphisms of the plane. This means, intuitively, that continuous deformations and reflections of the spatial structure do not affect satisfaction of the property. For example, the property "the intersection of regions P and Q is connected" is a topological property. On the other hand, the property "P lies above Q" is *not* topological. In this section we present a result of [20] showing that topological properties of semi-algebraic regions in a spatial database can be completely summarized by a finite first-order structure, called the *topological invariant* of the database. Moreover, the topological invariant of a semi-algebraic database can be constructed very efficiently – in NC (polylogarithmic time using polynomially many processors).

We use the following model for spatial databases. We assume given an infinite set **names** (consisting of names of regions). A spatial database *vocabulary* is a finite subset *Reg* of **names**. A *structure I* over a vocabulary *Reg* is a mapping from *Reg* to subsets of $\mathbf{R}^2$. For each $r \in Reg$, $I(r)$ provides a set of points called the *extent of r*. We generally refer to a set of points in the plane as a *region*. In practice, each $I(r)$ is finitely specified, although this may be transparent to the user. All regions considered here are compact, and specified by a disjunction of conjunctions of polynomial inequalities with rational coefficients (such regions are usually referred to as *semi-algebraic*). A region is *connected* if its boundary is connected. In the following, the term *region* will be used in the restricted manner just described, unless otherwise specified.

As discussed earlier, we are interested here in topological properties of spatial structures. Two structures I, J over a spatial vocabulary *Reg* are *topologically equivalent* iff there exists a bijection $\lambda : \mathbf{R}^2 \to \mathbf{R}^2$ such that both λ and λ^{-1} are continuous and for each $r \in Reg$, $\lambda(I(r)) = J(r)$. A property τ of spatial structures is *topological* if it is invariant under homeomorphisms, that is for all topologically equivalent structures I, J

over a given vocabulary, I satisfies τ iff J satisfies τ.

Topological invariants. In [20] it was shown that one can efficiently compute from a given semi-algebraic spatial structure I a finite relational structure $top(I)$ called the *topological invariant of I*, that describes completely the topological properties of I. We briefly describe the construction of the invariant and the results.

The invariant is constructed from a *maximum topological cell decomposition* of the spatial structure. A *topological cell decomposition* of I is a partition of $\mathbf{R}^2$ into finitely many subsets called *cells*, such that for every homeomorphism λ of $\mathbf{R}^2$, if $\lambda(I) = I$ then $\lambda(c)$ is a cell for every cell c. It is easily verified that for each spatial structure I there exists a unique maximal (in terms of number of cells) topological cell decomposition. The maximum topological cell decomposition can be constructed from a semi-algebraic spatial structure in NC, using results on *cell complexes* obtained in [4], [18]. We summarize their approach, slightly adapted to our context. Given a semi-algebraic spatial structure I over a vocabulary Reg, a *sign assignment* is a mapping $\sigma : Reg \rightarrow \{o, -, \partial\}$, and the *sign class* of σ is the set $I^\sigma = \bigcap_{r \in Reg} r^{\sigma(r)}$, where r^o is the interior of r, r^∂ is the boundary, and r^- is the exterior. A *cell complex* for I is a partition of $\mathbf{R}^2$ into finitely many, non-empty, pairwise disjoint regions, called *cells*, such that:

1. each cell is homeomorphic to $\mathbf{R}^0, \mathbf{R}^1$ or $\mathbf{R}^2 - \{\text{a finite set of points}\}$. The *dimension* of a cell is defined in the obvious manner.

2. the closure of each cell is a union of other cells;

3. each cell is included in some sign class I^σ.

It is shown in [18] that a cell complex can be constructed from a given semi-algebraic spatial structure in NC. One can further show that the maximum topological cell decomposition can be constructed from the cell complex obtained in [18], and the overall complexity remains NC.

The topological invariant for a spatial structure I is build up from the maximum topological cell decomposition for I. Cells of dimension 0,1 and 2 are called vertices, edges, and faces, respectively. The topological invariant associated to spatial structures over a vocabulary Reg is a finite structure consisting of the following relations (their meaning is explained intuitively):

1. Unary relations *Vertex, Edge, Face*, and *Exterior-face* providing the cells of dimension 0, 1, 2, and a distinguished face of dimension 2 called the *exterior face*.

2. *Endpoints* is a ternary relation providing endpoint(s) for edges.

3. *Face-Edge* is a binary relation providing, for each face (including the exterior cell), the edges on its boundary.

4. *Face-Vertex* is a binary relation providing, for each face (including the exterior cell), the vertices adjacent to it.

5. For each region name $p \in Reg$, a unary relation p providing the set of cells contained in region p.

6. *Orientation* is a 5-ary relation providing the clockwise and counter-clockwise orientation of edges incident to each vertex. More precisely, $(\hookleftarrow, v, e_1, e_2, e_3) \in Orientation$ iff v is a vertex, e_1, e_2, e_3 are edges of faces incident to v, and e_2 lies between e_1 and e_3 in the clockwise order on the incidents cells of v, and $(\hookrightarrow, v, e_1, e_2) \in Orientation$ iff v is a vertex, e_1, e_2, e_3 are cells incident to v, and e_2 lies between e_1 and e_3 in the counterclockwise order on the incident cells of v.

Let $inv(Reg)$ denote the above vocabulary and let top denote the mapping associating to each spatial structure I over Reg its topological invariant over $inv(Reg)$.

The main result on the topological invariant is the following.

Theorem 5.1 *Let* Reg *be a spatial vocabulary.*
(i) *The mapping top associating to each spatial structure over* Reg *its topological invariant is computable in polynomial time (and* NC*) with respect to the size of the representation of* I.
(ii) *For all spatial structures* I, J *over* Reg, I *and* J *are topologically equivalent iff* $top(I)$ *and* $top(J)$ *are isomorphic.*

How can topological invariants be used? Suppose that a topological query is posed against the spatial database. In principle, the query can be answered by another query posed against the topological invariant. Since the topological invariant is in most cases much simpler than the full spatial data, this strategy is likely to be more efficient. In order for this to work, topological queries in the spatial query language need to be *effectively translated* into queries in some query language for topological invariants. We present an example that provides some intuition into the difficulties involved in translating FO sentences on the spatial representation into FO sentences on the invariant. Consider the query on schema $Reg = \{P, Q\}$:

"Regions P and Q intersect only on their boundaries."

Clearly, this is a topological property. It can be expressed in FO with order over the reals by the sentence:

(†) $\forall x \forall y[(P(x,y) \wedge Q(x,y)) \rightarrow$
        $(boundary_P(x,y) \wedge boundary_Q(x,y))],$

where $boundary_P(x,y)$ (and similarly $boundary_Q(x,y)$) is the formula:

$$P(x,y) \wedge \forall x_1 \forall y_1 \forall x_2 \forall y_2[(x_1 < x < x_2 \wedge y_1 < y < y_2) \rightarrow$$
$$\exists x' \exists y'(x_1 < x' < x_2 \wedge y_1 < y' < y_2 \wedge \neg P(x',y')].$$

Clearly, the same property can be expressed by the FO sentence over $inv(Reg)$:

(‡) $\forall u[(P(u) \wedge Q(u)) \rightarrow (Vertex(u) \vee Edge(u))].$

However, how to get from (†) to (‡) is mysterious. The difficulty is to algorithmically extract the topological meaning of a sentence like (†) that uses non-topological statements involving reals and $<$. This problem is considered in [22]. It is solved for the case when Reg contains a single region name (however, the region can be quite general); it has been recently shown by Grohe and Segoufin that the translation is not possible with several region names.

It is natural to wonder if these techniques can be extended beyond dimension two. Unfortunately, the picture is largely negative. The existence of a topological invariant for 3-dimensional semi-algebraic databases implies a positive answer to an open problem in Knot Theory: the existence of an invariant for topologically equivalent knots [7]. In dimension four (and higher) there is no finite topological invariant, because topological equivalence itself is undecidable. This is shown by adapting the proof of an undecidability result on topological equivalence of manifolds [19]. The latter proof is by reduction of the word problem for finitely generated groups to isomorphism of the fundamental groups of two topological spaces, which in turn is equivalent to their being homeomorphic.

In summary, we have shown how to produce, from a semi-algebraic spatial database which is conceptually infinite, a finite structure that precisely summarizes the topological properties of the original structure. This example is typical of applications where structures that are in principle infinite are turned into finite structures that can be efficiently manipulated.

6   From Finite Back to Infinite?

We lastly consider a surprising idea: is it useful to sometimes model objects which are finite as infinite structures? Admittedly, this is not altogether new: after all, computers, which in practice are always finite, are better modeled by a Turing machine with infinite tape than by a finite-state machine. In this section we pose the same question about another well-known finite object: the World Wide Web.

In [3], we propose a model that views the Web as an *infinite* labeled graph. We believe this captures the intuition that exhaustive exploration of the Web is –or will soon become– prohibitively expensive. The infiniteness assumption can be viewed as a convenient metaphor, much like Turing machines with infinite tapes are useful abstractions of computers with finite (but potentially very large) memory. As a consequence of the infiniteness assumption, exhaustive exploration of the Web is penalized in our model by a non-terminating computation. Thus, our model draws a sharp distinction between exhaustive exploration of the Web and more controlled types of computation.

Note that this approach is fundamentally different from other attempts to model infinite data (e.g., [14], [15]) which focus on finitely representable structures. In contrast, we do not assume the Web is finitely represented. Instead, we view it as a possibly non-recursive infinite labeled graph which can never be entirely explored. Intuitively, a node in the graph is an abstraction of a Web page, and the labeled edges represent html links between pages. The links allow navigating through the Web, in hypertext style. Our model leads to a focus on querying and computation where exploration of the Web is controlled.

We consider queries which are mappings whose domain consists of pairs (o, I) where I is an infinite labeled graph and o is a designated starting node of I. For each pair (o, I), the result of the query consists of a subset of the nodes of I.

We begin by exploring the notion of computable query in the context of the Web. Our model is along the lines of the computable queries of Chandra and Harel [6]. We briefly describe a machine model of computation on the Web that we call a Web machine. This works much like a Turing machine, but takes as input an infinite string and may produce an infinite answer. Based on the Web machine, we can define the notions of *finite computability* and *eventual computability* of queries. The latter notion arises from the fact that infinite answers to queries are allowed. A query is *finitely computable* if its answer is always finite and computable by a halting Web machine.

A query is *eventually computable* if there is a Web machine, possibly non-terminating, which eventually outputs each node in the answer to the query (a node cannot be retracted once it is output).

Examples   The notions of finitely computable and eventually computable queries are illustrated by the following queries on input (o, I):

1. Finitely computable:

   - Find the nodes reachable from o by a path labeled $a.b.c$ (an a-labeled edge, followed by a b-labeled edge, followed by a c-labeled edge).

   - Find the nodes o' such that there is a path of length at most k from o to o'.

   - Find all nodes lying on a cycle of length at most 3 which contains o.

2. Eventually computable with possibly infinite answers (so not finitely computable):

   - Find the nodes reachable from o.

   - Find the nodes referencing o.

   - Find the nodes belonging to a cycle.

3. Eventually computable with finite answers, but not finitely computable:

   - Find the nodes on the shortest cycle containing o.

   - Find the node(s) at the shortest distance from o that reference o.

4. Not eventually computable:

   - Find all nodes that do not belong to a cycle.

   - Find all nodes which are not referenced by any other node.

   - Output o if and only if all nodes reachable from o have non-nil references[2].

---

[2]Nil references can be modeled by references to a special node named *nil*.

In particular, it is clear from the above examples that finitely computable and eventually computable properties are not closed under complement.

We next describe informally two specialized machine models that capture directly the main styles of computing used on the Web: browsing and searching. The *browser* machine model allows for navigational exploration of the Web. The *browse/search* machine additionally allows searching in the style of search engines.

The idea underlying the browser machine is to access the Web navigationally, by following references starting from the source node o. A browser machine has an infinite browsing tape, an infinite work tape, and a right-infinite output tape. It is equipped with a finite state control which includes a special state called *expand*. The computation of the machine on input (o, I) is as follows. Initially, the browsing tape contains the encoding of the source node o. If the *expand* state is reached at any point in the computation and the browsing tape contains the encoding of some node o', this is replaced on the browsing tape by an encoding of all nodes in I with incoming edges from o'.

Obviously, browser machines have limited computing ability, since they can only access the portion of I reachable from the source node o. However, this is an intuitively appealing approach for controlling the computation. The following result confirms the central role of this style of computation in the context of the Web.

Theorem 6.1 *Each finitely computable Web query is finitely computable by a browser machine.*

We next augment browser machines with a search mechanism. The search is essentially a selection operation such as: (i) select all edges with label "Department"; (ii) select all edges with label A and node 556 as destination; and (iii) select all edges. In general, a search triggers an eventually computable subquery, whose result may be infinite. This leads to the problem of integrating non-terminating subcomputations into the computation of a query. We adopt the following model.

A *browse/search machine* is a browser machine augmented with a right-infinite search-answer tape and a separate search-condition tape. There is a distinguished *search* state. The computation of the machine is non-deterministic. A search is triggered by writing a selection operation on the search-condition tape, then entering the search state. The search-answer tape functions similarly to the answer tape of an eventually computable query. Answers to previously triggered searches arrive on the search-answer tape at arbitrary times and in arbitrary order. More precisely, suppose the set of selections triggered up to some given point in the computation is $\{\sigma_1, \ldots, \sigma_n\}$. In any subsequent move of the machine, a (possibly empty) finite subset of the answers to some of the σ_i's is appended to the search-answer tape. This is non-deterministic. The order in which answers are produced is arbitrary. Each tuple in the answer to σ_i is prefixed by σ_i (everything is encoded in the obvious way). It is guaranteed that all answers to a triggered search will be eventually produced. However, note that there is generally no way to know at a given time if all answers to a particular search have been obtained. The rest of the computation occurs as in the browser machine. A Web query q is finitely computable by a browse/search machine if there exists a browse/search machine W such that *each* computation of W on input (o, I) halts and produces an encoding of $q(o, I)$ on the answer tape[3]. The definition of query eventually computable by a browse/search machine is analogous.

What is the power of browse/search machines? This is elucidated by the following result.

Theorem 6.2   (i) *A Web query is eventually computable if and only if it is eventually computable by a browse/search machine.*

(ii) *A Web query is finitely computable if and only if it is finitely computable by a browse/search machine (if and only if it is finitely computable by a browse machine).*

We close with a brief discussion of the ability of query languages to express queries on the Web. We consider in [3] the classical languages FO, Datalog, and Datalog¬ (see [1] for a presentation of these languages). The questions of interest for each language are the following: (i) Are the queries in the language finitely computable or eventually computable? (ii) Which fragments of each language can be implemented by browsers and which by a combination of browsing and searching? We provide syntactic restrictions that guarantee computability by browsers or by browse/search machines in FO and Datalog$^{(\neg)}$.

One of the interesting results of [3] is with respect to negation. The "positive" fragment of FO is eventually computable. The addition of recursion yields no problem. However, negation brings trouble, and some simple FO queries are not eventually computable. The Datalog¬ languages

---

[3]However, it should be clear that a browse/search machine that uses the search feature in a non-trivial way cannot terminate.

yield some surprises: the standard semantics, stratified and well-founded [24], are ill-suited for expressing eventually computable queries, whereas the more procedural inflationary semantics [2], [16] turns out to be naturally suited to express such queries, and thus has a fundamental advantage over the first two semantics.

7 Conclusion

The relationship between the finite and the infinite raises fundamental issues in the context of logic. We examined the impact of the finiteness assumption, and noted the failure of many classical results when structures are restricted to be finite, such as the Compactness theorem. We discussed two very effective and elegant alternative techniques for proving expressiveness results on finite structures: Ehrenfeucht-Fraissé games and 0-1 laws. We also illustrated ways to finitely represent information of interest about infinite structures, using the example of topological invariants of two-dimensional semi-algebraic regions. Conversely, we argued that in some situations it is beneficial to model finite objects as infinite structures.

References

[1] S. Abiteboul, R. Hull, V. Vianu, *Foundations of Databases*, Addison-Wesley, Reading-Massachusetts, 1995.

[2] S. Abiteboul, V. Vianu, Procedural and declarative database update languages, *Proc. ACM Symp. on Principles of Database Systems*, 1988, 240–250.

[3] S. Abiteboul, V. Vianu, Queries and computation on the Web, *Proc. Intern. Conf. on Database Theory*, Springer-Verlag, 1997.

[4] M. Ben-Or, D. Kozen, J. Reif, The complexity of elementary algebra and geometry, *J. of Computer and System Science*, **32** (1986), 251–264.

[5] A. Blass, Y. Gurevich, D. Kozen, A zero-one law for logic with a fixed point operator, *Information and Control*, **67** (1985), 70–90.

[6] A. K. Chandra, D. Harel, Computable queries for relational data bases, *Journal of Computer and System Sciences*, **21**, 2 (1980), 156–178.

[7] R. Crowell, R. Fox, *Introduction to Knot Theory*, Springer-Verlag, GTM vol. 57, 1963.

[8] R. A. DiPaola, The recursive unsolvability of the decision problem for a class of definite formulas, *J. of the ACM*, **16**, 2 (1969), 324–327.

[9] A. Ehrenfeucht, An application of games to the completeness problem for formalized theories, *Fund. Math.*, **49** (1961), 129–141.

[10] H. Enderton, *A Mathematical Introduction to Logic*, Academic Press, 1972.

[11] R. Fagin, Probabilities on finite models, *Journal of Symbolic Logic*, **41**, 1 (1976), 50–58.

[12] R. Fraissé, Sur les classifications des systèmes de relations, *Publ. Sci. Univ. Alger*, **I:1** (1954).

[13] Y. V. Glebskiĭ, D. I. Kogan, M. I. Liogonkiĭ, V. A. Talanov, Range and degree of realizability of formulas in the restricted predicate calculus, *Kibernetika*, **2** (1969), 17–28.

[14] D. Harel, T. Hirst, Completeness results for recursive data bases, *Proc. ACM Symp. on Principles of Database Systems*, 1993, 244–252.

[15] P. Kanellakis, G. Kuper, P. Revesz, Constraint query languages, *Proc. ACM Symp. on Principles of Database Systems*, 1990, 299–313.

[16] P. G. Kolaitis, C.H. Papadimitriou, Why not negation by fixpoint? *Proc. ACM Symp. on Principles of Database Systems*, 1988 231–239.

[17] P. Kolaitis, M. Vardi, 0-1 laws for infinitary logics, *Information and Computation*, **98** (1992), 258–294.

[18] D. Kozen, C-K. Yap, Algebraic cell decomposition in NC, *Proc. IEEE Symp. on Foundations of Computer Science*, 1985, 515–521.

[19] A. A. Markov, Unsolvability of the problem of homeomorphy, *Proc. Intern. Congress of Mathematics*, 1958, 300–306 (in Russian).

[20] C. H. Papadimitriou, D. Suciu, V. Vianu, Topological queries in spatial databases, *J. of Computer and System Sciences*, **58** (1999), 29–53.

[21] R. Rado, Universal graphs and universal functions, *Acta Arith.*, **9** (1964), 331–340.

[22] L. Segoufin, V. Vianu, Querying spatial databases via topological invariants, *Proc. ACM Symp. on Principles of Database Systems*, 1998.

[23] V. A. Talanov, V. V. Knyazev, The asymptotic truth value of infinite formulas, *All-union Seminar on Discrete Mathematics and Its Applications*, 1984, 56–61.

[24] A. Van Gelder, K. A. Ross, J. S. Schlipf, The well-founded semantics for general logic programs, *Proc. ACM Symp. on Principles of Database Systems*, 1988, 221–230.

AUTHOR INDEX

Other titles in the DMTCS series: